Effect of the Modification of Catalysts on the Catalytic Performance

Effect of the Modification of Catalysts on the Catalytic Performance

Editors

Florica Papa
Anca Vasile
Gianina Dobrescu

MDPI • Basel • Beijing • Wuhan • Barcelona • Belgrade • Manchester • Tokyo • Cluj • Tianjin

Editors

Florica Papa
Surface Chemistry and Catalysis
Romanian Academy
"Ilie Murgulescu" Physical Chemistry Institute
Bucharest
Romania

Anca Vasile
Chemical Kinetics
"Ilie Murgulescu" Institute of Physical Chemistry of the Romanian Academy
Bucharest
Romania

Gianina Dobrescu
Surface Chemistry and Catalysis
"Ilie Murgulescu" Institute of Physical Chemistry of the Romanian Academy
Bucharest
Romania

Editorial Office
MDPI
St. Alban-Anlage 66
4052 Basel, Switzerland

This is a reprint of articles from the Special Issue published online in the open access journal *Catalysts* (ISSN 2073-4344) (available at: www.mdpi.com/journal/catalysts/special_issues/Modification_Catalytic_Performance).

For citation purposes, cite each article independently as indicated on the article page online and as indicated below:

LastName, A.A.; LastName, B.B.; LastName, C.C. Article Title. *Journal Name* **Year**, *Volume Number*, Page Range.

ISBN 978-3-0365-6727-3 (Hbk)
ISBN 978-3-0365-6726-6 (PDF)

© 2023 by the authors. Articles in this book are Open Access and distributed under the Creative Commons Attribution (CC BY) license, which allows users to download, copy and build upon published articles, as long as the author and publisher are properly credited, which ensures maximum dissemination and a wider impact of our publications.
The book as a whole is distributed by MDPI under the terms and conditions of the Creative Commons license CC BY-NC-ND.

Contents

About the Editors vii

Florica Papa, Anca Vasile and Gianina Dobrescu
Effect of the Modification of Catalysts on the Catalytic Performance
Reprinted from: *Catalysts* 2022, 12, 1637, doi:10.3390/catal12121637 1

Monica Raciulete, Crina Anastasescu, Florica Papa, Irina Atkinson, Corina Bradu and Catalin Negrila et al.
Band-Gap Engineering of Layered Perovskites by Cu Spacer Insertion as Photocatalysts for Depollution Reaction
Reprinted from: *Catalysts* 2022, 12, 1529, doi:10.3390/catal12121529 5

Gabriela Petcu, Elena Maria Anghel, Elena Buixaderas, Irina Atkinson, Simona Somacescu and Adriana Baran et al.
Au/Ti Synergistically Modified Supports Based on SiO_2 with Different Pore Geometries and Architectures
Reprinted from: *Catalysts* 2022, 12, 1129, doi:10.3390/catal12101129 33

Veronica Bratan, Anca Vasile, Paul Chesler and Cristian Hornoiu
Insights into the Redox and Structural Properties of CoOx and MnOx: Fundamental Factors Affecting the Catalytic Performance in the Oxidation Process of VOCs
Reprinted from: *Catalysts* 2022, 12, 1134, doi:10.3390/catal12101134 53

Silviu Preda, Polona Umek, Maria Zaharescu, Crina Anastasescu, Simona Viorica Petrescu and Cătălina Gîfu et al.
Iron-Modified Titanate Nanorods for Oxidation of Aqueous Ammonia Using Combined Treatment with Ozone and Solar Light Irradiation
Reprinted from: *Catalysts* 2022, 12, 666, doi:10.3390/catal12060666 81

Md. Abu Hanif, Jeasmin Akter, Young Soon Kim, Hong Gun Kim, Jae Ryang Hahn and Lee Ku Kwac
Highly Efficient and Sustainable ZnO/CuO/g-C_3N_4 Photocatalyst for Wastewater Treatment under Visible Light through Heterojunction Development
Reprinted from: *Catalysts* 2022, 12, 151, doi:10.3390/catal12020151 101

Yulin Li, Ping She, Rundong Ding, Da Li, Hongtan Cai and Xiufeng Hao et al.
Bimetallic PdCo Nanoparticles Loaded in Amine Modified Polyacrylonitrile Hollow Spheres as Efficient Catalysts for Formic Acid Dehydrogenation
Reprinted from: *Catalysts* 2021, 12, 33, doi:10.3390/catal12010033 119

Tong Zhang, Wenge Qiu, Hongtai Zhu, Xinlei Ding, Rui Wu and Hong He
Promotion Effect of the Keggin Structure on the Sulfur and Water Resistance of Pt/CeTi Catalysts for CO Oxidation
Reprinted from: *Catalysts* 2021, 12, 4, doi:10.3390/catal12010004 133

Gianina Dobrescu, Florica Papa, Razvan State, Monica Raciulete, Daniela Berger and Ioan Balint et al.
Modified Catalysts and Their Fractal Properties
Reprinted from: *Catalysts* 2021, 11, 1518, doi:10.3390/catal11121518 149

Wei Song, Ran Zhao, Lin Yu, Xiaowei Xie, Ming Sun and Yongfeng Li
Enhanced Catalytic Hydrogen Peroxide Production from Hydroxylamine Oxidation on Modified Activated Carbon Fibers: The Role of Surface Chemistry
Reprinted from: *Catalysts* **2021**, *11*, 1515, doi:10.3390/catal11121515 . . . . **165**

Zexiang Chen, Meiqing Shen, Chen Wang, Jianqiang Wang, Jun Wang and Gurong Shen
Improvement of Alkali Metal Resistance for NH_3-SCR Catalyst Cu/SSZ-13: Tune the Crystal Size
Reprinted from: *Catalysts* **2021**, *11*, 979, doi:10.3390/catal11080979 . . . . **183**

About the Editors

Florica Papa

Dr. Florica Papa is a senior researcher in the Department of Surface Chemistry and Catalysis at "Ilie Murgulescu" Physical Chemistry Institute of the Romanian Academy Bucharest, Romania. The most significant research directions are material chemistry, heterogeneous catalysis, and photocatalysis. Her research activities are related to synthesis of mono and bimetallic nanoparticles with controlled morphology for applications in environmental depollution, such as selective catalytic and photocatalytic reduction of nitrates, photodegradation of persistent organic pollutants, photocatalytic oxidation of ammonia nitrogen with ozone in water. Another area of her research interests is the oxidation of C1–C4 aliphatic hydrocarbons on simple and doped oxides; oxidative coupling of methane on rare earth oxides; and the impact of the fractal dimension on the catalytic performance. She was project director for 3 national grants and work team member for more than 25 national and international projects. She is coauthor of 54 ISI articles, 2 book chapters, and 4 patents.

Anca Vasile

Dr. Anca Vasile is senior chemistry researcher in the Department of Chemical Kinetics at "Ilie Murgulescu" Institute of Physical Chemistry of the Romanian Academy (Bucharest, Romania). Dr. Vasile research deals with the synthesis, testing and evaluation of catalytic materials for applications in environmental depollution, mainly focused on heterogeneous catalysis, photocatalysis and adsorption. Her interest in the research activity is centered on: kinetics of gas–solid interaction; lower olefin (C3–C4) oxidation on multicomponent oxide catalysts; semiconductor properties of oxide catalysts; dynamics of the lattice oxygen in oxide catalysts for selective oxidation catalysis; synthesis of supported metallic nanoparticles; and catalytic removal of organic/inorganic compounds from water, photocatalytic reduction of nitrates from aqueous solutions.

Gianina Dobrescu

Dr. Gianina Dobrescu is working as senior physics researcher at Romanian Academy, "Ilie Murgulescu" Physical Chemistry Institute, Bucharest, Romania. Research interest: application of fractal theory in surface science, such as adsorption mechanism on fractal surfaces, applying fractal theory to characterize surfaces, modeling adsorption on fractal surfaces, computing fractal dimension from micrographs (TEM, SEM, AFM, STM), experimental adsorption isotherms and scattering experiments (light scattering, X-Ray scattering, neutron scattering), growth surfaces, and computation of time and spatial scaling exponents.

Editorial

Effect of the Modification of Catalysts on the Catalytic Performance

Florica Papa, Anca Vasile * and Gianina Dobrescu *

"Ilie Murgulescu" Institute of Physical-Chemistry of the Romanian Academy, 202 Spl. Independentei, 060021 Bucharest, Romania
* Correspondence: avasile@icf.ro (A.V.); gdobrescu@icf.ro (G.D.)

1. Introduction

Changing the composition and structure of a catalyst to obtain a positive impact on its performance is challenging. Therefore, the optimization of the surface and the bulk properties (electronic or physical structure) offers a strategy for the development of advanced catalysts used to be used in the global challenges of energy conversion and environmental protection [1–3] (Figure 1).

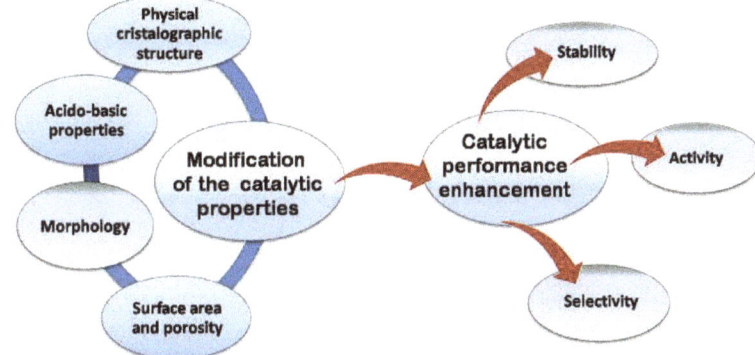

Figure 1. The impact of catalytic properties modification on the catalytic performances.

Catalyst performance plays a significant role in the catalytic processes and is expressed in terms of selectivity, activity, and stability (resistance to deactivation and regeneration capacity). Successful catalyst development depends on several factors, including the preparation method, the interaction between the active phase and support, the structural and physicochemical properties of the active metal or support, and the metal precursor used in the preparation [4].

Starting from the premises shown above that underline the special importance of this essential field in catalysis and catalytic materials, we felt honored to receive the invitation to be guest editors of this Special Issue. Following the excellent collaboration with Ms. Assistant Editor Mia Zhang in 2021 and with Ms. Assistant Editor Maeve Yue, starting in 2022, as well as following the special work undertaken by the entire editing team, 10 articles could be successfully published, including two reviews. Lastly, we are grateful to all of the authors of these publications for their excellent research work and for their contribution to this Special Issue.

2. This Special Issue

The purpose of the Special Issue, "Effect of Catalyst Modification on Catalytic Performance", was to gather a collection of papers that present new strategies for modifying catalysts that facilitate the establishment of composition–performance and structure–performance relationships. In what follows, we present some conclusions from the works published in this volume.

Raciulete et al. [5] developed a multi-step ion-exchange methodology for the fabrication of $Cu(LaTa_2O_7)_2$ lamellar architectures by exchanging Rb^+ with a much smaller Cu^{2+} spacer in the $RbLaTa_2O_7$ host to achieve photocatalysts capable of wastewater depollution. Cu-modified layered perovskites exhibited enhanced photocatalytic activity compared to the $RbLaTa_2O_7$ host. The superior photocatalytic activity of CuLTO-800R was attributed to its narrow band gap and photogenerated–carriers separation. Petcu et al. [6] presented the synthesis and characterization of new photocatalysts obtained through the immobilization of titanium and gold on the supports with various porous structures (micropores—zeolite Y, micro and mesopores—hierarchical zeolite Y, smaller mesopores—MCM-48 and larger mesopores—KIT-6). The effects of porous structure and surface properties on TiO_2 dispersion, crystal structure, the nature of the interaction between support–titanium species and Au-TiO_2, respectively, were studied. The authors further investigated the influence of the support properties, the presence of TiO_2 and Au species, and their interaction with amoxicillin photodegradation under UV and visible light irradiation. In another paper [7] pristine titanate nanorods were modified with iron to improve the light absorption and separation of photogenerated charges and to favor ammonia photodegradation under solar light irradiation. The morphological and structural characterizations (SEM, XRD, XRF, UV-Vis, H_2-TPR, NH_3-TPD, PL, PZC) of the studied catalysts were correlated with their activity on ammonia degradation with ozone- and photo-assisted oxidation. Preda et al. [7] optimized the aqueous ammonia oxidation process to obtain a high ammonia conversion and increase selectivity to gaseous nitrogen-containing products. In reference [8], the authors focused on the preparation of a ZnO/CuO/g-C_3N_4 (ZCG) nanocomposite through an efficient co-crystallization method, performed a photocatalytic test for the treatment of dye-containing wastewater followed by visible light irradiation and evaluated the material's durability and recycling capability. The outstanding photocatalytic performance was observed due to heterojunction formation among the g-C_3N_4, CuO-NPs, and ZnO-NPs compounds, which minimized the photogenerated e^--h^+ pair recombination and increased the electron flow rate. Li et al. [9] modified polyacrylonitrile hollow nanospheres (HPAN) derived from the polymerization of acrylonitrile in the presence of polystyrene emulsion (as template) using surface amination with ethylenediamine (EDA), using them as a support for loading Pd or PdCo nanoparticles. The authors demonstrated that the prepared PdCo nanoparticles supported on the surface of aminated polyacrylonitrile hollow nanospheres (EDA-HPAN) could be used as a highly active and stable catalyst for the dehydrogenation of formic acid. Zhang et al. [10] investigated the promotion of the Keggin structure to the sulfur and water resistance of Pt/CeTi catalysts for CO oxidation. They prepared Pt catalysts using cerium titanium composite oxide (CeTi), ammonium molybdophosphate with Keggin structure-modified CeTi (Keg-CeTi), and molybdophosphate without Keggin structure-modified CeTi (MoP-CeTi) as supports. Their research revealed that the high SO_2 and H_2O resistance of Pt/Keg-CeTi in CO oxidation was related to its stronger surface acidity, surface cerium and molybdenum species reduction, and lower SO_2 adsorption and transformation compared to Pt/CeTi and Pt/MoP-CeTi. In reference [11], an easy synthetic process for H_2O_2 generation was described, and a new comprehension of the conception and mechanistic examination of metal-free N- and O-doped carbon materials were also provided. The content and type of surface functional groups were improved by treating the PAN-based ACF with concentrated mixed acid in different volume ratios. The selectivity and yield of H_2O_2 in the reaction on modified activated carbon fiber (ACF) catalysts were well interconnected with the amounts of pyrrolic/pyridone nitrogen (N5) and desorbed carboxyl–anhydride groups from the ACF surface. The possible reaction pathway over the ACF catalysts promoted by N5 was also shown. In order to improve the alkali metal resistance of commercial catalyst Cu/SSZ-13 for ammonia-selective catalytic reduction (NH_3-SCR) reaction, Chen et al. [12] developed a simple method to synthesize Cu/SSZ-13 with a core–shell-like structure. Cu/SSZ-13, with a crystal size of 2.3 µm, exhibited excellent resistance to Na poisoning. They investigated physical structure characterization (XRD, BET, SEM, NMR) and chemical acidic distribution (H_2-TPR, UV-Vis, Diethylamine-TPD, pyridine-DRIFTs, EDS). Bratan et al. [13] presented in their review the main factors affecting the catalytic performances of CoOx and MnOx metal-oxide catalysts concerning the total oxidation of hydrocarbons. The catalytic behavior of the studied oxides was discussed and could be closely related to their redox properties, nonstoichiometric, defective structure, and lattice oxygen mobility. It was emphasized that controlling the structural and textural properties of the studied metal oxides, such as the specific surface area and specific morphology, plays an important role in catalytic applications. Dobrescu et al. [14] reviewed scientific results related to oxide catalysts, such as lanthanum cobaltites and ferrites with perovskite structure, and nanoparticle catalysts (such as Pt, Rh, Pt-Cu, etc.), emphasizing their fractal properties and the influence of their fractal modification on both the fractal and catalytic properties. They discussed some methods used to compute the fractal dimensions of the catalysts (micrograph fractal analysis and the adsorption isotherm method) and computed catalysts' fractal dimensions, underlining that increasing the fractal dimensions of the catalysts is of significant demand in heterogeneous catalysis.

In conclusion, the Special Issue, "Effect of the Modification of Catalysts on the Catalytic Performance", should be of great interest to all researchers involved in this scientific area regarding the synthesis and characterization of various catalysts or catalytic materials; catalytic performance (activity and selectivity); synergetic effect; the modification of catalysts or suitable promoters that are added to modify the catalyst structure, improve stability, or enhance the catalytic reactions, enabling better activity or selectivity; and the reaction mechanism and kinetic parameters (reaction rate and activation energy).

Author Contributions: Conceptualization, writing—review and editing A.V., G.D. and F.P. All authors have read and agreed to the published version of the manuscript.

Conflicts of Interest: The authors declare no conflict of interest.

References

1. Wang, C.; Wang, Z.; Mao, S.; Chen, Z.; Wang, Y. Coordination environment of active sites and their effect on catalytic performance of heterogeneous catalysts. *Chin. J. Catal.* **2022**, *43*, 928. [CrossRef]
2. Wang, S.; Wang, J.; Zhu, M.; Bao, X.; Xiao, B.; Su, D.; Li, H.; Wang, Y. Molybdenum-Carbide-Modified Nitrogen-Doped Carbon Vesicle Encapsulating Nickel Nanoparticles: A Highly Efficient, Low-Cost Catalyst for Hydrogen Evolution Reaction. *J. ACS* **2015**, *137*, 15753. [CrossRef] [PubMed]
3. Deng, J.; Li, M.; Wang, Y. Biomass-derived carbon: Synthesis and applications in energy storage and conversion. *Green Chem.* **2016**, *18*, 4824–4854. [CrossRef]
4. Védrine, J.C. Metal oxides in heterogeneous oxidation catalysis: State of the art and challenges for a more sustainable world. *ChemSusChem* **2019**, *12*, 577. [CrossRef] [PubMed]
5. Raciulete, M.; Anastasescu, C.; Papa, F.; Atkinson, I.; Bradu, C.; Negrila, C.; Eftemie, D.-I.; Culita, D.C.; Miyazaki, A.; Bratan, V.; et al. Band-Gap Engineering of Layered Perovskites by Cu Spacer Insertion as Photocatalysts for Depollution Reaction. *Catalysts* **2022**, *12*, 1529. [CrossRef]
6. Petcu, G.; Anghel, E.M.; Buixaderas, E.; Atkinson, I.; Somacescu, S.; Baran, A.; Culita, D.C.; Trica, B.; Bradu, C.; Ciobanu, M.; et al. Au/Ti Synergistically ModifiedSupports Based on SiO_2 with Different Pore Geometries and Architectures. *Catalysts* **2022**, *12*, 1129. [CrossRef]
7. Preda, S.; Umek, P.; Zaharescu, M.; Anastasescu, C.; Petrescu, S.V.; Gîfu, C.; Eftemie, D.-I.; State, R.; Papa, F.; Balint, I. Iron-Modified Titanate Nanorods for Oxidation of Aqueous Ammonia Using Combined Treatment with Ozone and Solar Light Irradiation. *Catalysts* **2022**, *12*, 666. [CrossRef]
8. Hanif, M.A.; Akter, J.; Kim, Y.S.; Kim, H.G.; Hahn, J.R.; Kwac, L.K. Highly Efficient and Sustainable ZnO/CuO/g-C_3N_4 Photocatalyst for Wastewater Treatment under Visible Light through Heterojunction Development. *Catalysts* **2022**, *12*, 151. [CrossRef]
9. Li, Y.; She, P.; Ding, R.; Li, D.; Cai, H.; Hao, X.; Jia, M. Bimetallic PdCo Nanoparticles Loaded in Amine Modified Polyacrylonitrile Hollow Spheres as Efficient Catalysts for Formic Acid Dehydrogenation. *Catalysts* **2022**, *12*, 33. [CrossRef]

10. Zhang, T.; Qiu, W.; Zhu, H.; Ding, X.; Wu, R.; He, H. Promotion Effect of the Keggin Structure on the Sulfur and Water Resistance of Pt/CeTi Catalysts for CO Oxidation. *Catalysts* **2022**, *12*, 4. [CrossRef]
11. Song, W.; Zhao, R.; Yu, L.; Xie, X.; Sun, M.; Li, Y. Enhanced Catalytic Hydrogen Peroxide Production from Hydroxylamine Oxidation on Modified Activated Carbon Fibers: The Role of Surface Chemistry. *Catalysts* **2021**, *11*, 1515. [CrossRef]
12. Chen, Z.; Shen, M.; Wang, C.; Wang, J.; Wang, J.; Shen, G. Improvement of Alkali Metal Resistance for NH3-SCR Catalyst Cu/SSZ-13: Tune the Crystal Size. *Catalysts* **2021**, *11*, 979. [CrossRef]
13. Bratan, V.; Vasile, A.; Chesler, P.; Hornoiu, C. Insights into the Redox and Structural Properties of CoOx and MnOx: Fundamental Factors Affecting the CatalyticPerformance in the Oxidation Process of VOCs. *Catalysts* **2022**, *12*, 1134. [CrossRef]
14. Dobrescu, G.; Papa, F.; State, R.; Raciulete, M.; Berger, D.; Balint, I.; Ionescu, N.I. Modified Catalysts and Their Fractal Properties. *Catalysts* **2021**, *11*, 1518. [CrossRef]

Article

Band-Gap Engineering of Layered Perovskites by Cu Spacer Insertion as Photocatalysts for Depollution Reaction

Monica Raciulete [1], Crina Anastasescu [1], Florica Papa [1,*], Irina Atkinson [1], Corina Bradu [2,3], Catalin Negrila [4], Diana-Ioana Eftemie [1], Daniela C. Culita [1], Akane Miyazaki [5], Veronica Bratan [1], Jeanina Pandele-Cusu [1], Cornel Munteanu [1], Gianina Dobrescu [1], Alexandra Sandulescu [1] and Ioan Balint [1,*]

[1] "Ilie Murgulescu" Institute of Physical Chemistry of the Romanian Academy, 202 Spl. Independentei, 060021 Bucharest, Romania
[2] PROTMED Research Center, University of Bucharest, 91–95 Spl. Independentei, 050107 Bucharest, Romania
[3] Faculty of Biology, University of Bucharest, 91–95 Spl. Independenței, 050095 Bucharest, Romania
[4] National Institute of Material Physics, 077125 Magurele, Romania
[5] Department of Chemical and Biological Sciences, Faculty of Science, Japan Women's University, 2-8-1 Mejirodai, Bunkyo-ku, Tokyo 112-8681, Japan
* Correspondence: frusu@icf.ro (F.P.); ibalint@icf.ro (I.B.)

Abstract: A multi-step ion-exchange methodology was developed for the fabrication of $Cu(LaTa_2O_7)_2$ lamellar architectures capable of wastewater depollution. The (001) diffraction line of $RbLaTa_2O_7$ depended on the guest species hosted by the starting material. SEM and TEM images confirmed the well-preserved lamellar structure for all intercalated layered perovskites. The UV–Vis, XPS, and photocurrent spectroscopies proved that Cu intercalation induces a red-shift band gap compared to the perovskite host. Moreover, the UV–Vis spectroscopy elucidated the copper ions environment in the Cu-modified layered perovskites. H_2-TPR results confirmed that Cu species located on the surface are reduced at a lower temperature while those from the interlayer occur at higher temperature ranges. The photocatalytic degradation of phenol under simulated solar irradiation was used as a model reaction to assess the performances of the studied catalysts. Increased photocatalytic activity was observed for Cu-modified layered perovskites compared to $RbLaTa_2O_7$ pristine. This behavior resulted from the efficient separation of photogenerated charge carriers and light absorption induced by copper spacer insertion.

Keywords: copper spacer; layered perovskite; phenol; photocatalysis; simulated solar irradiation

1. Introduction

Layered perovskites represent an emerging class of materials whose applications are constantly rising in various fields spanning from dielectrics [1], superconductivity [2], or luminescence [3] to catalysis and photocatalysis [4,5]. The development of stable, low-cost, and high-efficiency photocatalysts that utilize solar energy could solve many environmental concerns [6]. Among various semiconductors, Dion-Jacobson (DJ)-type layered perovskites are effective photocatalysts for water splitting under solar light irradiation [7]. Their interlamellar space allows for achieving stronger light absorption with more efficient charge carrier separation. In this way, the photocatalytic performance of the new composites dramatically increases.

The class of DJ perovskites possesses a composition of $M[A_{n-1}B_nO_{3n+1}]$, where M is an alkali metal, A is a lanthanide, B is a d^0 transitional metal, and n is the number of BO_6 octahedra. For example, the 2D blocks of $RbLaTa_2O_7$ are composed of corner-shared TaO_6 octahedra, interleaved with 12 coordinate La^{3+} cations, and separated by a Rb^+ interlayer cation. An essential feature of these materials consists of their ability to accommodate a variety of spacers into the inorganic matrix while the structure of perovskite blocks remains unchanged. Due to the layered structure flexibility, their physicochemical properties are tailored through ion exchange, pillaring, exfoliation, or restacking methods [8]. Compounds such as clay minerals, graphite, and transition metal dichalcogenides are known to accommodate guest species in their interlayer space. Therefore, a unique nanoarchitecture with attractive properties is obtained compared to those of the parental materials. Layered perovskites are amenable to alkali-proton exchange reactions to obtain solid acid phases. Protons located between galleries can be further exchanged with various molecules (alcohols, amines, phosphorous, alcohols, and carboxylic acids) leading to new organic-inorganic hybrid compounds [9,10]. Exfoliation into ultrathin nanosheets is feasible for the fabrication of nanostructured devices [11] or used as building blocks for novel materials [12]. It is worth noting that the size and configuration of the spacer cation influence the physicochemical properties of the photocatalysts [13]. The recent work by Minich et al. [14] revealed that the interlayer water plays a crucial role in the formation of n-alkylamines intercalated perovskite-like bismuth titanate $H_2K_{0.5}Bi_{2.5}Ti_4O_{13} \cdot H_2O$. Fan et al. [15] demonstrated that (Pt, TiO_2) intercalated $HLaNb_2O_7$ possesses superior photocatalytic H_2 evolution compared to unmodified TiO_2. Wu and coworkers [16] fabricated $Fe_2O_3/HLaNb_2O_7$ through an intercalation route, which showed an enhanced hydrogen production rate compared to individual oxides.

An approach to improve the charge separation efficiency in layered perovskites refers to noble metal intercalation at specific interlayer sites. As an example, RuO_2- [17] and Pt- [18] intercalated $KCa_2Nb_3O_{10}$ possess high photocatalytic performance for water splitting. Unfortunately, the high cost, the limited availability, as well as the non-uniform distribution of these guests restrict their practical application. Regarding the photocatalysis field, TiO_2 has served as a benchmark of light-related photocatalytic processes. However, its large band gap, the fast carrier recombination, or the particle agglomerations are factors that limit the TiO_2 efficiency. To overcome these drawbacks, searching for innovative functional materials for (i) expanding the light-harvesting wavelength range and (ii) suppressing the quick recombination of photogenerated charge carriers is highly desirable.

Copper oxide p-type semiconductors have become a promising alternative in the field of catalysis, photovoltaics, and energy storage applications [19]. These materials possess outstanding properties, such as narrow band gap, non-toxicity, and low price. CuO has a band gap of 1.21–2.1 eV and a monoclinic crystal structure, while Cu_2O has a band gap of 2.2–2.9 eV and a cubic crystal structure [20]. Alternatively, the interlayer gallery of the layered perovskites provides a space-confined environment for the exchangeable ions.

One strategy for developing visible-light-driven photocatalysts is to create a new valence band (VB) located at more negative potentials than O 2p orbitals, giving rise to a narrower band gap [21]. Recently, new visible-light-driven H_2- and O_2- evolving photocatalysts have been developed by Ag^+ and Cu^+ alkali-ion exchanges in various wide-band-gap perovskites [22]. Similarly, Cu(I)-ion-exchanged $K_2La_2Ti_3O_{10}$ photocatalysts demonstrated enhanced activity for H_2 production with a response up to 620 nm [23]. Therefore, engineering the interlayer spacing of lamellar materials is a reliable route to improve photocatalytic performance.

Phenolic compounds are a class of ubiquitous contaminants frequently present in wastewater due to their employment in industry and human activities [24]. Alternative approaches have been developed over the last few decades to remove persistent organic pollutants [25]. The photocatalytic mineralization process remains one of the most attractive technologies for the degradation of recalcitrant contaminants contained in water sources into harmless CO_2 and H_2O. Its efficiency is due to reactive oxygen species generated by the photocatalyst during light exposure that degrade the hazardous compounds.

We report a versatile route to fabricate a novel $Cu(LaTa_2O_7)_2$ layered architecture through the modification of the $RbLaTa_2O_7$ host via protonation-amine intercalation steps. One aim of this work was to explore the nature and role of copper species (interlayer and/or surface) on the physico-chemical properties of the $Cu(LaTa_2O_7)_2$ matrix. The ultimate goal was to study the photocatalytic degradation of phenol under simulated solar irradiation over the new Cu(LaTa$_2$O$_7$)$_2$ assemblies, depending on their thermal treatments.

2. Results and Discussion

2.1. Crystal Structure and Morphology

The XRD patterns for the pristine layered perovskite (RbLTO), proton-exchanged form (HLTO), n-butylamine-derived perovskite (BuALTO), copper-intercalated perovskite (CuLTO), copper-intercalated perovskite calcined at 500 °C (CuLTO-500C), and copper-intercalated perovskite reduced at 800 °C (CuLTO-800R) were collected in Figure 1. The lattice parameters and the phase composition of the synthesized materials are listed in Table 1.

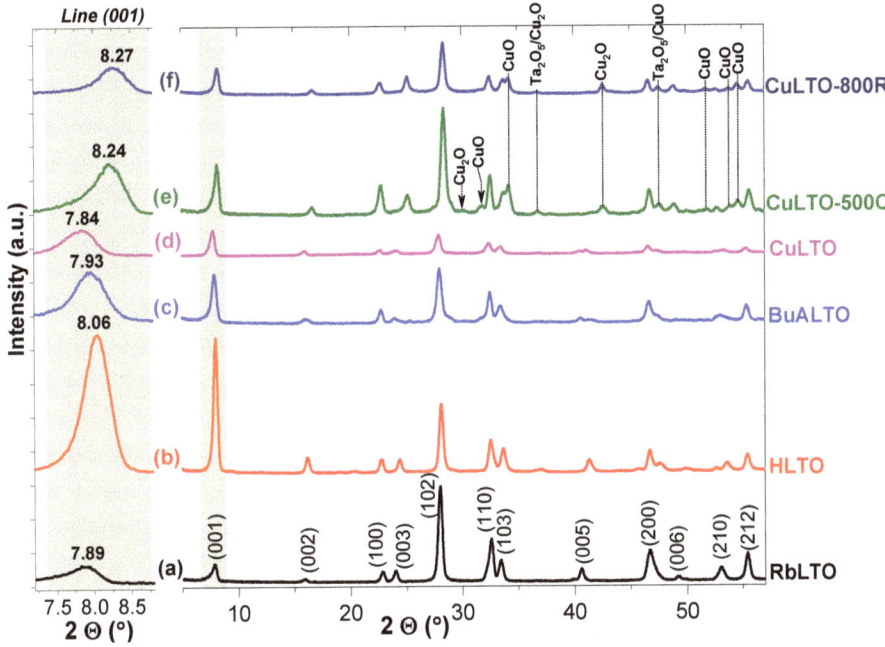

Figure 1. XRD patterns of (**a**) RbLTO, (**b**) HLTO, (**c**) BuALTO, (**d**) CuLTO, (**e**) CuLTO-500C, (**f**) CuLTO-800R. In brackets are given the Miller indices of the RbLaTa$_2$O$_7$ phase.

Table 1. Lattice parameters, d-spacing, gallery height, and phase composition for the RbLTO host and its guest-intercalated products.

Sample	Lattice Parameters		d-Spacing for l$_{(001)}$, (Å)	Gallery Height (Å)	Phase Composition
	a (Å)	c (Å)			
RbLTO	3.881	11.121	11.185	3.69	RbLaTa$_2$O$_7$
HLTO	3.883	10.924	10.951	3.45	HLaTa$_2$O$_7$
BuALTO	3.889	11.113	11.113	3.61	(C$_4$H$_{11}$N)LaTa$_2$O$_7$
CuLTO	3.888	11.023	11.228	3.73	Cu(LaTa$_2$O$_7$)$_2$
CuLTO-500C	3.880	10.718	10.719	3.22	Cu(LaTa$_2$O$_7$)$_2$, Cu$_2$O, CuO, Ta$_2$O$_5$
CuLTO-800R	3.884	10.676	10.676	3.18	Cu(LaTa$_2$O$_7$)$_2$, Cu$_2$O, CuO, Ta$_2$O$_5$

The XRD pattern of the RbLTO solid shows the RbLaTa$_2$O$_7$ structure with tetragonal symmetry (P4/*mmm*, cell parameters a = 3.881 and c = 11.121 Å, PDF card no. 01-089-0389). Acid treatment of pristine produced the HLaTa$_2$O$_7$ phase (PDF card no. 01-081-1194), in accordance with previously reported data for proton-exchanged compounds. The protonation step left the a parameter unchanged, while the (001) diffraction line was shifted towards higher 2Θ angles (8.06°). This fact indicates a smaller c parameter for the HLTO solid. Subsequent treatment of the proton-exchanged perovskite with *n*-butylamine (*n*-C$_4$H$_9$-NH$_2$) leads to a shift of the (001) reflection at a lower 2Θ angle (7.93°) in BuALTO solid. This shift is a consequence of the amine's intercalation into the protonated perovskite's interlayer spaces. Onward, the intercalation of the Cu^{2+} spacer into the BuALTO solid resulted in a further shift of the (001) line to lower angles (2Θ = 7.84°). This observation indicates that copper was introduced in between the [LaTa$_2$O$_7$]$^-$ layers. The benefit of using the *n*-butylamine as an intercalation precursor is that of forming a complex with the metallic ion. From this standpoint, [Cu(NH$_2$-C$_4$H$_9$)$_4$(H$_2$O)$_2$]$^{2+}$ can be considered as a vector to incorporate copper into perovskite layers.

The XRD patterns of both CuLTO-500C and CuLTO-800R reveal the preservation of the stratified structure but also the presence of distinctive lines attributed to Cu$_2$O (PDF card no. 01-078-2076), CuO (PDF card no. 01-089-5895), and Ta$_2$O$_5$ (PDF card no. 00-018-1304) crystalline phases. Figure 2B shows that either the calcination or reducing treatments of CuLTO solid resulted in the dramatic shortening of the *d*-distance. Thus, values of 10.719 Å for CuLTO-500C and 10.676 Å for CuLTO-800C were observed in comparison to the pristine layered perovskite.

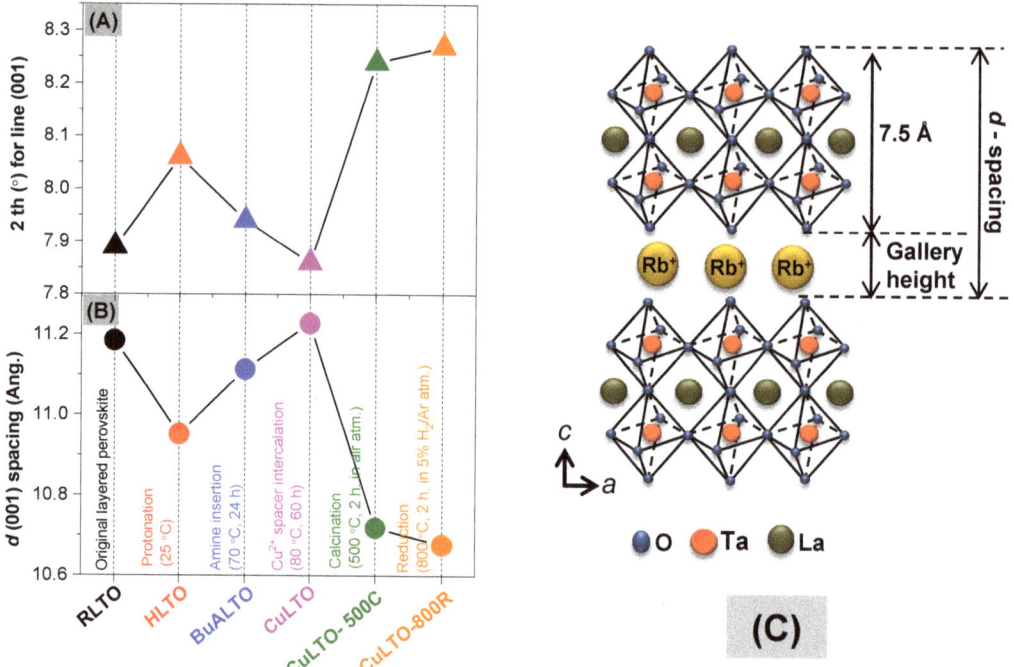

Figure 2. Evolution of the (**A**) 2-theta and (**B**) *d*-spacing of the RbLTO host and its guest-intercalated spacers. The layered perovskite structure of RbLaTa$_2$O$_7$ is schematically given in (**C**).

The nature of the spacer influences the overall structure of the newly assembled compound as it can distort the perovskite from its ideal structure through the tilting of the oxygen octahedra, the shifting of the layers, or Jahn-Teller distortions [26]. The shortening of the c-axis of the CuLa$_2$Ti$_3$O$_{10}$ compound was observed by Hyeon et al. after the solid was heated up to 700 °C [27]. The authors attributed this contraction to the flattening of the CuO$_4$ unit toward quasi-square planar coordination. Ogawa and Kuroda [28] reported that the alkyl-ammonium ions intercalated layered silicates are aligned parallel to the silicate sheets. Similarly, Sasaki et al. [29] indicated two modes of accommodation of the alkyl chain into the titanate layered structure. For a chain length up to the carbon number of three (C$_3$), a nearly parallel arrangement has been suggested, while for C$_5$ or more, the interlayer distance expanded linearly with an increase in the chain length. These authors claimed that in the case of butylamine, both types of arrangement coexisted (flat or perpendicular). In another study regarding layered titanate-intercalated pyridine, two values of the d-spacing were reported: (i) a value of 3 Å when pyridine was inserted flat into the interlayer space and (ii) 5.8 Å when pyridine was intercalated perpendicular to the surface [30]. The chain length of the butylamine has a value of 7.6 Å, and its alkyl chain usually tilts or overlaps [31].

Collectively, our results appear to be consistent with the literature review. We suppose that n-butylamine was inserted in between the layers of the BuALTO solid in such a way, probably in a flat orientation, that it does not significantly expand the d_{001} spacing.

Figure 2A displays the variation of the d-spacing and 2-theta angle for the (001) line diffraction over the synthesized materials attributed to the RbLaTa$_2$O$_7$ host structure (Figure 2B).

The gallery height was calculated by the difference between the d values derived from the (001) reflection peaks of all the solids and the perovskite layer thickness (~7.5 Å) [16]. Table 1 gives the interlayer heights of the RbLTO host and its guest-intercalated spacers. The results underline that the gallery height of the solids is spacer-related.

The chemical reactions that take place after each synthesis step are described in Equations (1)–(4):

$$Rb_2CO_3 + La_2O_3 + 2Ta_2O_5 \rightarrow 2RbLaTa_2O_7 + CO_2\uparrow \text{ (Layered host synthesis)} \quad (1)$$

$$RbLaTa_2O_7 + HNO_3 \rightarrow HLaTa_2O_7 + RbNO_3 \text{ (Step I: Protonation)} \quad (2)$$

$$HLaTa_2O_7 + C_4H_9\text{-}NH_2 \rightarrow (C_4H_9\text{-}NH_3)LaTa_2O_7 + 1/22\ H_2O \text{ (Step II: n-Butylamine insertion)} \quad (3)$$

$$4[(C_4H_9\text{-}NH_3)(LaTa_2O_7)] + 2Cu(NO_3)_2 \rightarrow [Cu(NH_2\text{-}C_4H_9)_4(H_2O)_2](LaTa_2O_7)_2 + Cu(LaTa_2O_7)_2 + 2NO_2 + 2H_2O \quad (4)$$
$$\text{(Step III: Copper intercalation)}$$

The methodology of replacing Rb$^+$ with Cu^{2+} in the RbLaTa$_2$O$_7$ interlayer host proceeds in three steps (Scheme 1). The first step involves the exchange of the interlayer rubidium cation with H$^+$ via protonation (step I, Equation (2)). The next stage consists of the expansion of the galleries of proton-exchanged layered perovskite by introducing a spacer such as n-butylamine (step II, Equation (3)). The role of the alkylamine is not only to increase the interlayer distance but also to form a [Cu-amine] complex, thus favoring the intercalation of Cu into the perovskite host (step III, Equation (4)).

Thermogravimetric (TG) analyses additionally confirm the various spacers inserted into the RbLaTa$_2$O$_7$ host. The TG curves of the HLTO, BuALTO, CuLTO, CuLTO-500C, and CuLTO-800R materials showed several stages of mass losses as a function of temperature. The obtained results are listed in Table S1 from supporting information. For comparison, the TG curve of RbLTO pristine is also given in Figure 3A, while its description was presented elsewhere [32]. The first peak centered at 134 °C is due to the elimination of adsorbed water, while the second peak at 305 °C corresponds to a partial dehydration of lanthanum oxide. After 600 °C, take place the elimination of CO$_2$.

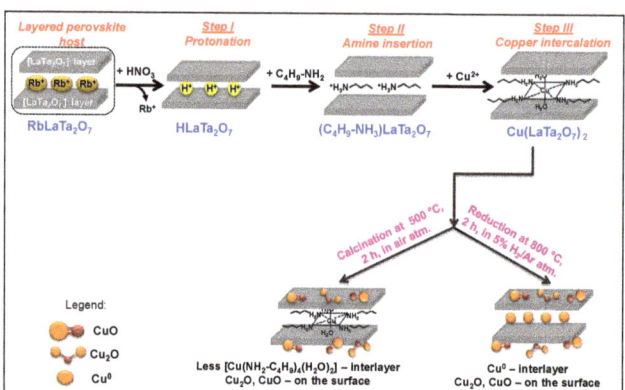

Scheme 1. Schematic pathway of Cu(LaTa$_2$O$_7$)$_2$ synthesis followed by the two different thermal treatments.

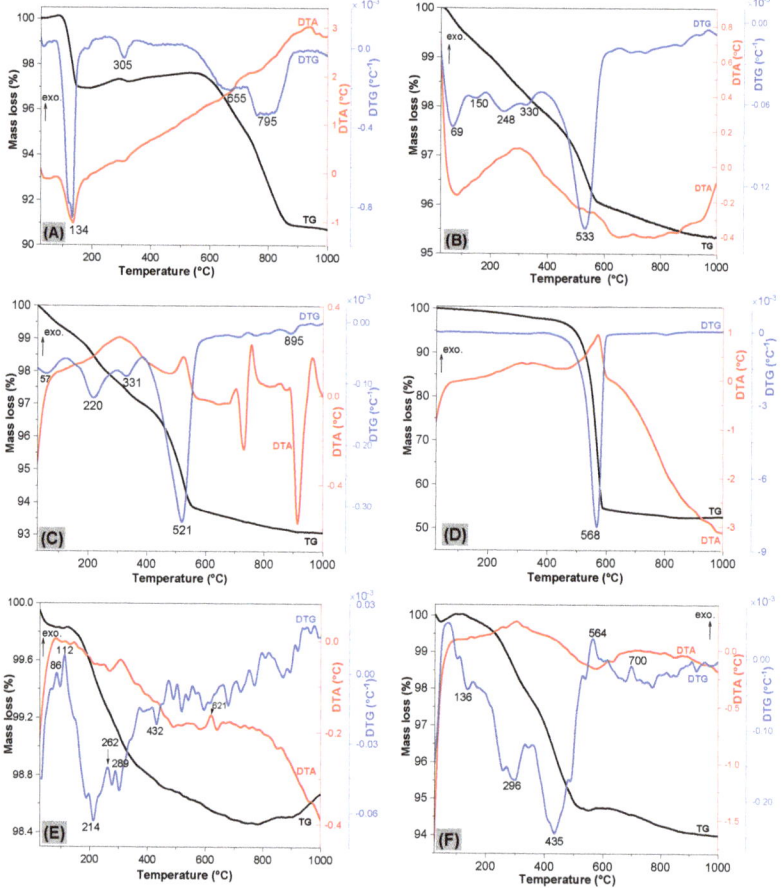

Figure 3. Thermogravimetric curves (TG, DTG, DTA) of (**B**) HLTO, (**C**) BuALTO, (**D**) CuLTO, (**E**) CuLTO-500C, and (**F**) CuLTO-800R, measured in an air atmosphere with a heating rate of 10 °C·min^{-1}. The TG curve of the RbLTO host is given for comparison in (**A**) [32].

For the HLTO sample, (Figure 3B) water elimination from 25 to 380 °C in four steps is associated with the removal of the adsorbed water. The experimental value of the measured weight loss (2.00%) was close to the theoretical value of 2.93%. The calculated value corresponds to 1 mole H_2O per $[LaTa_2O_7]^-$ unit formula of protonated perovskite. The mass loss of 1.37% (peak at 533 °C) is due to the loss of the interlayer proton as water, according to Equation (5). The last step on the TG curve (650–1000 °C) indicates negligible mass loss (0.32%), showing that the deprotonation process is completed up to 650 °C.

$$HLaTa_2O_7 \rightarrow LaTa_2O_{6.5} + 1/2\, H_2O \qquad (5)$$

The theoretical weight loss should be 1.48% if all Rb^+ cations replaced by protons are located in the interlayer space. According to these data, a protonation degree of 93% has been achieved in the HLTO solid. The successful protonation step was also confirmed by ion chromatography showing that 99.2% Rb^+ cations were replaced by H^+ during the acid treatment of the $RbLaTa_2O_7$ host with nitric acid. This result is consistent with reports in the literature for other related protonated layered perovskites.

Thermal analysis attests that n-butylamine was intercalated and helps to quantify the amount of the inserted amine. The first mass loss (25–124 °C) of the BuALTO (Figure 3C) is attributed to the elimination of adsorbed water. In the temperature range of 124–1000 °C, the total weight loss of 6.36% is assigned to the elimination of amine, which takes place in multiple steps. In the study of Geselbracht et al. [33], the total mass loss of the hexylamine-intercalated $HCa_2Nb_{3-x}Ta_xO_{10}$ has been calculated in the domain of 25–900 °C. In the research work on octylamine-intercalated layered vanadium oxide, the authors ascribed the weight loss at low-temperature to the release of free amine molecules, while the higher-temperature one was attributed to the combustion of interlayer amine [34]. For this study, the complete decomposition of the inserted n-butylamine takes place in multiple steps. The peak at 521 °C observed on the TG curve of the BuALTO solid certifies that the butylamine is located in the interlayer position. Its experimental mass loss was determined to be 3.61%. The calculated weight loss for the amine total decomposition should be 10.64%, while a lower experimental value was found. It implies that only 33.92% of the butylamine was located in between the $[LaTa_2O_7]^-$ layers.

After the insertion of the Cu^{2+} spacer (Figure 3D), the metal-ligand complex $[Cu(NH_2\text{-}C_4H_9)_4(H_2O)_2]$ is formed. The large mass loss of 43.39% observed for CuLTO is indicative of [Cu-amine] complex decomposition with the elimination of gaseous NO_2, CO_2, and NH_3 compounds, Equation (6).

$$\left[\begin{array}{c} H_9C_4\text{-}H_2N \diagdown \overset{H_2O}{\underset{|}{Cu}} \diagup NH_2\text{-}C_4H_9 \\ H_9C_4\text{-}H_2N \diagup \underset{H_2O}{\overset{|}{}} \diagdown NH_2\text{-}C_4H_9 \end{array}\right]^{2+} \xrightarrow{Temp.} CuO + 2NO_2\uparrow + 4CO_2\uparrow + 4NH_3\uparrow + 2H_2O \qquad (6)$$

The TG curves of the CuLTO-500C and CuLTO-800R samples (Figure 3E,F) show different behavior compared to the HLTO, BuALTO, and CuLTO layered perovskites. Various exothermic effects observed on the DTA curves of both thermally treated materials could signify a possible phase transition. Indeed, the presence of copper phases on CuLTO-500C and CuLTO-800R was subsequently probed by XRD analysis (Figure 1).

SEM images of RbLTO, HLTO, BuALTO, and Cu-modified layered perovskites are shown in Figure 4A–E. The well-preserved lamellar structure of the starting material is observed for all studied architectures. The synthesized Rb- and H- forms do not show significant particle size differences. On the other hand, the particles of BuA- and Cu-intercalated products tend to form lamellar agglomerates with a smaller size as compared to RbLTO's original layered perovskite. The transmission electron microscopy (TEM) images further confirm that both n-butylamine- and Cu-based layered perovskites (Figure 5A–F) retain their stratified structures.

Figure 4. SEM images of (**A**) RbLTO, (**B**) HLTO, (**C**) BuALTO, (**D**) CuLTO, (**E**) CuLTO-500C, and (**F**) CuLTO-800R materials.

The elemental mapping images (Figure 6) show that in the BuALTO and CuLTO compounds, La, Ta, and O are homogeneously distributed within each sample. No presence of rubidium was observed. This observation demonstrated that the cation exchange took place. The presence of a copper peak in the CuLTO novel architecture (Figure 6B) demonstrates the successful Cu intercalation process.

Table 2 reports the surface chemical composition of RbLTO, CuLTO, CuLTO-500C, and CuLTO-800R perovskites as determined by XPS analysis. The results illustrate that the surface composition of the solids changes, depending on the applied thermal treatment (as fresh, calcination, and reduction). Notably, the bulk of the Cu-modified perovskites is constantly preserved as a $Cu(LaTa_2O_7)_2$ crystalline phase. This observation is confirmed by the XRD patterns (Figure 1).

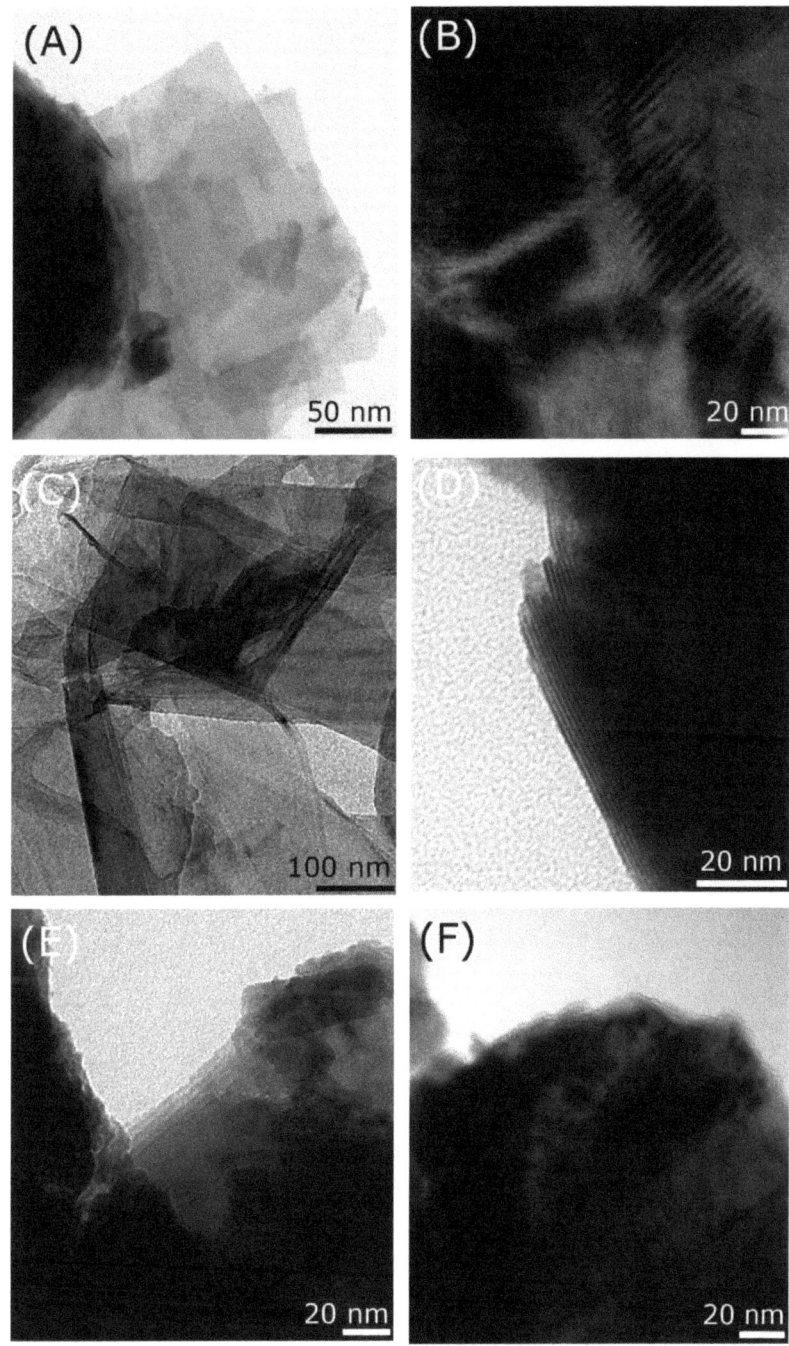

Figure 5. TEM images of (**A**,**B**) BuALTO, (**C**,**D**) CuLTO, (**E**) CuLTO-500C, and (**F**) CuLTO-800R.

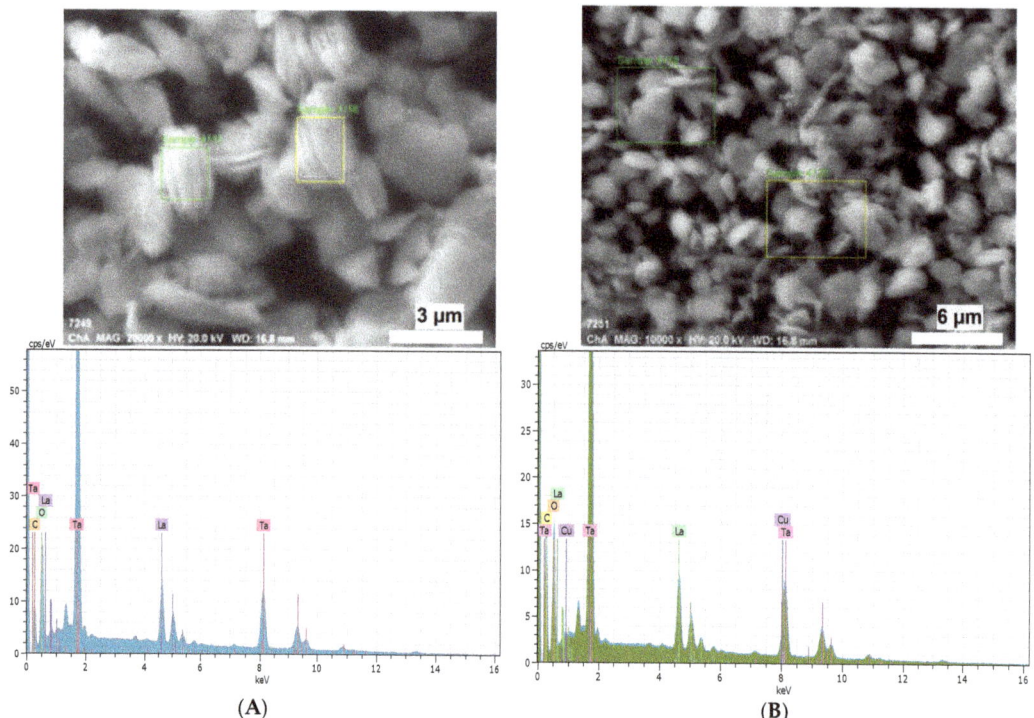

Figure 6. SEM-EDS elemental mapping images of (**A**) BuALTO and (**B**) CuLTO lamellar structures for O, La, Ta, and Cu elements.

Table 2. Surface chemical composition and atomic ratio determined by XPS analysis for RbLTO host and its guest-intercalated products.

Sample	Chemical Analysis of the Products (at. %)						Atomic Ratio
	Rb (%)	C (%)	O (%)	La (%)	Ta (%)	Cu (%)	Cu/(La + Ta)
RbLTO	10.0	39.6	36.5	5.4	8.5	0.0	0.00
CuLTO	0.0	67.0	32.9	0.0	0.0	0.1	0.00
CuLTO-500C	0.0	33.3	58.4	1.5	5.6	1.3	0.18
CuLTO-800R	0.0	39.3	45.1	3.1	8.9	3.5	0.29

The Cu/(La + Ta) surface atomic ratio of Cu-intercalated layered perovskites is slightly decreased compared to the theoretical value of the $RbLaTa_2O_7$ host [i.e., Rb/(La + Ta) = 0.33]. Note that there is a coverage of the CuLTO surface by the organic component.

2.2. Optical Absorption of Samples by UV–Vis Spectroscopy

The optical properties of the RbLTO host and its modified compounds (HLTO, BuALTO, CuLTO, CuLTO-500C, and CuLTO-800R) were studied by UV–Vis spectroscopy (Figure 7A). Moreover, this is a helpful tool to understand the coordination environment of Cu^{2+} ions contained in the lamellar space of the perovskite host.

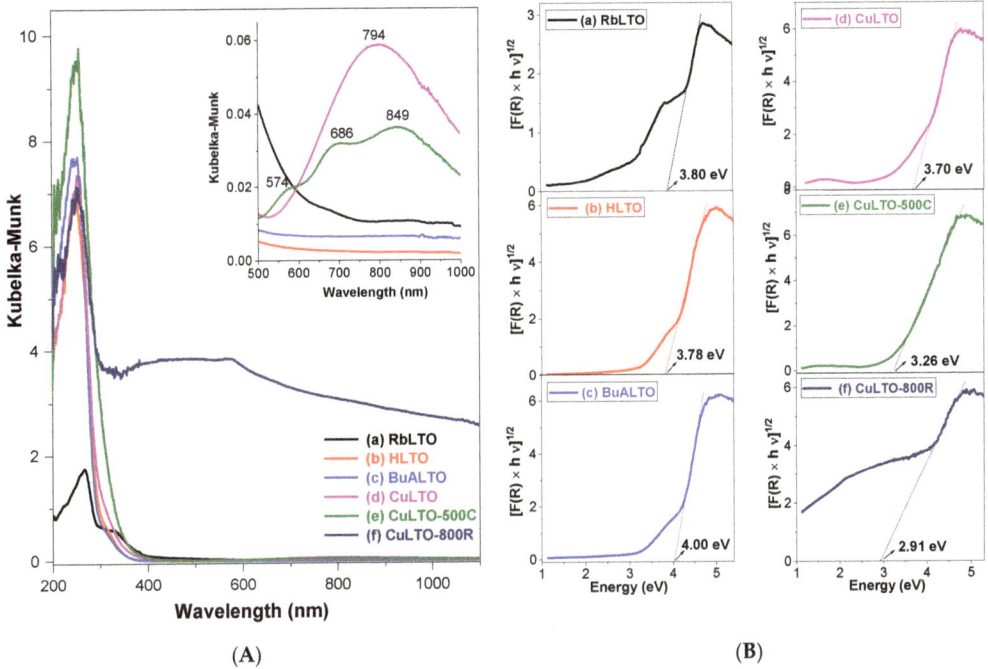

Figure 7. (**A**) UV–Vis absorption spectra and (**B**) Tauc plot of the (a) RbLTO, (b) HLTO, (c) BuALTO, (d) CuLTO, (e) CuLTO-500C, and (f) CuLTO-800R.

RbLTO displays an absorption edge at 268 nm with a small shoulder at about 331 nm, corresponding to the electron excitation of the O 2p valence band to the Ta 5d conduction band. The intercalation of various spacers into the perovskite host layers induces an increase in light absorption. An absorption edge at about 252 nm is observed for the H$^+$-, n-butylamine-, and Cu^{2+}-intercalated perovskites. Notably, copper insertion determines significant changes in UV–Vis spectra in the visible absorption domain (inset of Figure 7A). The CuLTO shows an absorption hump between 520 and 950 nm assigned to the d–d electronic transition of Cu^{2+} ions [35]. After calcination at 500 °C, three absorption edges at 573, 686, and 849 nm were perceived. This observation indicates that the copper environment has changed for CuLTO-500C. A reduction at 800 °C leads to a broader and less intense band between 400 and 580 nm in the visible domain.

Different authors claimed that if Cu^{2+} ions are in perfect octahedral coordination, the d–d transition band will appear between 750–800 nm. If Cu^{2+} cations are in a distorted octahedral (nearly square planar) configuration, this band will have a blue shift to 600–750 nm. Researchers assigned the broadband between 400 and 450 nm to the charge transfer band for either single- or double-O-bridged copper pairs [36,37].

In line with previous literature studies, it appears that for the CuLTO solid, the Cu^{2+} ion is in perfect octahedral coordination, confidently attributed to the complexion of copper ions with the butyl-amino ligands [38]. After calcination at 500 °C (CuLTO-500C sample), the presence of two Cu^{2+} species could be observed: (i) copper surrounded by amino-ligand and (ii) copper surrounded by oxygen due to surface CuO$_x$ species. A reduction at 800 °C (i.e., CuLTO-800R sample) leads to considerable increases in light absorption compared to the CuLTO and CuLTO-500C solids, which may be due to the charge transfer band for either the Cu-O-Cu or ⟨Cu(O)₂Cu⟩ pairs and the metallic copper [36,37].

The optical band gap energies (Eg) of all photocatalysts are presented in Figure 7B. The Eg was estimated by Tauc curves corresponding to indirect band transitions. The Eg values of all layered perovskites follow the sequence: CuLTO-800R (2.91 eV) < CuLTO-500C (3.26 eV) < CuLTO (3.70 eV) < HLTO (3.78 eV) < RbLTO (3.80 eV) < BuALTO (4.00 eV). Overall, the RbLaTa$_2$O$_7$ host can accommodate various molecules in between layers, enhancing light absorption. The significant decrease in CuLTO-500C and CuLTO-800R band gaps impacts the photocatalytic reaction and the selectivity, as will be seen later.

2.3. Tauc Plot Derived from Photocurrent Spectroscopy (PCS)

A similar method to UV–Vis spectroscopy to estimate the optical band gap of semiconductors is photocurrent spectroscopy. For this purpose, a Tauc plot of $(J_{ph}/\Phi \times h\upsilon)^n$ versus wavelength was represented, where J_{ph} is the photocurrent, Φ is the photon flux, and n is $\frac{1}{2}$ due to indirect transition [39]. A straight line was fitted at the lower energy region to determine the optical band gap of the layered materials. Figure 8A illustrates the allowed indirect band gap Tauc plots of photoanodes using the normalized photocurrent response. The estimated band gap energies of RbLTO, CuLTO, CuLTO-500C, and CuLTO-800R are 3.51, 3.44, 3.06, and 2.88, respectively. These values imply that the insertion of the copper spacer into the RbLaTa$_2$O$_7$ host leads to band-gap shrinking.

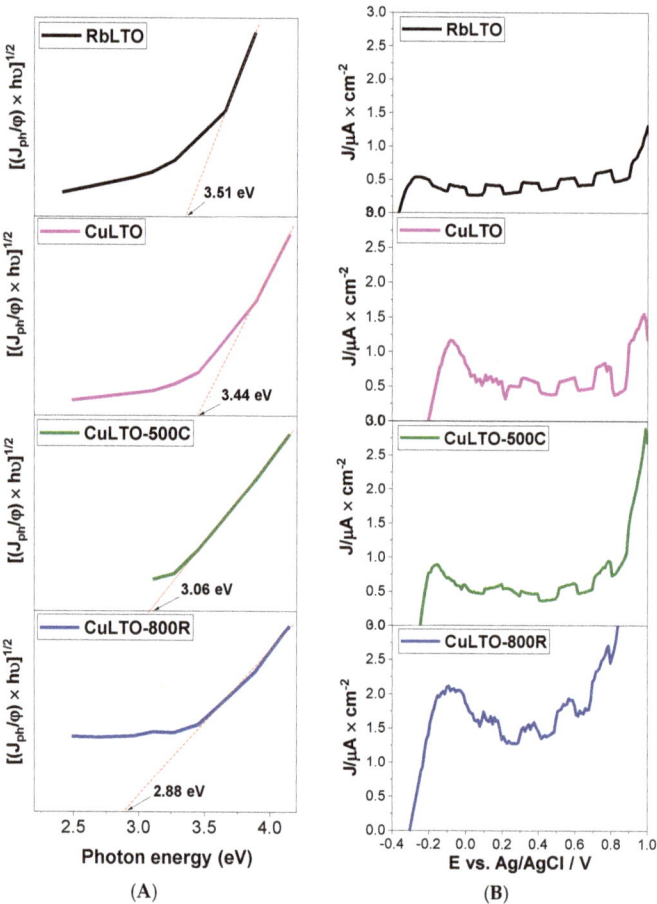

Figure 8. (**A**) Allowed indirect band gap Tauc plots of RbLTO, CuLTO, CuLTO-500C, and CuLTO-800R photoanodes and (**B**) LSV plots under chopped AM 1.5 simulated solar light.

The linear sweep voltammetry (LSV) plots under chopped AM 1.5 simulated solar irradiation over RbLTO, CuLTO, CuLTO-500C, and CuLTO-800R photoanodes are compared in Figure 8B. The unmodified RbLTO exhibits the lowest photocurrent density. The LSV response increases for the Cu-modified layered perovskites, while the highest on/off photocurrent response appears on CuLTO-800R. All subjected materials showed a clear characteristic of n-type semiconductors.

2.4. FTIR Absorption Spectra of the Layered Perovskites

The functional groups in RbLaTa$_2$O$_7$ original layered perovskite and its spacer-modified compounds were studied by FTIR spectroscopy (Figure 9). In the RbLTO host, the band corresponding to Ta-O-Ta was identified at 550 cm^{-1}. The strong peaks at 613 and 651 cm^{-1} are due to Ta-O asymmetric stretching vibrations (ν_{as} Ta-O), while the band at 903 cm^{-1} represents the Ta-O symmetric stretching vibration (ν_s Ta-O) of TaO$_6$ octahedra [40]. Rb$^+$/H$^+$ exchange determines one broad band at around 3379 cm^{-1} and a second one at 1638 cm^{-1} corresponding to the asymmetric stretching (ν_{as}O-H) and bending mode (δ_sO-H) of hydroxyl groups. The band at 934 cm^{-1} is related to hydroxyl groups of protonated materials. The protonation step induces a shift to the lower wavenumber of all vibrations related to Ta-O. The introduction of the *n*-butylamine spacer induces important changes in the FTIR spectra of the BuALTO solid. Thus, this sample displays new vibrations allocated to the CH$_2$ asymmetric stretching band (2964 cm^{-1}), the CH$_2$ symmetric stretching band (2853 cm^{-1}), and the bending vibrations of the CH$_2$ groups (1443–1416 cm^{-1}). These bands are abbreviated as ν_{as}CH$_2$, ν_sCH$_2$, and δCH$_2$, respectively. Additionally, signals of the N-H stretching vibration (3026–3171 cm^{-1}) and the C-N bond (1149–1205 cm^{-1}) are observed [41] confirming the insertion of amine in between the [LaTa$_2$O$_7$]$^-$ layers. Absorbance below 1000 cm^{-1} reveals two broad bands at 903 and 804 cm^{-1} associated with N-H out-of-plane vibration, specific to primary amines [42]. Notably, the band at 903 cm^{-1} overlaps over the symmetric vibration of Ta-O.

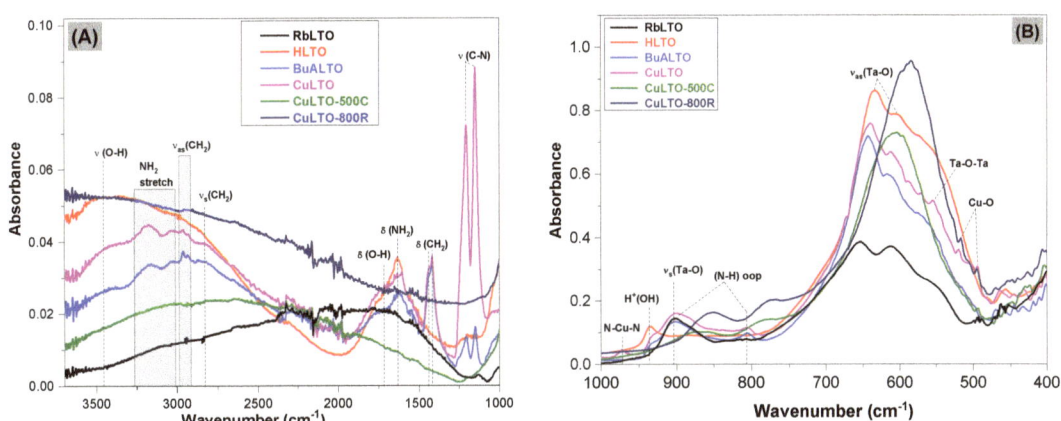

Figure 9. FTIR spectra at the (**A**) high-wavenumber and (**B**) low-wavenumber regions for (a) RbLTO, (b) HLTO, (c) BuALTO, (d) CuLTO, (e) CuLTO-500C, and (f) CuLTO-800C.

After Cu^{2+} insertion (CuLTO sample), the signal related to the C-N stretching vibration (1149–1205 cm^{-1}) increases significantly. Alternatively, a new small band develops around 973 cm^{-1} attributed to N-Cu-N stretching vibration [43]. As a matter of fact, the positively charged Cu^{2+} ions act as a Lewis acid while the (*n*-C$_4$H$_9$-NH$_3^+$) ligand with one pair of electrons acts as a Lewis base, and consequently, it has a great tendency to form [Cu(NH$_2$-C$_4$H$_9$)$_4$(H$_2$O)$_2$]$^{2+}$ complex. Undeniably, the TG analysis of CuLTO confirms the presence of the [copper-amine] complex. Moreover, the UV–Vis spectra certify the existence of Cu^{2+} in a perfect octahedron configuration due to the interlayer [Cu-amine] complex. The insertion of spacers determines a redshift of the asymmetric TaO_6 octahedra with 26 cm^{-1} for HLTO, 14 cm^{-1} for BuALTO, and 19 cm^{-1} for CuLTO, respectively, as compared to the RbLTO host. In order to verify the occurrence of interactions between the layered materials and the amine, the FTIR spectrum of neat butylamine is given in the supporting information, Figure S1. It is noticed that the band at 1079–1133 cm^{-1} for both the BuALTO and CuLTO solids is shifted towards higher frequencies compared to neat C_4-amine.

After the different thermal treatments (i.e., calcination/reduction) were applied to a CuLTO fresh sample, all IR bands disappeared in the wavenumbers domain higher than 1000 cm^{-1}. A close view of the FTIR fingerprint domain (Figure 9B) shows the occurrence of new bands at 851 cm^{-1} for CuLTO-500C and 774 cm^{-1} for CuLTO-, compared to the BuALTO and CuLTO samples. It implies that the copper environment has changed in the thermally treated solids. Additionally, the lattice vibration peaks referred to the asymmetric stretching mode of Ta-O converged to one band at 602 cm^{-1} for CuLTO-500C and 585 cm^{-1} for CuLTO-800R. This fact could be attributed to a rearrangement of the interlayer species accompanied by the disordering of the $[LaTa_2O_7]^-$ sheets.

The literature data covering Cu-based materials showed that the bands positioned at 578 cm^{-1} and 435 cm^{-1} are due to υ(Cu-O) [44]. Gopalakrishnan et al. [45] indicate that the Cu-O vibration of Cu_2O nanoparticles appears at about 618 cm^{-1}. Morioka et al. [46] observed small peaks at 428, 503, and 609 cm^{-1} but it has been unclear whether these bands corresponded to CuO and/or Cu_2O. In our preparation, two small bands appear (i) at 495 cm^{-1} for CuLTO, CuLTO-500C, and CuLTO-800R and (ii) at about 516 cm^{-1} for the CuLTO and CuLTO-800R solids. Therefore, this observation witnessed the successful copper spacer insertion into the perovskite structure into the above-stated materials.

2.5. Temperature-Programmed Reduction (TPR) Measurements

To further confirm the successful insertion of a Cu^{2+} spacer into the perovskite galleries, the optical properties were corroborated with those of the H_2-TPR and XPS studies. We focused in particular on the copper chemical state in the assembled Cu-based layered perovskites. The H_2-TPR profiles for CuLTO, CuLTO-500C, and CuLTO-800C catalysts are shown in Figure 10A. The reduction of CuO_x takes place via sequences (Equations (7) and (8)):

$$2CuO + H_2 \rightarrow Cu_2O + H_2O \qquad (7)$$

$$Cu_2O + H_2 \rightarrow 2Cu + H_2O \qquad (8)$$

All Cu-modified photocatalysts show two distinct regions in the temperature ranges of 50–450 °C and 450–800 °C, respectively. The CuLTO solid displays three separate reduction peaks, with the maxima occurring at 220, 313, and 536 °C, respectively. The first two low-temperature peaks represent the reduction of Cu^{2+} to Cu^+ and Cu^+ to Cu^0 species situated on the catalyst surface [47]. The third large and intense peak is ascribed to the reduction of copper species, (i.e., Cu^{2+} to Cu^0) located between the perovskite layers [48–50]. Similar results were found by Nestroinaia et al. [51] for Ni-intercalated layered double hydroxide, where the peak between 400–550 °C was attributed to the reduction in Ni^{3+} existing inside brucite-like layers. Hence, the TPR peak at 536 °C of the CuLTO catalyst confidently attests that copper intercalated between the $[LaTa_2O_7]^-$ layers. The calcination treatment over the Cu-fresh solid shifts the reduction peaks towards higher temperature values. Thus, in the TPR profile of the CuLTO-500C, a low-temperature broadened peak is observed along with a high-temperature distinctive one. The broad peak deconvoluted into two peaks was attributed to Cu^{2+}/Cu^+ (280 °C) and Cu^+/Cu^0 (448 °C) surface species. The shoulder positioned at 545 °C corresponded to a one-step reduction of Cu^{2+}/Cu^0 interlayer species. The CuLTO-800R material shows a broad peak over the studied temperature range, deconvoluted in four peaks. The low-temperature region (peaks at 194 and 371 °C) indicates the two-step reducibility of surface Cu^{2+}. The third peak, centered at 532 °C, is assigned to interlayer copper while the peak at about 650 °C is ascribed to a partial reduction of some tantalum species in this sample. Previous reports indicated that the reduction of Ta_2O_5 requires temperatures of around 700 °C [52,53]. This means that the existence of metallic copper in CuLTO-800R solid could promote the reduction of tantalum species at lower temperatures. Generally, the surface Cu^+/Cu^0 redox couple is beneficial for photocatalytic applications relevant to the degradation of organic molecules [54].

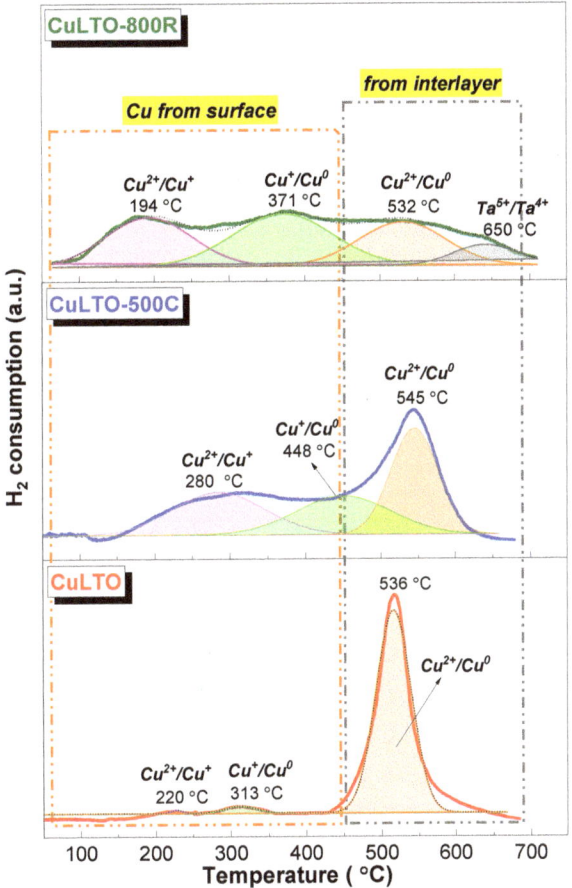

Figure 10. H_2-TPR profiles of CuLTO, CuLTO-500C, and CuLTO-800R catalysts.

The corresponding H_2 consumption values of Cu-intercalated layered perovskites, expressed in µmoles·g^{-1}, are reported in Table 3. For the CuLTO solid, the H_2 consumption by copper localized on the surface of the catalyst is quite small, confirming that the highest amount of copper is located in the interlayer. This observation is in line with the XPS data of CuLTO, which suggests that Cu^{2+} is present as bulk species rather than surface ones, as copper has a weak XPS signal (discussed later in the manuscript). In the case of the CuLTO-500C sample, a higher H_2 consumption value was observed for the copper localized on the surface, accompanied by a slight decrease in the H_2 uptake owing to the interlayer species. For the CuLTO-800R solid, only 66.58 µmoles·g^{-1} of H_2 consumption was calculated for the copper localized into perovskite-like layers. This observation indicates that the reduction process enhanced the migration of interlayer copper particles into highly dispersed surface Cu^+/Cu^0. The Cu^+/Cu^0 redox couple facilitates the separation of photogenerated carriers, thus improving the photocatalytic reaction [55].

Table 3. H$_2$ consumptions in H$_2$-TPR experiments.

Catalyst	H$_2$ Consumption (μmoles·g^{-1})			Total H$_2$ Consumption (μmoles·g^{-1})
	Cu from Surface		Cu from Interlayer	
	Cu^{2+}/Cu$^+$	Cu$^+$/Cu^0	Cu^{2+}/Cu^0	
CuLTO	11.64	44.25	662.16	718.05
CuLTO-500C	149.92	140.88	466.61	757.41
CuLTO-800R	91.37	88.54	66.58	246.49

Figure S2 shows the schematic copper species distribution on the surface and into the interlayer of the modified catalysts. Doubtless, in the CuLTO photocatalyst, most of the copper is localized between the galleries, and only a small amount of Cu remains on the catalyst surface. This observation is associated with TG analysis (Figure 3D), which shows that an interlayered [Cu-amine] complex formed in the above-stated sample. When CuLTO was air-calcined at 500 °C, copper species were located both on the surface and interlayer galleries. Indeed, the XRD pattern (Figure 1) validates the appearance of supplementary CuO$_x$ reflexions in CuLTO-500C. For the CuLTO-800R sample, the H$_2$-TPR study certifies that most of the copper species are emplaced on the catalyst surface, and only a small amount of copper remained in the perovskite layers. Based on these results, a few conclusions are necessary. Between the perovskite galleries, the metal-amine complex bridges the oxygen from the [TaO$_6$] octahedra. It may be possible that between the layers, there is also Cu^{2+} that comes from the Cu(LaTa$_2$O$_7$)$_2$ crystalline phase. After the calcination process, the decomposition of Cu(LaTa$_2$O$_7$)$_2$ to CuO$_x$ occurs, which migrates onto the catalyst surface. During this process, the decomposition of the [Cu-amine] complex from the interlayer also takes place. When the material is thermally treated at 800 °C in slightly reducing conditions, a higher migration of copper on the catalyst surface occurs.

2.6. XPS Measurements

X-ray photoelectron spectroscopy (XPS) analysis was carried out to further investigate the oxidation state and chemical composition of the newly assembled architectures. Figure 11A shows the XPS spectra of the Cu 2p$^{3/2}$ emission line for the CuLTO, CuLTO-500C, and CuLTO-800R solids. The C 1 s, O 1 s, La 3 d, and Ta 4 f emission lines of the Cu-modified layered perovskites are given in the supporting information, Figures S3 and S4.

The photoelectron profile of the Cu 2p$^{3/2}$ region for the CuLTO fresh solid was very weak. In fact, the H$_2$-TPR results confirmed that most of the Cu entities are located in the interlayer position for this sample, while the penetration depth for XPS detection is only several nanometers. Notice that the butylamine covered the surface of the above-mentioned sample. This is the reason no detectable copper peak can be observed on its XPS spectra. In CuLTO-500C material, there is a peak at a binding energy of 933.6 eV and a weak shake-up satellite at 943.6 eV. According to literature reports, the binding energy (BE) of Cu 2p$^{3/2}$ related to Cu^{2+} ions ranges between 932.8–933.6 eV (with shake-up satellite features at 940–945 eV) [56,57]. The obtained spectra agree with the literature data, confirming the presence of Cu^{2+} as a major species in the CuLTO-500C solid. After the treatment of CuLTO at 800 °C in a reductive atmosphere, the peak for Cu^{2+} disappeared, and one single peak at 932.8 eV was detected. The results reveal that in the CuLTO-800R solid, copper predominantly existed as Cu$^+$ and Cu^0 species. In line with TPR studies, the reducing conditions lead to a migration of Cu^{2+} from the interlayer space towards the surface of the catalyst in the form of Cu$^+$ and metallic copper.

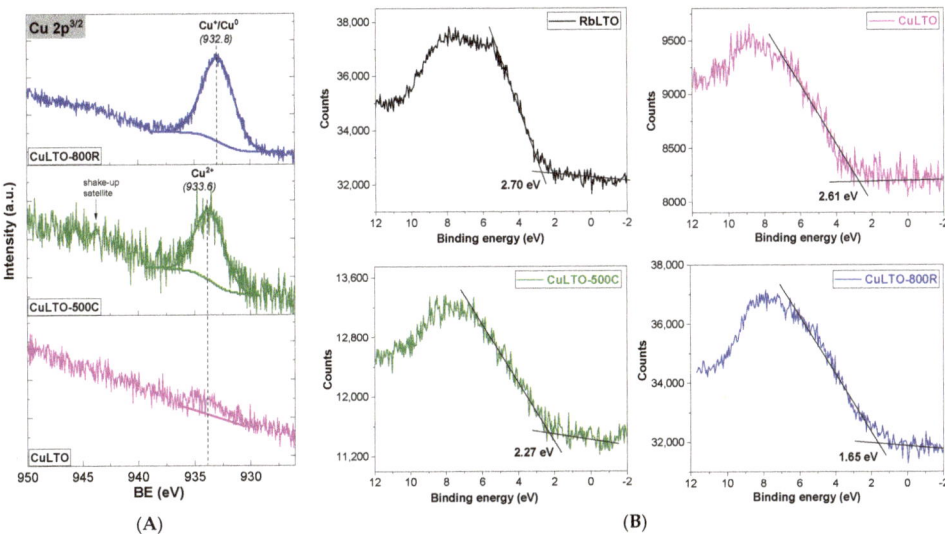

Figure 11. (**A**) XPS spectra for the Cu2p$^{3/2}$ emission line of (a) CuLTO, (b) CuLTO-500C, and (c) CuLTO-800R materials, and (**B**) valence band of unmodified RbLTO and Cu-modified perovskites.

The valence band (VB) spectra of the RbLTO, CuLTO, CuLTO-500C, and CuLTO-800R layered materials were also investigated by the XPS technique (Figure 11B). The VB values of the four samples were obtained by the linear extrapolation of the leading edge to the extended baseline of the VB spectra. Undeniably, copper insertion into the perovskite layered structures leads to the narrowing of the valence band compared to unmodified RbLTO material.

The band gap energy values derived from the UV–Vis, XPS, and photocurrent spectroscopies were compared in Figure 12. The Eg calculated by the three experimental techniques undoubtedly proves that copper intercalation into perovskite layers induces a red band-gap shift.

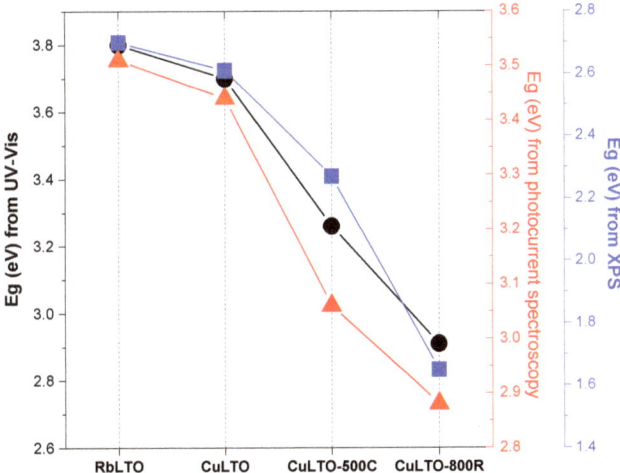

Figure 12. Comparison of the band gap values calculated from UV–Vis, XPS, and photocurrent spectroscopies for the RbLTO, CuLTO, CuLTO-500C, and CuLTO-800R materials.

3. Evaluation of Photocatalytic Activity

The physical features of the synthesized layered perovskites are expected to show different photocatalytic behavior. With this aim, the photodegradation of phenol (Ph) under simulated solar irradiation over the RbLTO, HLTO, CuLTO, CuLTO-500C, and CuLTO-800R photocatalysts was investigated as a model reaction. Table 4 gives the conversion of phenol after 4 h of reaction and the specific surface area over the studied photocatalysts.

Table 4. Photocatalytic results of the tested photocatalysts.

Photocatalyst	Phenol Conversion (%) [1]	SSA (m$^2 \cdot$g^{-1})
RbLTO	7.6	1.5
HLTO	8.5	2.6
CuLTO	13.0	4.1
CuLTO-500C	14.9	4.9
CuLTO-800R	16.6	3.4

[1] Phenol conversion after 4 h of reaction time (experimental conditions: 110 mL of 50 mg·L^{-1} phenol aqueous solution, 0.05 g catalyst, T = 18 °C, light source AM 1.5).

The conversion of phenol over an unmodified RbLTO photocatalyst was about 7.6% after 4 h of light irradiation. After Rb$^+$ replacement by H$^+$, the catalytic performance increased, reaching 8.5% of phenol conversion for the HLTO photocatalyst. The insertion of Cu^{2+} ions in between the [LaTa$_2$O$_7$]$^-$ slabs influenced the phenol conversion and its selectivity. The CuLTO photocatalyst improves its activity up to 13.0% phenol conversion after 4 h of reaction time. Additionally, the copper spacer has a beneficial role in the layered perovskite structure by (a) increasing its specific surface area from 1.5 m$^2 \cdot$g^{-1} for the RbLTO host up to 4.1 m$^2 \cdot$g^{-1} for CuLTO and (b) enhancing the photocatalytic performance for phenol photodegradation. The different thermal treatments applied to the CuLTO fresh solid determine distinct behavior for the photodegradation of phenol. Accordingly, the air-calcination of the CuLTO-500C catalyst led to an increase in phenol conversion by up to 14.9%. On the other hand, the reduced CuLTO-800R catalyst determines the highest conversion of phenol (16.6%) among all studied photocatalysts.

The photocatalytic reaction follows pseudo-first-order kinetics according to the equation $\ln(C/C_0) = -kt$, where C and C_0 designate the phenol concentration at time 0 and time t, and k is the apparent rate constant. Figure 13A shows the plot of $\ln(C_0/C)$ versus the irradiation time for the photocatalytic degradation of phenol over the subjected catalysts. The order of the apparent rate constant k values after 1 h of reaction (Figure 13B) follows the sequence: RbLTO (0.032 min^{-1}) < HLTO (0.063 min^{-1}) < CuLTO (0.091 min^{-1}) < CuLTO-500C (0.111 min^{-1}) < CuLTO-800R (0.133 min^{-1}). These results demonstrate that the introduction of a copper spacer promotes an increase in the reaction rate. The enhanced activity of CuLTO-800R compared to the RbLTO host was favored by its narrow band gap (3.16 eV), as determined from UV–Vis, XPS, and photocurrent spectroscopies. Additionally, the H$_2$-TPR and XPS measurements confirmed the presence of a Cu$^+$/Cu^0 redox couple localized on the CuLTO-800R catalyst surface, which is responsible for its higher activity.

Very interesting results were perceived for the distribution of the reaction products shown in Figure 13C. Therefore, hydroquinone (HQ) and 1,2-dihydroxybenzene (1,2-DHBZ) were identified as major intermediate products over the RbLTO, HLTO, and CuLTO photocatalysts. Contrarily, both CuLTO-500C and CuLTO-800R photocatalysts lead to the formation of benzoquinone (BQ) as a product intermediate. These results imply that the reaction mechanism is distinct in the Cu-intercalated layered perovskites. Figure 13D shows the catalytic efficiency of phenol mineralization over the RbLTO, HLTO, CuLTO, CuLTO-500C, and CuLTO-800R studied catalysts. The enhanced Cu-modified photocatalysts' activity compared to the RbLTO host is attributed to a better charge separation at the layer/modified interfaces and improved light harvesting ability due to band-gap shrinking. Notice that the CuLTO-800R solid was the most effective photocatalyst regarding the efficiency of phenol mineralization, with 2.82 µmoles·h^{-1} of CO$_2$ and 1.78 µmoles·h^{-1} of

H_2, respectively. The stability/reusability tests were performed with the catalyst displaying the best performances for the degradation of phenol under identical reaction conditions. The catalyst was recovered after centrifugation, dried, and used for the next run. After three cycling processes (Figure 13E), the degradation efficiency of phenol over CuLTO-800R remained satisfactory. In order to understand the morphological evolution of CuLTO-800R after three photocatalytic runs, SEM and TEM analyses were carried out (Figure S5). Morphological investigation revealed no significant change in the shape of the spent CuLTO-800R catalyst after the reusability test. The good stability of the CuLTO-800R material may be attributed to the more homogeneous dispersion of the Cu atoms within the $[LaTa_2O_7]^-$ network. These observations demonstrate the long-term stability of the CuLTO-800R material.

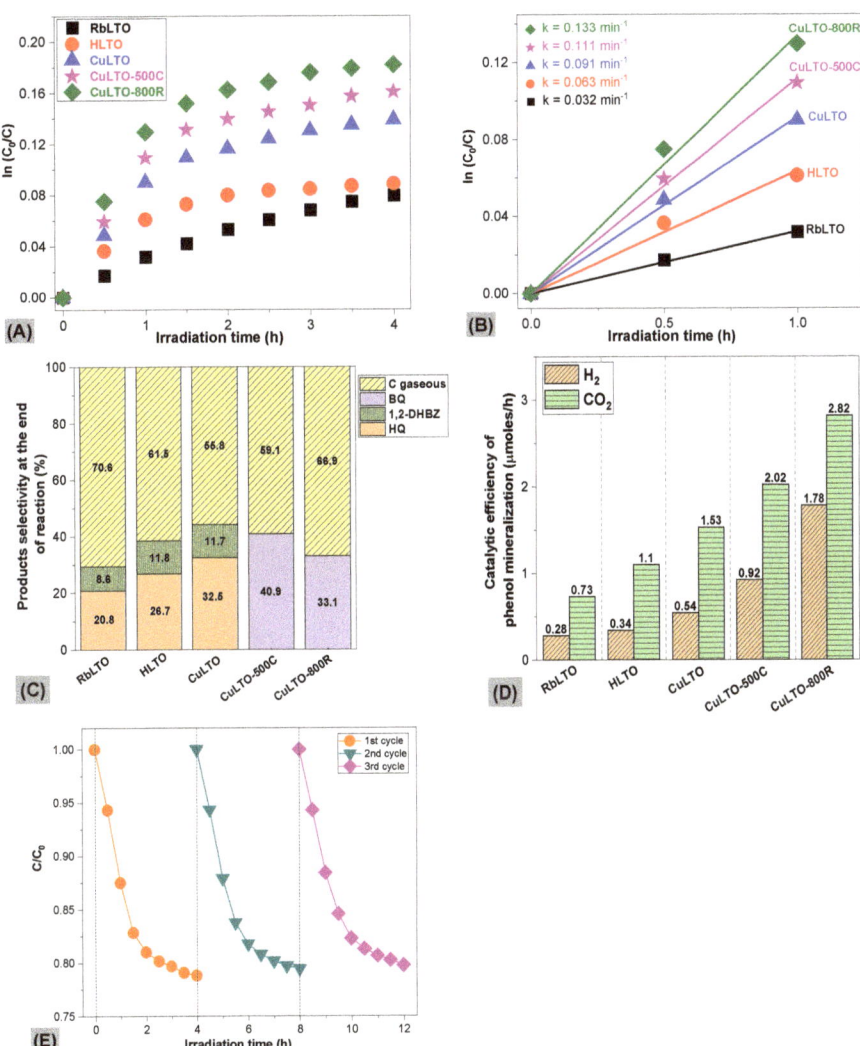

Figure 13. Graphical representation of the (**A**) pseudo-first-order kinetic plot on photodegradation of phenol, (**B**) apparent rate constant k (in min^{-1}), (**C**) the product selectivity at the end of the reaction, (**D**) the catalytic efficiency of phenol mineralization, and (**E**) the stability of the CuLTO-800R photocatalyst.

The benefit of copper ion insertion was relevant for the Cu-modified photocatalysts, where the specific surface area (SSA) of CuLTO was 4.1 $m^2 \cdot g^{-1}$ (Table 4). This value is about three times larger than that of the RbLaTa$_2$O$_7$ host (1.5 $m^2 \cdot g^{-1}$). Both thermally treated photocatalysts displayed higher SSA against the perovskite host with values of 4.9 $m^2 \cdot g^{-1}$ for CuLTO-500C and 3.4 $m^2 \cdot g^{-1}$ for CuLTO-800R, respectively. Doubtless, the insertion of various spacers into a layered compound provides original routes for tuning the physicochemical properties of the newly obtained materials.

In order to understand the mechanism, reactive oxygen species (ROS) generation under simulated solar irradiation was investigated according to the previously reported procedure [58]. Figure 14 displays the photogeneration of (A) a hydroxyl radical and (B) a superoxide anion over the CuLTO, CuLTO-500C, and CuLTO-800R photocatalysts. The hydroxyl radical (•OH) occurrence was evaluated based on photoluminescence (PL) emission peaking around 450 nm due to the umbelliferone formation, a coumarine derivative product obtained in the presence of photogenerated hydroxyl radicals. Figure 14A is indicative of the lack of •OH in the investigated photocatalytic systems, with no significant peaks being perceived around 450 nm. The superoxide anion (•O$_2^-$) formation under the simulated solar irradiation was evaluated from its interaction with XTT sodium salt, resulting in formazan production, with a characteristic peak appearing in the UV–Vis spectra. According to Figure 14B, only the CuLTO catalyst proves to generate •O$_2^-$. For the aforementioned sample, a broad band ranging from 420 to 550 nm arises after 20 min of irradiation.

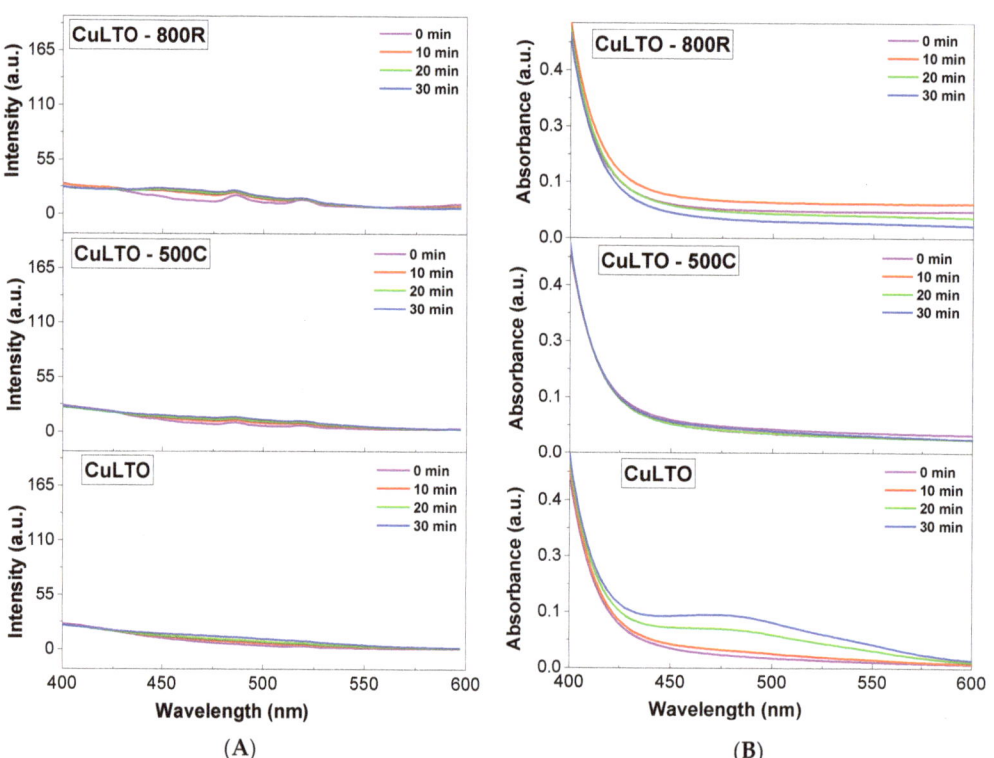

Figure 14. Photogeneration of (**A**) hydroxyl radical (•OH) and (**B**) superoxide anion (•O$_2^-$) over the CuLTO, CuLTO-500C, and CuLTO-800R photocatalysts.

Correlating ROS generation with photocatalytic activity, it is concluded that the reaction mechanism involves the generation of photoinduced electron-hole pairs. For the CuLTO catalysts, the superoxide anion ($\bullet O_2^-$) consumes the electrons (e^-) generated by irradiation, decreasing the photocatalytic activity. The reaction mechanism of the CuLTO-500C and CuLTO-800R photocatalysts is also produced through e^-/h^+, but the absence of $\bullet O_2^-$ is beneficial for the reaction of mineralization. A possible reaction mechanism of Cu-modified perovskites is advanced in Figure 15A.

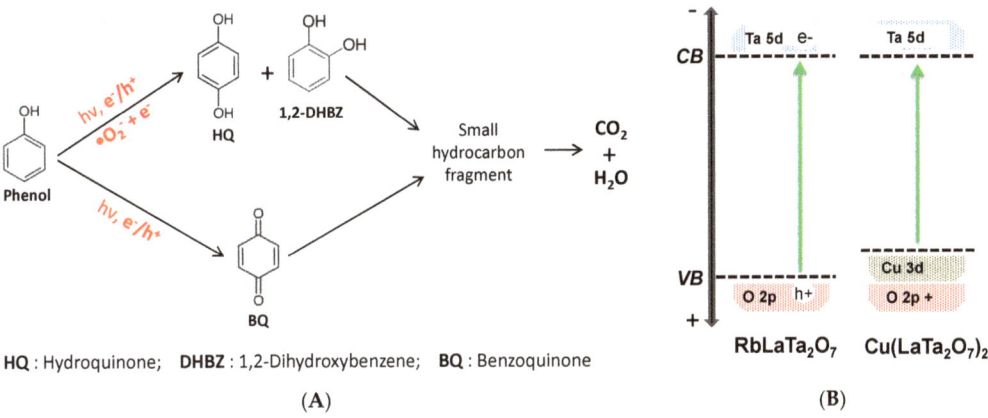

Figure 15. Schematic representation of (A) phenol photocatalytic pathways over Cu-intercalated perovskites, and (B) possible electronic structures of unmodified RbLaTa$_2$O$_7$ and Cu(LaTa$_2$O$_7$)$_2$-modified perovskites.

The photocatalytic degradation of phenol under simulated solar irradiation is an effective example of the degradation of persistent water contaminants [59]. Due to its abundance and low cost, copper is an affordable alternative to expensive noble metals (Au, Ag), widely used in various applications. The p-type CuO$_x$ possesses remarkable properties such as catalytic and antibacterial activities, optoelectronic properties, and high stability [60]. Further, perovskite with a layered structure, such as Bi$_2$WO$_6$, [61,62] demonstrated excellent activity in the mineralization of pollutants under light irradiation due to the transfer of electrons to the surface of the photocatalyst along the layered network. The recombination of charge carriers in such perovskites was depressed by the electron transfer to the layered host.

The advantages of the proposed Cu-modified layered perovskites can be envisaged as a "nanoreactor" where the reactant is confined within a layered perovskite structure in a restricted area, limiting the migration and recombination of photogenerated carriers. From the perspective of the electronic structure of the RbLaTa$_2$O$_7$ perovskite, its valence band (VB) is composed of O 2p orbitals, while the Ta 5d orbitals occupy the conduction band (CB). The Rb and La cations do not directly contribute to a band formation and solely build up the crystal structure of the perovskite. The symmetry of the metal-oxygen octahedral/tetrahedral coordination and the VB and CB structures also play crucial roles in determining the photocatalytic activity [63]. A successful approach to raise the valence band and decrease the band-gap size of the mixed oxides is to combine the elements with d^{10} (i.e., Cu$^+$, Ag$^+$) and d^0 (i.e., Nb^{5+}, Ta^{5+}) electron configurations. In these compounds, the band-gap excitations take place from a metal-to-metal charge transfer (MMCT) transition between electron-donating d^{10} and electron-accepting d^0 configurations [64]. As an example, Cu$_5$Ta$_{11}$O$_{30}$ and Cu$_3$Ta$_7$O$_{19}$ perovskites displayed a narrower band gap compared to unmodified layered oxides due to an MMCT between the Cu 3d and Ta 5d orbitals [65]. Zhang et al. [66] proposed a practical strategy to fine-tune the energy band structure of new oxynitrides by forming solid solutions between two perovskite-type materials. These authors reported that the VB of the synthesized materials was formed by N 2p with minor O 2p, while the CB was composed of Ti 3d in the majority and O 2p, La 5d, and N 2p in the minority. Jiang and co-workers [67] found that the electronic structure of γ-$Cu_3V_2O_8$ consisted of O 2p states while the bonding Cu 3d and V 3d states were observed to lie deeper within the valence band. The CB was found to be dominated by unoccupied Cu 3d orbitals, with V 3d states residing at higher energies.

When comparing our results with the findings in prior studies, a similar conclusion was reached. Here, it is postulated that the insertion of a copper spacer modifies the valence band of Cu-based photocatalysts. The narrow band gap of Cu-intercalated perovskites could be attributed to their valence band consisting of a Cu 3d and O 2p orbitals mixture. A possible electronic structure of the $RbLaTa_2O_7$ host and its modified $Cu(LaTa_2O_7)_2$ perovskite are proposed in Figure 15B. The current literature on the photocatalytic degradation of phenol admitted that during the reaction, hydroquinone, catechol, and p-benzoquinone are identified as major intermediates. Other reaction intermediate products (chloro-hydroquinone, 4-chlorocatechol, and resorcinol) can eventually be converted into acetylene, maleic acid, CO, and CO_2. Different mechanisms have been proposed for the photodegradation of organic pollutants, where the hydroxyl radical (•OH), the superoxide radical (•O_2^-), and the hole (h^+) are reviewed as the main reactive species [68]. Trapping experiments carried out on a phosphorus-doped carbon-supported Cu_2O composite [69] indicated that the •O_2^- radical was the major active species, while the hydroxyl radical played a minor role in the phenol photodegradation. Therefore, the intercalation of copper between the interlayer spaces of perovskite will serve as a promoter for charge separation between electrons and holes, leading to enhancements in the photocatalytic degradation of pollutants. This study shows the great versatility of RbLaTa-based layered perovskite for producing assembled materials with different properties by only controlling the spacer.

4. Materials and Methods

4.1. Synthesis of Catalytic Materials

Preparation of $RbLaTa_2O_7$ layered perovskite host. $RbLaTa_2O_7$ powder was synthesized by the solid-state route as reported in a previous paper [70]. The stoichiometric amounts of Rb_2CO_3, La_2O_3, and Ta_2O_5 with a 50% molar excess of Rb_2CO_3 were treated in an air atmosphere at 1200 °C for 18 h with one intermediate grinding. The obtained white solid was washed with water and air-dried at 110 °C overnight. This sample was designated as RbLTO.

Preparation of $HLaTa_2O_7$ protonated perovskite. In a typical proton exchange reaction, the starting layered perovskite (1 g) was dispersed in an aqueous solution of 3M HNO_3 (100 mL) and stirred at room temperature for one week, with the daily renewal of the acid solution according to ref. [71]. This solid was labeled as HLTO.

Amine-intercalated layered perovskite. In order to intercalate copper species in between interlayers of the perovskite, the expansion of its interlamellar gallery is needed. Therefore, the $HLaTa_2O_7$ protonated layered perovskite (0.5 g) was dispersed in a 30 mL n-butylamine (BuA)-water mixture (1/1, v/v) and heated at 70 °C for 24 h. The white suspension was filtered, washed with acetone and water, and then air-dried at 110 °C overnight. The obtained solid was designated as BuALTO.

Assembling of $Cu(LaTa_2O_7)_2$ novel architecture. For the copper intercalation procedure, the amine-treated powder (0.2 g) was dispersed in an aqueous solution of $Cu(NO_3)_2 \cdot 4H_2O$ (100 mL, 0.057 M) and kept at 80 °C for 60 h. The resultant mixture was washed with water and air-dried overnight at 110 °C. This solid was designated as CuLTO.

To further investigate the nature and localization of the copper spacer in the novel layered architecture, the CuLTO fresh solid was subjected to different thermal treatments, as follows: (i) calcination in an air atmosphere at 500 °C for 2 h (sample labeled as CuLTO-500C) and (ii) reduction at 800 °C in a 5% H_2/Ar atmosphere for 2 h (sample denoted as CuLTO-800R).

4.2. Characterization of Photocatalysts

The RbLaTa$_2$O$_7$ original layered perovskite and its corresponding intercalated-guest spacers were characterized by various techniques. X-ray diffraction (XRD) patterns were obtained on a Rigaku Corporation Ultima IV diffractometer, Tokyo, Japan (monochromatized Cu Kα (λ = 0.15418 nm). The average crystallite size was calculated using the Debye-Scherrer equation. The X-ray photoelectron microscopy (XPS) experiments were carried out on a SPECS spectrometer, Berlin, Germany with a PHOIBOS 150 analyzer, using non-monochromatic Al Kα radiation (1486.7 eV). The charge compensation was realized by a flood gun of Specs FG15/40 type. The acquisition was operated at a pass energy of 20 eV for the individual spectral lines and 50 eV for the extended spectra. SEM pictures were collected with a high-resolution microscope, an FEI Quanta3D FEG device, Brno, Czech Republic, at an accelerating voltage of 5 kV, in high-vacuum mode with an Everhart-Thornley secondary electron (SE) detector coupled with EDX (energy dispersive X-ray) analysis. TEM micrographs were obtained on a JEM-1400 apparatus, (JEOL Ltd., Tokyo, Japan) operated at 100 kV. The Rb leaching during the protonation step was checked by ion chromatography (Dionex ICS 900, Sunnyvale, CA, USA) through the dosing of Rb$^+$ cations. Nitrogen BET-specific surface areas were measured at 77 K with a Micromeritics ASAP 2020 instrument, Norcross, GA, USA. Thermo-differential analyses (TG/DTA/DTG) were recorded with Mettler Toledo TGA/SDTA 851^e apparatus (Greifensee, Switzerland) in an air atmosphere using an alumina crucible between 25–1000 °C at a heating rate of 10 °C·min^{-1}. Infrared spectra were obtained in transmission mode using a JASCO spectrophotometer, Tokyo, Japan. UV–Vis spectra were acquired on a Perkin Elmer Lambda 35 spectrophotometer, Shelton, CT, USA equipped with an integrating sphere. The reflectance was converted to absorption using the Kubelka-Munk function. The optical band gap of the samples was calculated according to the formula, E_g (eV) = 1240/λ (wavelength in nm). The temperature-programmed reduction measurements (H$_2$-TPR) were performed using a CHEMBET-3000 Quantachrome Instrument, Boynton Beach, FL, USA equipped with a thermal conductivity detector (TCD). In a typical experiment, the fresh sample (0.050 g) was heated up to 800 °C at the constant rate of 10 °C·min^{-1} of the 5 vol.% H$_2$/Ar reduction gas and a flow rate of 70 mL·min^{-1}. The hydrogen consumption was estimated from the area of the recorded peaks. The calibration of the TCD signal was performed by injecting a known quantity of hydrogen (typically 50 µL) into the carrier gas (Ar). The experimentally obtained peak surface (mV·s) was thus converted into micromoles of hydrogen. The photoelectrochemical measurements were carried out in an electrochemical cell equipped with a quartz window using a three-electrode configuration. All measurements were performed on a Zahner IM6 potentiostat, Zahner-Elektrik GmbH, Kronach-Gundelsdorf, Germany. A deaerated solution of 0.1 M Na$_2$SO$_4$ was used as an electrolyte. The counter electrode was a high-surface Pt wire, and the reference electrode was Ag/AgCl. The working electrode (geometric area of ~3.5 cm^2) was prepared by the deposition of the interest powder (0.010 g) onto the transparent conductive (TCO) glass (Solaronix, Aubonne, Switzerland). Linear sweep voltammetry (LSV) was carried out at a 10 mV/s rate under chopped AM 1.5 simulated solar light (Peccell-L01, Yokohama, Japan).

4.3. Photocatalytic Degradation of Phenol under Simulated Solar Irradiation

Photocatalytic test. The photocatalytic degradation experiments were performed in a batch photoreactor thermostated at 18 °C (Scheme 2). The AM 1.5 light beam was provided by a solar light simulator (Peccell-L01, Yokohama, Japan) equipped with a 150 W short-arc Xe lamp (1000·W·m^{-2}). The photocatalyst powder (0.05 g) was suspended in 110 mL of 50 mg·L^{-1} aqueous phenol solution. During tests, the Ar carrier gas (10 mL·min^{-1}) was continuously purged into the reaction solution and passed through a refrigerant cooled to −5 °C. This cooler helped to remove the liquid vapors from entering the GC equipment. The photoreactor was provided with a quartz window of 4.5 × 4.5 cm^2 for light irradiation. Prior to each experiment, the suspension was kept in dark conditions for 30 min under stirring to attain the adsorption-desorption equilibrium. Subsequently, the reaction vessel was exposed to simulated solar light for 4 h of reaction time.

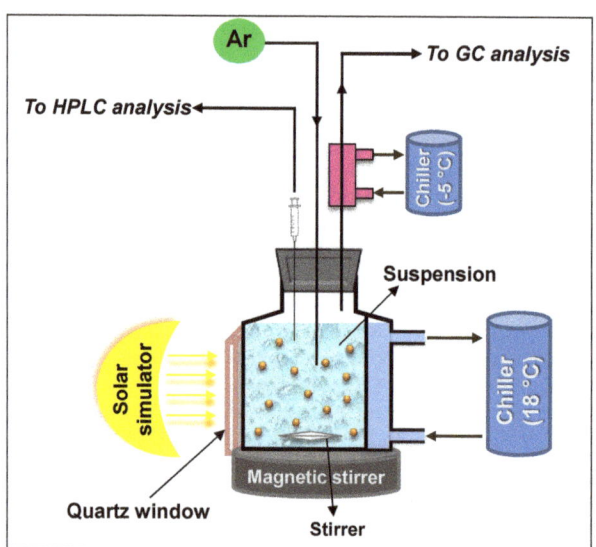

Scheme 2. Schematic diagram of the experimental setup used for the photocatalytic degradation of phenol under simulated solar light.

Analytical methods. Phenol and intermediate ring compounds (i.e., hydroquinone, benzoquinone, 1,2-dihydroxibenzene) were identified and quantified by high-performance liquid chromatography (HPLC) (Waters, Alliance e2659, Milford, MA, USA) equipped with a UV–Vis detector (Waters, model 2489) (λ = 273 nm) and a Kromasil 100 5-C18 column. Aliquots of 2 mL of the reaction mixture were collected at 30 min time intervals, filtered through 0.22 µm Q-Max membrane filter, and analyzed by HPLC. The mobile phase was a mixture of Milli-Q water and methanol (50/50, v/v) applied in the isocratic elution program. The flow rate of the mobile phase was 1 mL·min^{-1}, and the injection volume was set to 2 µL. The evolved gases were analyzed every 30 min using online gas chromatography (Buck Scientific, Norwalk, CT, USA) equipped with a TCD detector. The H_2 and O_2 were separated and quantified on Molecular Sieve 5 Å, whereas CO_2 was on the Haysep column.

Reactive oxygen species (ROS) generation. For the hydroxyl radical (•OH) trapping, an amount of 0.001 g catalyst was suspended in 10 mM coumarin (Merck, Darmstadt, Germany) solution and exposed to simulated solar irradiation (Peccel-L01 Solar Simulator, Yokohama, Japan). The formation of a fluorescent product from coumarin interaction with the photogenerated hydroxyl radicals is investigated with a Carry Eclipse fluorescence spectrometer (Agilent Technologies, Santa Clara, CA, USA) for λ_{exc} = 330 nm. Superoxide anion (•O_2^-) monitoring was performed by suspending 0.003 g of catalyst in a 3 mM solution of 2,3-Bis(2-methoxy-4-nitro-5-sulfophenyl)-2H-tetrazolium-5-carboxanilide (XTT sodium salt). The XTT formazan appears as a result of XTT interaction with the photogenerated •O_2^- and generates a broad peak around 470 nm depicted with a UV–Vis spectrophotometer (Analytik Jena Specord 200 Plus, Jena, Germany).

5. Conclusions

A multi-step ion-exchange methodology has been exploited by exchanging Rb^+ with a much smaller Cu^{2+} spacer in the $RbLaTa_2O_7$ host to achieve photocatalysts capable of wastewater depollution. The XRD patterns revealed that the interlayer distance of the lamellar perovskites was spacer-related. The UV-Vis, XPS, and photocurrent spectroscopies demonstrated band-gap narrowing after copper intercalation. The H_2-TPR results indicated that the Cu species located on the surface were reduced at a lower temperature while those from the interlayer occurred at higher temperature ranges. Cu-modified layered perovskites exhibited enhanced photocatalytic activity compared to the $RbLaTa_2O_7$ host. Experiments proved that the reaction mechanism over Cu-intercalated perovskites was produced via the generation of photoinduced e^-/h^+ pairs. The superior photocatalytic activity of CuLTO-800R was attributed to its narrow band gap and photogenerated-carriers separation. Cu-based architectures are promising materials because of their stable perovskite-like slabs and flexible interlayer galleries that facilitate modification.

Supplementary Materials: The following supporting information can be downloaded at: https://www.mdpi.com/article/10.3390/catal12121529/s1. Figure S1: FTIR spectra of neat n-butylamine; Figure S2: Copper species distribution in CuLTO, CuLTO-500C, and CuLTO-800R catalysts; Figure S3: XPS spectra of (a) RbLTO, (b) CuLTO, (c) CuLTO-500C, and (d) CuLTO-800R layered architectures for C 1s and O 1s emission lines; Figure S4: XPS spectra of (a) RbLTO, (b) CuLTO, (c) CuLTO-500C, and (d) CuLTO-800R layered architectures for La 3d and Ta 4f emission lines; Figure S5: (A) SEM and (B) TEM images of CuLTO-800R photocatalyst after the stability/reusability test; Table S1: Mass loss at different steps of TG curves of all layered perovskite materials. References [32,71,72] are cited in the supplementary materials.

Author Contributions: Conceptualization, I.B.; methodology, I.B., F.P. and M.R.; validation, I.B. and F.P.; investigation, M.R., C.A., F.P., I.A., C.N., D.-I.E., D.C.C., A.M., V.B., J.P.-C., C.M., G.D. and A.S.; resources, I.B. and F.P.; writing—original draft preparation, M.R.; writing—review and editing, I.B. and F.P.; supervision, I.B.; project administration, I.B., F.P. and C.B.; funding acquisition, C.B. and F.P. All authors have read and agreed to the published version of the manuscript.

Funding: This research was funded by the MINISTRY OF RESEARCH, INNOVATION, AND DIGITIZATION, CNCS/CCCDI-UEFISCDI, grant number PN-III-P2-2.1-PTE-2019-0222, within PNCDI III.

Data Availability Statement: All data are available upon reasonable request from the authors.

Conflicts of Interest: The authors declare no conflict of interest. The funders had no role in the design of the study; in the collection, analyses, or interpretation of data; in the writing of the manuscript; or in the decision to publish the results.

References

1. Fang, M.; Kim, C.H.; Mallouk, T.E. Dielectric Properties of the Lamellar Niobates and Titanoniobates $AM_2Nb_3O_{10}$ and $ATiNbO_5$ (A = H, K, M = Ca, Pb), and Their Condensation Products $Ca_4Nb_6O_{19}$ and $Ti_2Nb_2O_9$. *Chem. Mater.* **1999**, *11*, 1519–1525. [CrossRef]
2. Liu, Y.; Mao, Z.-Q. Unconventional superconductivity in Sr_2RuO_4. *Phys. C Supercond. Appl.* **2015**, *514*, 339–353. [CrossRef]
3. Ida, S.; Ogata, C.; Eguchi, M.; Youngblood, W.J.; Mallouk, T.E.; Matsumoto, Y. Photoluminescence of Perovskite Nanosheets Prepared by Exfoliation of Layered Oxides, $K_2Ln_2Ti_3O_{10}$, $KLnNb_2O_7$, and $RbLnTa_2O_7$ (Ln: Lanthanide Ion). *J. Am. Chem. Soc.* **2008**, *130*, 7052–7059. [CrossRef] [PubMed]
4. Hu, Y.; Mao, L.; Guan, X.; Tucker, K.A.; Xie, H.; Wu, X.; Shi, J. Layered perovskite oxides and their derivative nanosheets adopting different modification strategies towards better photocatalytic performance of water splitting. *Renew. Sustain. Energy Rev.* **2020**, *119*, 109527. [CrossRef]
5. Oshima, T.; Yokoi, T.; Eguchi, M.; Maeda, K. Synthesis and photocatalytic activity of $K_2CaNaNb_3O_{10}$, a new Ruddlesden-Popper phase layered perovskite. *Dalton Trans.* **2017**, *46*, 10594–10601. [CrossRef] [PubMed]
6. Atri, S.; Tomar, R. A Review on the Synthesis and Modification of Functional Inorganic-Organic-Hybrid Materials via Microwave-Assisted Method. *ChemistrySelect* **2021**, *6*, 9351–9362. [CrossRef]
7. Machida, M.; Yabunaka, J.-I.; Kijima, T. Synthesis and Photocatalytic Property of Layered Perovskite Tantalates, $RbLnTa_2O_7$ (Ln = La, Pr, Nd, and Sm). *Chem. Mater.* **2000**, *12*, 812–817. [CrossRef]
8. Sanjaya Ranmohotti, K.G.; Josepha, E.; Choi, J.; Zhang, J.; Wiley, J.B. Topochemical Manipulation of Perovskites: Low-Temperature Reaction Strategies for Directing Structure and Properties. *Adv. Mater.* **2011**, *23*, 442–460. [CrossRef]
9. Wang, B.; Dong, X.; Pan, Q.; Cheng, Z.; Yang, Y. Intercalation behavior of n-alkylamines into an A-site defective layered perovskite $H_2W_2O_7$. *J. Solid State Chem.* **2007**, *180*, 1125–1129. [CrossRef]
10. Uppuluri, R.; Gupta, A.S.; Rosas, A.S.; Mallouk, T.E. Soft chemistry of ion-exchangeable layered metal oxides. *Chem. Soc. Rev.* **2018**, *47*, 2401–2430. [CrossRef] [PubMed]
11. Funatsu, A.; Taniguchi, T.; Tokita, Y.; Murakami, T.; Nojiri, Y.; Matsumoto, Y. Nd^{3+}-doped perovskite nanosheets with NIR luminescence. *Mater. Lett.* **2014**, *114*, 29–33. [CrossRef]

12. Timmerman, M.A.; Xia, R.; Le, P.T.P.; Wang, Y.; ten Elshof, J.E. Metal Oxide Nanosheets as 2D Building Blocks for the Design of Novel Materials. *Chem. A Eur. J.* **2020**, *26*, 9084–9098. [CrossRef] [PubMed]
13. Li, F.; Xie, Y.; Hu, Y.; Long, M.; Zhang, Y.; Xu, J.; Qin, M.; Lu, X.; Liu, M. Effects of alkyl chain length on crystal growth and oxidation process of two dimensional tin halide perovskites. *ACS Energy Lett.* **2020**, *5*, 1422–1429. [CrossRef]
14. Minich, I.A.; Silyukov, O.I.; Gak, V.V.; Borisov, E.V.; Zvereva, I.A. Synthesis of organic-inorganic hybrids based on perovskite-like bismuth titanate $H_2K_{0.5}Bi_{2.5}Ti_4O_{13} \cdot H_2O$ and *n*-alkylamines. *ACS Omega* **2020**, *5*, 8158–8168. [CrossRef]
15. Fan, L.; Wei, Y.; Cheng, Y.; Huang, Y.; Hao, S.; Wu, J. Preparation and photocatalytic properties of $HLaNb_2O_7/(Pt, TiO_2)$ perovskite intercalated nanomaterial. *Int. J. Hydrog. Energy* **2014**, *39*, 7747–7752. [CrossRef]
16. Wu, J.; Cheng, Y.; Lin, J.; Huang, Y.; Huang, M.; Hao, S. Fabrication and photocatalytic properties of $HLaNb_2O_7/(Pt, Fe_2O_3)$ pillared nanomaterial. *J. Phys. Chem. C* **2007**, *111*, 3624–3628. [CrossRef]
17. Ebina, Y.; Sakai, N.; Nagasaki, T. Photocatalyst of lamellar aggregates of RuO_x-loaded perovskite nanosheets for overall water splitting. *J. Phys. Chem. B* **2005**, *109*, 17212–17216. [CrossRef]
18. Oshima, T.; Wang, Y.; Lu, D.; Yokoi, T.; Maeda, K. Photocatalytic overall water splitting on Pt nanocluster-intercalated, restacked $KCa_2Nb_3O_{10}$ nanosheets: The promotional effect of co-existing ions. *Nanoscale Adv.* **2019**, *1*, 189–194. [CrossRef]
19. Jiang, C.; Moniz, S.J.A.; Wang, A.; Zhang, T.; Tang, J. Photoelectrochemical devices for solar water splitting-materials and challenges. *Chem. Soc. Rev.* **2017**, *46*, 4645–4660. [CrossRef]
20. Majumdar, D.; Ghosh, S. Recent advancements of copper oxide based nanomaterials for supercapacitor applications. *J. Energy Storage* **2021**, *34*, 101995. [CrossRef]
21. Wang, Q.; Domen, K. Particulate photocatalysts for light-driven water splitting: Mechanisms, challenges, and design strategies. *Chem. Rev.* **2020**, *120*, 919–985. [CrossRef] [PubMed]
22. Watanabe, K.; Iwashina, K.; Iwase, A.; Nozawa, S.; Adachi, S.-I.; Kudo, A. New Visible-Light-Driven H_2- and O_2-Evolving Photocatalysts Developed by Ag(I) and Cu(I) Ion Exchange of Various Layered and Tunneling Metal Oxides Using Molten Salts Treatments. *Chem. Mater.* **2020**, *32*, 10524–10537. [CrossRef]
23. Iwashina, K.; Iwase, A.; Kudo, A. Sensitization of wide band gap photocatalysts to visible light by molten CuCl treatment. *Chem. Sci.* **2015**, *6*, 687–692. [CrossRef]
24. Wysocka, I.; Kowalska, E.; Trzciński, K.; Łapiński, M.; Nowaczyk, G.; Zielińska-Jurek, A. UV-Vis-Induced Degradation of Phenol over Magnetic Photocatalysts Modified with Pt, Pd, Cu and Au Nanoparticles. *Nanomaterials* **2018**, *8*, 28. [CrossRef] [PubMed]
25. Cordova Villegas, L.G.; Mashhadi, N.; Chen, M.; Mukherjee, D.; Taylor, K.E.; Biswas, N. A Short Review of Techniques for Phenol Removal from Wastewater. *Curr. Pollut. Rep.* **2016**, *2*, 157–167. [CrossRef]
26. Alonso, J.A.; Martínez-Lope, M.J.; Casais, M.T.; Fernández-Díaz, M.T. Evolution of the Jahn−Teller Distortion of MnO_6 Octahedra in $RMnO_3$ Perovskites (R = Pr, Nd, Dy, Tb, Ho, Er, Y): A Neutron Diffraction Study. *Inorg. Chem.* **2000**, *39*, 917–923. [CrossRef]
27. Hyeon, K.-A.; Byeon, S.-H. Synthesis and Structure of New Layered Oxides, $M^{II}La_2Ti_3O_{10}$ (M = Co, Cu, and Zn). *Chem. Mater.* **1998**, *11*, 352–357. [CrossRef]
28. Ogawa, M.; Kuroda, K. Preparation of Inorganic-Organic Nanocomposites through Intercalation of Organoammonium Ions into Layered Silicates. *Bull. Chem. Soc. Jpn.* **1997**, *70*, 2593–2618. [CrossRef]
29. Sasaki, T.; Kooli, F.; Iida, M.; Michiue, Y.; Takenouchi, S.; Yajima, Y.; Izumi, F.; Chakoumakos, B.C.; Watanabe, M. A Mixed Alkali Metal Titanate with the Lepidocrocite-like Layered Structure. Preparation, Crystal Structure, Protonic Form, and Acid−Base Intercalation Properties. *Chem. Mater.* **1998**, *10*, 4123–4128. [CrossRef]
30. Sasaki, T.; Izumi, F.; Watanabe, M. Intercalation of Pyridine in Layered Titanates. *Chem. Mater.* **1996**, *8*, 777–782. [CrossRef]
31. Sun, C.; Peng, P.; Zhu, L.; Zheng, W.; Zhao, Y. Designed Reversible Alkylamine Intercalation-Deintercalation in the Layered Perovskite-Type Oxide $KCa_2Nb_3O_{10}$. *Eur. J. Inorg. Chem.* **2008**, *24*, 3864–3870. [CrossRef]
32. Raciulete, M.; Papa, F.; Culita, D.C.; Munteanu, C.; Atkinson, I.; Bratan, V.; Pandele-Cusu, J.; State, R.; Balint, I. Impact of $RbLaTa_2O_7$ Layered Perovskite Synthesis Conditions on their Activity for Photocatalytic Abatement of Trichloroethylene. *Rev. Roum. Chim.* **2018**, *63*, 821–828.
33. Geselbracht, M.J.; White, H.K.; Blaine, J.M.; Diaz, M.J.; Hubbs, J.L.; Adelstein, N.; Kurzman, J.A. New solid acids in the triple-layer Dion-Jacobson layered perovskite family. *Mater. Res. Bull.* **2011**, *46*, 398–406. [CrossRef]
34. Cheng, S.; Hwang, H.-D.; Maciel, G.E. Synthesis and pillaring of a layered vanadium oxide from V_2O_5 at ambient temperature. *J. Mol. Struct.* **1998**, *470*, 135–149. [CrossRef]
35. Castellini, E.; Bernini, F.; Sebastianelli, L.; Sainz-Díaz, C.I.; Serrano, A.; Castro, G.R.; Malferrari, D.; Brigatti, M.F.; Borsari, M. Interlayer-Confined Cu(II) Complex as an Efficient and Long-Lasting Catalyst for Oxidation of H_2S on Montmorillonite. *Minerals* **2020**, *10*, 510. [CrossRef]
36. Mokhtar, A.; Medjhouda, Z.A.K.; Djelad, A.; Boudia, A.; Bengueddach, A.; Sassi, M. Structure and intercalation behavior of copper II on the layered sodium silicate magadiite material. *Chem. Pap.* **2018**, *72*, 39–50. [CrossRef]
37. Ding, Z.; Frost, R.L. Study of copper adsorption on montmorillonites using thermal analysis methods. *J. Colloid Interface Sci.* **2004**, *269*, 296–302. [CrossRef] [PubMed]
38. Hakiki, A.; Kerbadou, R.M.; Boukoussa, B.; Zahmani, H.H.; Launay, F.; Pailleret, A.; Pillier, F.; Hacini, S.; Bengueddach, A.; Hamacha, R. Catalytic behavior of copper-amine complex supported on mesoporous silica SBA-15 toward mono-Aza-Michael addition: Role of amine groups. *J. Inorg. Organomet. Polym. Mater.* **2019**, *29*, 1773–1784. [CrossRef]

39. Chen, Z.; Deutsch, T.G.; Dinh, H.N.; Domen, K.; Emery, K.; Forman, A.J.; Gaillard, N.; Garland, R.; Heske, C.; Jaramillo, T.F.; et al. Photoelectrochemical Water Splitting. Incident photon-to-current efficiency and photocurrent spectroscopy. In *Photoelectrochemical Water Splitting*; Chen, Z., Dinh, H.N., Miller, E., Eds.; Briefs in Energy; Springer: New York, NY, USA, 2013; pp. 87–97. [CrossRef]
40. Ghosh, B.; Halder, S.; Sinha, T.P. Dielectric Relaxation and Collective Vibrational Modes of Double-Perovskites A_2SmTaO_6 (A = Ba, Sr and Ca). *J. Am. Ceram. Soc.* **2014**, *97*, 2564–2572. [CrossRef]
41. Kumar, D.V.R.; Kim, I.; Zhong, Z.; Kim, K.; Lee, D.; Moon, J. Cu(II)-alkyl amine complex mediated hydrothermal synthesis of Cu nanowires: Exploring the dual role of alkyl amines. *Phys. Chem. Chem. Phys.* **2014**, *16*, 22107–22115. [CrossRef]
42. Espeel, P.; Goethals, F.; Driessen, F.; Nguyen, L.-T.T.; Du Prez, F.E. One-pot, additive-free preparation of functionalized polyurethanes via amine-thiol-ene conjugation. *Polym. Chem.* **2013**, *4*, 2449–2456. [CrossRef]
43. Du, J.; Zhao, Y.; Chen, J.; Zhang, P.; Gao, L.; Wang, M.; Cao, C.; Wen, W.; Zhu, C. Difunctional Cu-doped carbon dots: Catalytic activity and fluorescence indication for the reduction reaction of p-nitrophenol. *RSC Adv.* **2017**, *7*, 33929–33936. [CrossRef]
44. Fang, L.; Dong, S.; Shi, L.; Sun, Q. Modification of $CuCl_2 \cdot 2H_2O$ by dielectric barrier discharge and its application in the hydroxylation of benzene. *New J. Chem.* **2019**, *43*, 12744–12753. [CrossRef]
45. Gopalakrishnan, K.; Ramesh, C.; Ragunathan, V.; Thamilselvan, M.; Thamilselvan, M. Antibacterial activity of Cu_2O nanoparticles on E Coli synthesized from Tridax Procumbens leaf extract and surface coating with polyaniline. *Dig. J. Nanomater. Biostruct.* **2012**, *7*, 833–839.
46. Morioka, T.; Takesue, M.; Hayashi, H.; Watanabe, M.; Smith, R.L. Antioxidation properties and surface interactions of polyvinylpyrrolidone-capped zerovalent copper nanoparticles synthesized in supercritical water. *ACS Appl. Mater. Interfaces* **2016**, *8*, 1627–1634. [CrossRef] [PubMed]
47. Torre-Abreu, C.; Ribeiro, M.F.; Henriques, C.; Delahay, G. The role of Si/Al ratio, copper content and co-cation. *Appl. Catal. B Environ.* **1997**, *14*, 261–272. [CrossRef]
48. Wang, Z.; Yan, X.; Bi, X.; Wang, L.; Zhang, Z.; Jiang, Z.; Xiao, T.; Umar, A.; Wang, Q. Lanthanum-promoted copper-based hydrotalcites derived mixed oxides for NO_x adsorption, soot combustion and simultaneous NO_x-soot removal. *Mater. Res. Bull.* **2014**, *51*, 119–127. [CrossRef]
49. Gervasini, A.; Bennici, S. Dispersion and surface states of copper catalysts by temperature-programmed-reduction of oxidized surfaces (s-TPR). *Appl. Catal. A Gen.* **2005**, *281*, 199–205. [CrossRef]
50. Zeng, Y.; Wang, T.; Zhang, S.; Wang, Y.; Zhong, Q. Sol-gel synthesis of $CuO-TiO_2$ catalyst with high dispersion CuO species for selective catalytic oxidation of NO. *Appl. Surf. Sci.* **2017**, *411*, 227–234. [CrossRef]
51. Nestroinaia, O.V.; Ryltsova, I.G.; Lebedeva, O.E. Effect of synthesis method on properties of layered double hydroxides containing Ni(III). *Crystals* **2021**, *11*, 1429. [CrossRef]
52. Wachs, I.E.; Briand, L.E.; Jehng, J.-M.; Burcham, L.; Gao, X. Molecular structure and reactivity of the group V metal oxides. *Catal. Today* **2000**, *57*, 323–330. [CrossRef]
53. Seisenbaeva, G.A.; Cojocaru, B.; Jurca, B.; Tiseanu, C.; Nedelec, J.-M.; Kessler, V.G.; Parvulescu, V.I. Mesoporous tantalum oxide photocatalyst: Structure and activity evaluation. *ChemistrySelect* **2017**, *2*, 421–427. [CrossRef]
54. Liu, M.; Qiu, X.; Hashimoto, K.; Miyauchi, M. Cu(II) nanocluster-grafted, Nb-doped TiO_2 as an efficient visible-light-sensitive photocatalyst based on energy-level matching between surface and bulk states. *J. Mater. Chem. A* **2014**, *2*, 13571–13579. [CrossRef]
55. Wang, D.; Pan, X.; Wang, G.; Yi, Z. Improved propane photooxidation activities upon nano Cu_2O/TiO_2 heterojunction semiconductors at room temperature. *RSC Adv.* **2015**, *5*, 22038–22043. [CrossRef]
56. Li, J.; Zeng, J.; Jia, L.; Fang, W. Investigations on the effect of Cu^{2+}/Cu^{1+} redox couples and oxygen vacancies on photocatalytic activity of treated $LaNi_{1-x}Cu_xO_3$ (x = 0.1, 0.4, 0.5). *Int. J. Hydrog. Energy* **2010**, *35*, 12733–12740. [CrossRef]
57. Li, F.; Zhang, L.; Evans, D.G.; Duan, X. Structure and surface chemistry of manganese-doped copper-based mixed metal oxides derived from layered double hydroxides. *Colloids Surf. A Physicochem. Eng. Asp.* **2004**, *244*, 169–177. [CrossRef]
58. Sandulescu, A.; Anastasescu, C.; Papa, F.; Raciulete, M.; Vasile, A.; Spataru, T.; Scarisoreanu, M.; Fleaca, C.; Mihailescu, C.N.; Teodorescu, V.S.; et al. Advancements on Basic Working Principles of Photo-Driven Oxidative Degradation of Organic Substrates over Pristine and Noble Metal-Modified TiO_2. Model Case of Phenol Photo Oxidation. *Catalysts* **2021**, *11*, 487. [CrossRef]
59. Sobczyński, A.; Duczmal, Ł.; Zmudziński, W. Phenol destruction by photocatalysis on TiO_2: An attempt to solve the reaction mechanism. *J. Molec. Catal. A Chem.* **2004**, *213*, 225–230. [CrossRef]
60. Rydosz, A. The use of copper oxide thin films in gas-sensing applications. *Coatings* **2018**, *8*, 425. [CrossRef]
61. Wu, L.; Bi, J.; Li, Z.; Wang, X.; Fu, X. Rapid preparation of Bi_2WO_6 photocatalyst with nanosheet morphology via microwave-assisted solvothermal synthesis. *Catal. Today* **2008**, *131*, 15–20. [CrossRef]
62. Tang, J.; Zou, Z.; Ye, J. Photocatalytic Decomposition of Organic Contaminants by Bi_2WO_6 Under Visible Light Irradiation. *Catal. Lett.* **2004**, *92*, 53–56. [CrossRef]
63. Yan, H.; Wang, X.; Yao, M.; Yao, X. Band structure design of semiconductors for enhanced photocatalytic activity: The case of TiO_2. *Prog. Nat. Sci. Mater. Int.* **2013**, *23*, 402–407. [CrossRef]
64. Dey, S.; Ricciardo, R.A.; Cuthbert, H.L.; Woodward, P.M. Metal-to-metal charge transfer in AWO_4 (A = Mg, Mn, Co, Ni, Cu, or Zn) compounds with the wolframite structure. *Inorg. Chem.* **2014**, *53*, 4394–4399. [CrossRef]
65. Palasyuk, O.; Palasyuk, A.; Maggard, P.A. Syntheses, optical properties and electronic structures of copper (I) tantalates: $Cu_5Ta_{11}O_{30}$ and $Cu_3Ta_7O_{19}$. *J. Solid State Chem.* **2010**, *183*, 814–822. [CrossRef]

66. Mao, L.; Cai, X.; Gao, H.; Diao, X.; Zhang, J. A newly designed porous oxynitride photoanode with enhanced charge carrier mobility. *Nano Energy* **2017**, *39*, 172–182. [CrossRef]
67. Jiang, C.-M.; Farmand, M.; Wu, C.H.; Liu, Y.-S.; Guo, J.; Drisdell, W.S.; Cooper, J.K.; Sharp, I.D. Electronic Structure, Optoelectronic Properties, and Photoelectrochemical Characteristics of γ-$Cu_3V_2O_8$ Thin Films. *Chem. Mater.* **2017**, *29*, 3334–3345. [CrossRef]
68. Xu, X.; Sun, Y.; Fan, Z.; Zhao, D.; Xiong, S.; Zhang, B.; Zhou, S.; Liu, G. Mechanisms for O_2^- and OH production on flowerlike $BiVO_4$ photocatalysis based on electron spin resonance. *Front. Chem.* **2018**, *6*, 64. [CrossRef]
69. Dubale, A.A.; Ahmed, I.N.; Chen, X.-H.; Ding, C.; Hou, G.-H.; Guan, R.-F.; Meng, X.; Yang, X.-L.; Xie, M.-H. A highly stable metal-organic framework derived phosphorus doped carbon/Cu_2O structure for efficient photocatalytic phenol degradation and hydrogen production. *J. Mater. Chem. A* **2019**, *7*, 6062–6079. [CrossRef]
70. Raciulete, M.; Papa, F.; Kawamoto, D.; Munteanu, C.; Culita, D.C.; Negrila, C.; Atkinson, I.; Bratan, V.; Pandele-Cusu, J.; Balint, I. Particularities of trichloroethylene photocatalytic degradation over crystalline $RbLaTa_2O_7$ nanowire bundles grown by solid-state synthesis route. *J. Environ. Chem. Eng.* **2019**, *7*, 102789. [CrossRef]
71. Raciulete, M.; Papa, F.; Negrila, C.; Bratan, V.; Munteanu, C.; Pandele-Cusu, J.; Culita, D.C.; Atkinson, I.; Balint, I. Strategy for Modifying Layered Perovskites toward Efficient Solar Light-Driven Photocatalysts for Removal of Chlorinated Pollutants. *Catalysts* **2020**, *10*, 637. [CrossRef]
72. Li, T.F.; Liu, J.J.; Jin, X.M.; Wang, F.; Song, Y. Composition-dependent electro-catalytic activities of covalent carbon-$LaMnO_3$ hybrids as synergistic catalysts for oxygen reduction reaction. *Electrochim. Acta* **2016**, *198*, 115–126. [CrossRef]

Article

Au/Ti Synergistically Modified Supports Based on SiO$_2$ with Different Pore Geometries and Architectures

Gabriela Petcu [1], Elena Maria Anghel [1], Elena Buixaderas [2], Irina Atkinson [1], Simona Somacescu [1], Adriana Baran [1], Daniela Cristina Culita [1], Bogdan Trica [3], Corina Bradu [4], Madalina Ciobanu [1] and Viorica Parvulescu [1,*]

[1] Institute of Physical Chemistry "Ilie Murgulescu" of the Romanian Academy, 202 Splaiul Independentei st., 060021 Bucharest, Romania
[2] Institute of Physics, Czech Academy of Sciences, Na Slovance 2, 18200 Prague, Czech Republic
[3] National Institute for Research & Development in Chemistry and Petrochemistry-ICECHIM, Spl. Independentei 202, 060021 Bucharest, Romania
[4] Research Center for Environmental Protection and Waste Management, University of Bucharest, Spl. Independentei 91-95, 050095 Bucharest, Romania
* Correspondence: vpirvulescu@icf.ro

Abstract: New photocatalysts were obtained by immobilization of titanium and gold species on zeolite Y, hierarchical zeolite Y, MCM-48 and KIT-6 supports with microporous, hierarchical and mesoporous cubic structure. The obtained samples were characterized by X-ray diffraction (XRD), N$_2$-physisorption, scanning and transmission electron microscopy (SEM/TEM), diffuse reflectance UV–Vis spectroscopy (DRUV-Vis), X-ray photoelectron spectroscopy (XPS), Raman and photoluminescence spectroscopy. The photocatalytic properties were evaluated in degradation of amoxicillin (AMX) from water, under UV (254 nm) and visible light (532 nm) irradiation. The higher degradation efficiency and best apparent rate constant were obtained under UV irradiation for Au-TiO$_2$-KIT-6, while in the visible condition for the Au-TiO$_2$-MCM-48 sample containing anatase, rutile and the greatest percent of Au metallic clusters were found (evidenced by XPS). Although significant values of amoxicillin degradation were obtained, total mineralization was not achieved. These results were explained by different reaction mechanisms, in which Au species act as e$^-$ trap in UV and e$^-$ generator in visible light.

Keywords: zeolite Y; hierarchical zeolite Y; 3D mesoporous silica; Au/Ti-supported catalysts; surface plasmon resonance effect; photocatalytic degradation of amoxicillin

1. Introduction

Active photocatalysts were obtained by supporting gold species, either on active (TiO$_2$) [1,2] or inert (SiO$_2$) [3] oxides. Activity of the Au-TiO$_2$ photocatalysts is a function of different parameters, such as electronic factors, gold and titania dispersion and interaction [4]. An important strategy to ensure highly dispersed gold species is selection of a suitable support with high surface area, enabling control of the supported particle size and agglomeration rate. Furthermore, it was reported [5] that TiO$_2$ presence on different supports is beneficial not only for photocatalytic activity but also in catalyst manufacturing by improving dispersion of the gold species. It was noticed that TiO$_2$ particles play a dual role in gold immobilization on silica support. Thus, TiO$_2$ particles, evenly dispersed into silica mesopores, act as a scaffold [6], preventing silica framework to collapse and there are an anchor able to interact with Au species. High dispersion of TiO$_2$ is a key factor because gold nanoparticles (AuNPs) interact preferentially with Ti sites. Thus, Au/Ti synergistically modified supports based on SiO$_2$, with different pore geometries and architectures, can enhance the photocatalytic performances under visible light conditions and represent a promising solution in environmental remediation area. The support porosity affects the size of immobilized TiO$_2$ particles due to the confinement effect inside the pores and accessibility of gold species to TiO$_2$ sites, leading to different interaction between titanium and gold as a function of the support nature. Great attention was paid to mesoporous silica due to its specific properties as high surface area and porosity, ordered porous structure and controllable pore size [7,8]. The previous studies [9–11] show that TiO$_2$ loading within SiO$_2$ support ensures modification of the charge carrier dynamics by reductions in recombination intensity and increases in charge carrier lifetime due to the impact with the support electric field. Thus, mesoporous silica is considered an ideal support for a high dispersion of metal on its surface, leading to achievement of catalytic materials with high performances [12]. Among the various ordered mesoporous silica, MCM-48 and KIT-6 materials are more suitable for photocatalytic applications due to their cubic arrangement and highly connected open porous networks, which provide a better access of light irradiation and reactants to the active sites [9,13]. In addition to that, the mesoporous silica with cubic structure favors the formation of a high amount of titania species with tetrahedral coordination, considered to be more photocatalytically active than octahedral coordinated TiO$_2$ [14]. In comparison with mesoporous silica support, a more homogeneous dispersion of TiO$_2$ was achieved using zeolites with microporous structure and high crystallinity degree [15]. However, in catalytic applications, zeolites have the disadvantage of limiting the pore access for reactants with larger molecules. In order to overcome this drawback, hierarchical zeolites were developed [16,17]. These materials preserve the advantages of zeolites as well as mesoporous silica materials, favoring the mass transfer of reactants. Further, mesoporous systems allow for an efficient adsorption/desorption process of the reactants/products [17]. The adsorption of reactants on a photocatalyst is a key step because photogenerated charge carriers have a short lifetime, so only those molecules adsorbed on the surface of the photocatalyst near the active sites can take part in the photocatalytic process.

The Au-TiO$_2$ system was intensively studied and used in wide applications as photocatalytic degradation of different pollutants (azo dyes, chloroaromatic compounds, phenol) from wastewater under visible or UV irradiation [18–20], hydrogen production [18] or photocatalytic reduction of CO$_2$ [12]. Associated with titanium, gold was supported on mesoporous silica materials as SBA-15 [21,22], MCM-41 [23] and microporous TS-1 zeolite [24]. However, from our knowledge, any comparative study has not been reported yet on the effects of support properties on Au-TiO$_2$ photocatalytic activity. No catalytic systems were obtained by immobilizing Au-TiO$_2$ on supports, such as zeolite Y, hierarchical zeolite Y and ordered cubic mesoporous silica (KIT-6 and MCM-48) supports. Therefore, in this work, new photocatalysts were obtained by immobilization of titanium and gold on the supports with various porous structure (micropores—zeolite Y, micro and mesopores—hierarchical zeolite Y, smaller mesopores—MCM-48 and larger mesopores—KIT-6). Zeolite Y exhibits a faujasite (FAU) framework with three-dimensional porous structure [25]. For comparison, two types of mesoporous silica support (MCM-48 and KIT-6), with 3D structure and smaller, respectively, larger mesopores, were selected. In addition to the porous structure, the surface properties were also varied by a change of the surfactant and by incorporation of aluminum in the silica network during the zeolite synthesis. The effects of porous structure and surface properties on TiO$_2$ dispersion, crystal structure, nature of interaction between support–titanium species and Au-TiO$_2$, respectively, were studied. Further, we investigated the influence of support properties, the presence of TiO$_2$ and Au species and their interaction on amoxicillin (AMX) photodegradation under UV and visible light irradiation as a model reaction for applications of these materials in degradation of organic compounds from contaminated water. Amoxicillin is a β-lactam antibiotic from the penicillin class with a broad spectrum and applications in veterinary and human medicine, hardly degradable and its removal from wastewater represents a real interest [26]. The effect of porous structures and adsorption on performances in photocatalytic degradation of amoxicillin (AMX) was also evidenced in previous studies [16,26] and formed the basis of the applications herein studied.

2. Results and Discussion

2.1. Properties of Supported Au/Ti Photocatalysts

X-ray patterns at wide angle of all samples, except that supported on KIT-6, present three distinct diffraction lines at 2θ ≈ 25.2, 37.8, 48.1°, indexed to (101), (004) and (200) crystal planes of anatase TiO_2 phase (Figure 1A). No reflections ascribed to TiO_2 were detected in the XRD pattern of the KTA sample. This may be the result of the high dispersion of titanium on KIT-6 mesoporous support surface with higher concentration of the silanol group. It has been shown that the ability of silica surface for chemical modification is determined by the content of silanol groups Si–OH [27]. The comparative study of two silicas (MCM-41, SBA-15) with hexagonal ordered mesoporous structure explained the decreasing of silanol density for MCM-41 by zeolitization of this silicate obtained in basic conditions. Moreover, the concentration of surface hydroxyl groups increased for MCM-48 compared to MCM-41 due to the disorder in its porous structure [28]. Therefore, we can consider that, among the mesoporous supports used, KIT-6 has a higher concentration of silanol on the surface, as it was demonstrated further by Raman spectroscopy (Figure S6a).

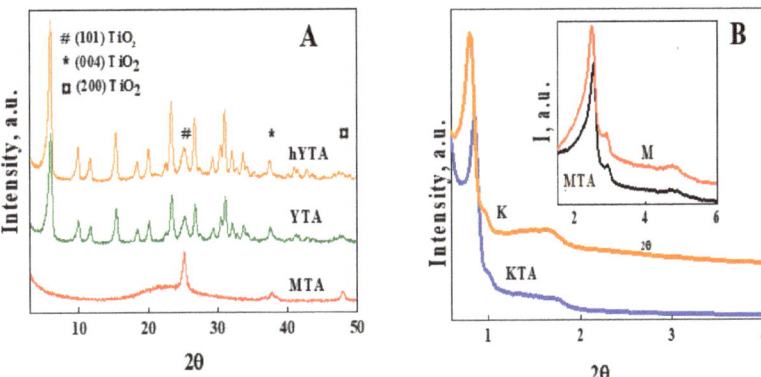

Figure 1. XRD patterns at wide angle (**A**) and low angle (**B**) of synthesized materials.

One can notice that gold-based compounds were not identified in the XRD patterns. The absence of gold reflections can be a consequence of the high dispersion of gold nanoparticles and their low content (<1%), below the limit of detection for XRD analysis [29]. The presence of Ti and Au in all the samples was confirmed by TEM-EDX data (Figure S1).

The XRD patterns of the synthesized photocatalysts showed that structure of the porous supports was insensitive to incorporation of the Ti and Au species. Wide-angle XRD patterns (Figure 1A) of photocatalysts supported on zeolite Y (YTA) and hierarchical zeolite Y (hYTA) have similar diffraction patterns with the corresponding supports [15]. Further, low-angle XRD patterns (Figure 1B) highlight the preservation of MCM-48 and KIT-6 ordered cubic mesoporous structure [7,13] for MTA and KTA samples.

TEM images of the porous host structure after immobilization of Ti and Au species are illustrated in Figure 2. Figure 2A shows the mesoporous cubic-ordered structure of KIT-6 and Figure 2B shows the crystalline structure of zeolite Y. The mesoporous structure of MCM-48 and hierarchical zeolite Y supports, respectively, the crystals from the last one, are illustrated in Figure S2.

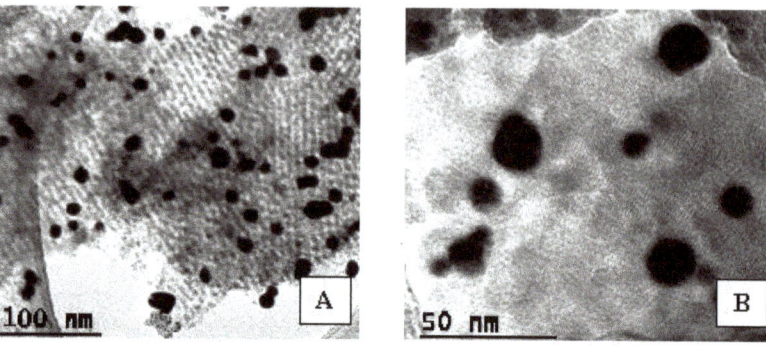

Figure 2. TEM images of KTA (**A**) and YTA (**B**) samples.

According to IUPAC classification, N_2 adsorption–desorption isotherms of Ti-Au photocatalysts (Figure 3) exhibit type IV isotherm with H1 hysteresis loop for materials based on KIT-6 and H3 for the ones based on MCM-48. Analogous N_2 adsorption–desorption isotherms with hysteresis loops of both supported catalysts and their corresponding mesoporous supports imply that shape and pore dimension were not altered by titanium and gold immobilization.

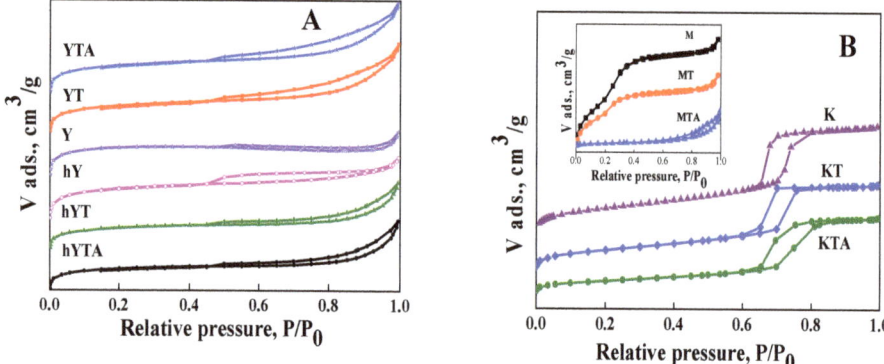

Figure 3. N_2 adsorption–desorption isotherms of the photocatalysts with zeolite Y supports (**A**) and mesoporous silica supports (**B**) MCM-48 (M) and KIT-6 (K).

Conversely, a hysteresis loop came out after TiO_2 immobilization on zeolite Y (Figure 3A), very likely due to mesopores formed among TiO_2 nanoparticles. After gold immobilization on MT, the N_2 adsorption–desorption isotherm does not maintain the steep inflection at the relative pressure near P/P_0 0.25–0.3 as a support isotherm (inseted isotherms in Figure 3B), indicating that metalic species are not immobilized in the pores as an ideally homogeneous layer [30], but rather form deposits at their entrance and out of them. However, titania and gold immobilization caused a more significant decrease in pore volume (Figure 3B) of the MCM-48 support (MT and MTA samples), as opposed to the other mesoporous supports used, e.g., KIT-6. This is due to smaller pore size of MCM-48 compared to KIT-6, as depictable in Table S1, for 10% titania loading. Our previous study on MCM-48-supported catalysts [13] pointed out a smaller decrease in pore volume in conditions of a lower TiO_2 loading (5%). The zeolite-supported catalysts show a combination of type I and IV isotherms with H3 and H4 hysteresis loop [14]. The absence of the adsorption hysteresis for MCM-48 and Ti-MCM-48 samples is a result of their uniform pore distribution. Values of surface areas and pore size after each impregnation step are summarized in Table S1. Further immobilization of gold caused a BET surface increase for the zeolite-supported samples in contrast with the mesoporous ones. The latter finding confirms gold immobilization on titania.

Morphology of the synthesized samples after titanium and gold immobilization is presented in Figure 4. No change in morphology was observed after the impregnations of supports. Thus, the preservation of octahedral morphology with smooth surface, specific to zeolite Y and no aggregates or any other observable defects in the structure of zeolite Y was evidenced. Further, SEM images of KTA and MTA samples revealed typical morphology of MCM-48 and KIT-6 mesoporous silica, respectively [13,31].

Figure 4. SEM images of the synthesized photocatalysts.

TEM images show a high dispersion of gold nanoparticles with diameters ranging from 5 to 24 nm for samples supported on KIT-6 and zeolite Y (Figure 2). Larger Au particles and association of Au-NPs (clusters) on hYT and MT supports are more obvious (Figure S2). These facts are in agreement with the XPS results from Table 1.

Table 1. XPS data: binding energies (BEs) and the quantitative assessment.

Sample	Binding Energy (eV)				Au Chemical Species rel.conc.			
	Au4f/2 Metallic nps	Au4f/2 Clusters	Au4f/2 Au$^+$	Au4f/2 Au^{3+}	Au Metallic nps	Clusters	Au^{1+}	Au^{3+}
YTA	83.3	84.3	85.3	86.7	44.7	18.8	18	18.6
hYTA	83.3	84.3	85.3	86.7	-	60.2	-	39.8
MTA	83.3	84.3	85.3	86.7	-	63.9	-	36.1
KTA	83.3	84.3	85.3	86.7	64.8	-	-	35.2

A higher content of Au clusters on the surface of the hYTA and MTA samples was depicted. XPS spectroscopy provides evidence for the low relative concentrations (<1%) of gold on the sample surface (<10 nm) as the photoelectron spectra are very noisy despite the large number of runs recorded during data acquisition. Moreover, the "band-like" shape of the Au4f spectra suggests the presence of different chemical species accommodated under the experimental spectra envelope.

XPS spectra for the synthesized samples presented in Figure S3 show four types of chemical species, as follows: Au metallic nanoparticles (4 nm), very small Au$^{\delta+}$ clusters as well as Au^{1+} and Au^{3+} oxidation states [32,33].

One can notice that a higher Au content favors the formation of additional species, such as: cluster type Au, Au^{1+} and Au^{3+} oxidation states. As XPS data showed (Table 1), different gold species were obtained as a result of their interaction with TiO$_2$ supported on different porous materials. In heterogeneous catalytic processes, a key factor that deserves great attention is the interaction between metallic species and oxide supports, the phenomenon known as strong metal–support interaction (SMSI), which implies an electron transfer between support and metal species immobilized. The SMSI phenomenon has an important role in photocatalyst stability and activity [34].

A close examination of the Au 4f spectral deconvolution suggests the influence of the synthesis procedure on the Au surface chemical species. In addition to the synthesis conditions, the nature of the surfactant leads to specific properties of the support (pore size and shape, volume and specific area, morphology, stability and surface properties) that influence the dispersion and nature of the titanium species and finally the gold ones. The dispersed titanium species represent a support for gold, their interaction being a strong one (SMSI). Thus, the samples hYTA and MTA whose supports were obtained using a cationic surfactant (TTAB and CTAB, respectively) display similar chemical species. Unlike these, the KTA sample (with support KIT-6 obtained using P123 nonionic surfactant) possesses the highest amount of metallic gold on the surface.

The surface chemistry of titanium, oxygen and silicon is assessed in Figure S4a–c. The two Ti2p peaks indicated that 4+ is the only oxidation state of titanium (Figure S4a). Ti^{4+} from Ti–O–Ti oligomers is represented in octahedral coordination (lower binding energy), while Ti^{4+} from the isolated Ti–O–Si sites is in tetrahedral (higher binding energy) coordination [35,36]. The ratio of these peak intensities indicates a small variation with titania loading. The lower concentration of TiO_2 on the surface of the KTA sample, accessible to XPS measurements, represents another argument regarding its homogeneous distribution on the silica surface as Ti-O-Si species. Although the Ti content is similar to that of the zeolite samples (YTA, hYTA), the highest percentage of TiO_2 on the surface was evidenced for the MTA sample (Figure S4b). This confirms a different interaction of titanium with the surface of the support. In the case of the latter support, the agglomerated oxide species predominate on the surface. Figure S5a–d show the deconvoluted O1s spectrum for all the samples under investigation. For KTA and MTA samples, a peak at ~533 eV is attributed to the O-Si bond from siliceous mesoporous supports (MCM-48, KIT-6), while the peak at a lower BE (~532.5 eV) is assigned to O-Si from zeolite supports. The oxygen peak located at ~530 eV indicates the presence of oxygen bonded to Ti in TiO_2 species with octahedral coordination. The OH groups adsorbed on the TiO_2/SiO_2 surface were detected at ~531.2 eV and in a range 533–534 eV, respectively. We emphasize that in the range 533–534 eV, traces of water adsorbed from the environment might be present on the porous silica surfaces. In order to distinguish between OH bonded as silanol in the sample lattice and the OH from adsorbed water, Raman spectroscopy was used (Figure S6c). For the KTA sample, the presence of both OH group types was evidenced; however, the narrow peak located at 3741 cm^{-1} suggests a higher content of OH from mesoporous silica lattice compared to zeolite Y, for which more adsorbed water is observed, or the YTA sample, which does not present silanol groups. Two distinct binding energies (BEs) were evidenced for Si2p XPS spectra (Figure S4b). The peak from higher BEs (103.2 eV) was assigned to Si-O from MCM-48 and KIT-6 mesoporous silica supports and the one from lower BEs (102.3 eV) to Si-O from zeolite supports. In the case of zeolite samples, the shift in peaks recorded by O1s and Si2p high-resolution XPS spectra is the result of Al^{3+} effect, which reduces the energy in the O-Si bond. Figure S4c shows the presence of aluminum (in a smaller amount than silicon) on the surface of the samples obtained using Y-type zeolite as a support.

Visible Raman spectra of the (Y/hY/K/M)TA powders are illustrated in Figure 5. The ultralow frequency peaks within 12–16 cm^{-1} (inset of the Figure 5) are due to surface vibrations in the particle pore structures [37–39]. Further, a densification process and, hence, enhancement in the nanoparticle sizes give rise to a lower-wavenumber shift in this peak. Consequently, the cubic KTA is expected to have smaller nanoparticle sizes of TiO_2 and/or bigger pore size than the other cubic sample, MTA. This is analogous to the low-angle XRD findings in Figure 1B where the lowest shifted peak (bellow 1°) due to d_{211} plane signifies bigger pore size in the KTA sample in contrast with MTA (peak located above 2°).

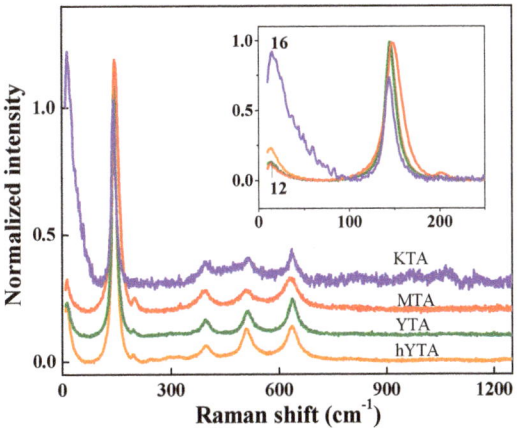

Figure 5. Raman spectra of the (K/M/Y/hY)TA catalysts (laser line of 514 nm).

The non-framework anatase modes (E_g located at 144, 197 and 640 cm^{-1}, B_{1g} at 400 and 519 cm^{-1} and A_{1g} at 507 cm^{-1} [40]) were identified in the Raman spectra of the (K/M/Y/hY)TA samples. Since the 197 cm^{-1} peak is not visible for the KTA sample, there is a phonon confinement effect typically present in the case of TiO_2 [41], for a grain size of only a few nanometers. Despite the strong Raman scattering of anatase, small spectral features of silica support are noticeable in the Raman spectrum of KTA_514 (Figures 5 and S6a) above 750 cm^{-1}. This is a consequence of the most pronounced plasmon resonant effect induced by gold [42] into the KTA catalyst, i.e., the 514 nm laser line used to excite sample is close to the ~550 nm absorption in Figure 6, as well as modification of the internal porosity of silica support by titania. The 144, 400 and 640 cm^{-1} modes of anatase are very sensitive to the crystallite size [43]. A wider peak at about 150 cm^{-1} in the case of the MTA might point out anatase nanoparticles smaller than 10 nm. Further, the presence of rutile in the MTA catalyst is confirmed by its shifted 633 cm^{-1} band [43] in Figure 5 and 616 cm^{-1} band in the UV-Raman spectrum in Figure S6. An enhanced 513 cm^{-1} band for hYTA shows the presence of oxygen motion in a plane bisecting T–O–T bonds (T represents Si and/or Al) [44] in zeolites. A tiny peak at 324 cm^{-1} could be attributable to ring breathing via Si−O−Si linkages [44] in zeolites.

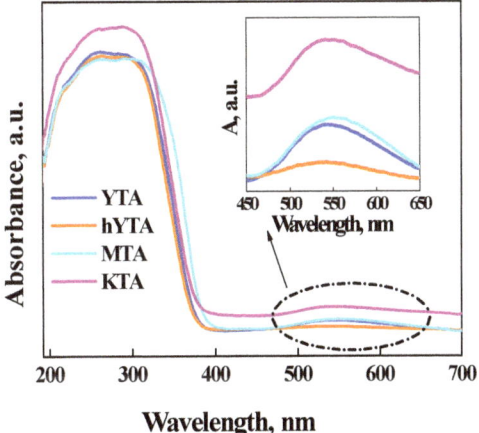

Figure 6. UV-Vis spectra of Ti-Au photocatalysts.

UV-Raman bands of the KA sample, obtained by immobilization of gold on KIT-6 support, are located at 385, 447, 492 and 603 cm^{-1} in Figure S6a, being assignable to six-, five-, four- and three-membered siloxane rings [45,46]. The latter bands are also named defect bands, D_1 and D_2 [46]. The next bands at 791 and 823 cm^{-1} belong to Si-O-Si linkages, while the 973 cm^{-1} band originates from stretching modes on the surface silanols, O_3Si-OH [45,47]. Thus, in comparison with the MTA sample, a higher silanol group content on KTA can be observed. The presence of the silanols is also supported by the high wavenumber band at 3741 cm^{-1} (Figure S6c), assignable to stretching of SiO-H linkages. Typically, UV-Raman investigations are used to remove a fluorescent background and identify possible highly dispersed titanium oxide and/or titanium ions in the micro- and mesoporous frameworks of supports [48]. Thus, loading of TiO_2 into mesoporous silica support should cause a diminishing of the 980 cm^{-1} band and/or shifting towards lower wavenumbers [47] due to gradual interaction with a Ti cation (Si-$O^{\delta-}$ ... $Ti^{\delta+}$) [49]. Since the intensity of the 971 cm^{-1} seems less affected in the case of the KTA spectrum, Ti atoms are very likely incorporated in the KIT-6 framework rather than interacting with surface silanol. No band at about 980 cm^{-1} but a wide band peaking up at 921 cm^{-1} is depicted for MTA in Figure S6a. Bands at 490, 542 and 1124 cm^{-1} are considered to belong to Si-O-Ti linkages [48,49]. It is obvious that UV-Raman is sensitive to the KIT-6 and MCM-48 supports of the KTA and MTA catalysts. Conversely, TiO_2 loading of the zeolite supports caused surface silanol interaction (see Figure S6b) meaning that the 970 cm^{-1} band vanishes. Non-framework anatase is noticeable in both visible and ultraviolet Raman spectra of the (M/K/Y/hY)TA catalysts.

UV-Vis diffuse reflectance spectra of the photocatalysts obtained by immobilization of Au-TiO_2 active species on micro/meso/micro and mesoporous supports are presented in Figure 6.

For all photocatalysts, a high absorption capacity was recorded in the UV domain, revealing a broad absorption band, resulting from overlapping of more absorbtion bands related to several titanium species. Three main absorbtion peaks can be observed for the synthesized samples around 220, 260 and 310 nm. The UV signal at arround 220 nm was associated to tetrahedral Ti species, while the absorption peak at around 260 nm was related to octahedral Ti-oxide species. The shoulder of absorption presented after 300 nm indicates the presence of a TiO_2 anatase phase on the support surface [14,21]. Furthermore, for the MTA sample, a red shift in the absorption bands was shown, corresponding to an increase in TiO_2 particle size, as evidenced using the Scherrer equation (Table S1). This behavior was explained by the quantum size effect that emerges for TiO_2 species with particle sizes <10 nm [15]. After Au immobilization, a specific band absorption in the visible domain, at around 550 nm, was observed. This weak signal recorded in the visible region is due to the surface plasmon resonance effect of gold nanoparticles [12]. This effect is determined by gold particles composition and size and their interaction with titania from the support [21]. These results are in agreement with those obtained by XPS spectroscopy (Table 1).

A better interaction between gold nanoparticles and material used as support leads to a higher stabilization with an increase in corresponding signal intensity [12]. The highest peak assigned to the plasmonic effect of gold nanoparticles was obtained for the photocatalyst supported on KIT-6. XPS results (Table 1) show the highest Au-NPs concentration on the surface for this sample. Gold nanoparticles have a better interaction with titanium species, being highly dispersed, as XRD results revealed. The intensity of the Au plasmonic effect decreased as follows: KTA > YTA ≥ MTA > hYTA. This variation can be very well correlated with the nature of the titanium (Figure S4a) and gold (Table 1) species on the surface. The highest intensity of the plasmonic effect for the KTA sample is due to a high dispersion of titania and, consequently, to the presence of the highest percent of Au metallic nanoparticles.

The band gap energy of photocatalysts was obtained using the Kubelka–Munk function by plotting $[F(R) \cdot h\nu]^{1/2}$ versus photon energy (eV). As shown in Table 2, a narrowing of the band gap was obtained after titanium immobilization on porous supports in comparison with bulk TiO_2 (3.2 eV) due to the high dispersion of titania as tetrahedral Ti-O-Si species [36,50]. The values of band gap energy depend on the physico-chemical properties of porous supports. Further, after gold immobilization, a widening of the band gap values for the all photocatalysis was evidenced. This behavior can be explained by the Burstein–Moss (BM) effect [51], as a result of plasmonic properties of gold nanoparticles.

Table 2. Band gap energy of photocatalysts.

Sample	YT	hYT	KT	MT
Eg (eV)	3.17	3.22	3.14	3.13
Sample	YTA	hYTA	KTA	MTA
Eg (eV)	3.20	3.26	3.19	3.17

Practically, under light irradiation, the conduction band of TiO_2 is filled by free electrons of Au nanoparticles. Therefore, since the conduction band of TiO_2 is blocked, the transitions of electrons from the valence band are allowed only at energy levels higher than the Fermi level. Therefore, the energy required for electron transitions must be large enough to allow the electron transfer to occur not just up to the conduction band, but up to a higher free energy level.

Photoluminescence (PL) spectra were measured to show the effect of active species and support on light-generated electrons and holes, since PL emission is a result of the recombination of the free carriers. The key factors for conjunction of the photoluminescence effect are considered the oxygen vacancies concentration [52,53]. These defects are trapping centers, which inhibit the recombination and simultaneously enable efficient joining electrons with the oxygen, which is why they are considered as active centers for the adsorption on the surface of water and OH− groups necessary for carrying out the photocatalytic reaction.

The photoluminescence results recorded after TiO_2 immobilization on different supports (Figure 7A) show the lowest emission for the MT sample with TiO_2 as a mixture of anatase and rutile crystalline species. This indicates that by supporting TiO_2 on MCM-48 mesoporous silica, the recombination of photogenerated charges was suppressed.

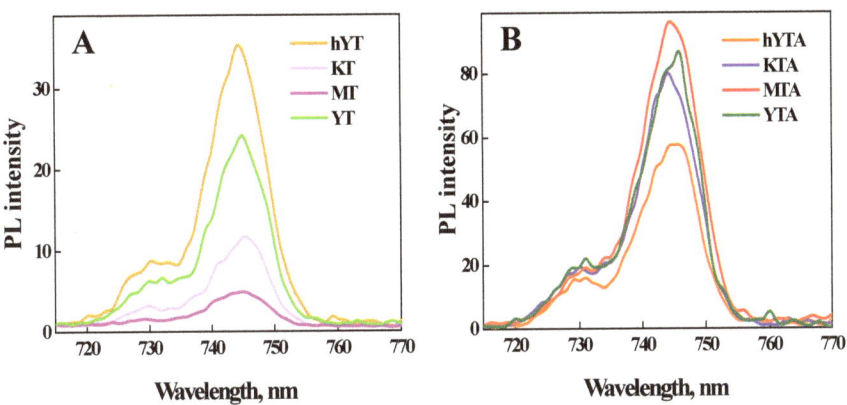

Figure 7. The PL spectra of Ti-modified materials (A) and Au-Ti photocatalysts (B).

For all gold-modified photocatalysts, the enhancement in PL emission was observed (Figure 7B), which can be explained by changes in the level of surface defects, due to the interaction between Au nanoparticles—TiO_2 with higher particle size and different numbers and types of surface defects. The reduction in surface defects enhances the electron number, available to recombine with the holes, leading to an enhancement in the near-band-edge emission [54]. A different number of oxygen vacancies could be created by TiO$_2$ immobilization, as a result of different interaction between TiO$_2$ and silica supports. The electron pair remained in the vacancy after oxygen defect formation at the interface of TiO$_2$/SiO$_2$ moves toward the neighboring Ti atoms, creating reduced Ti^{3+} and Ti^{2+} species.

2.2. Adsorption and Photocatalytic Activity

The AMX adsorption on the obtained materials was evaluated in dark conditions. The obtained results on the adsorption increasing in time are presented in Figure 8.

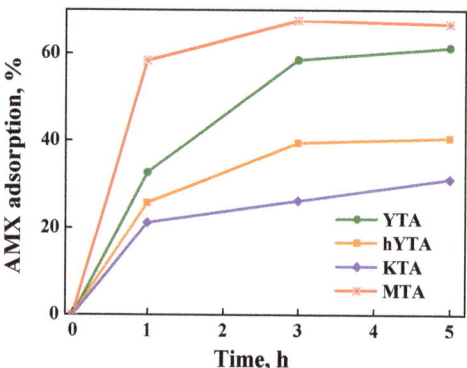

Figure 8. Adsorption capacity of photocalysts in amoxicillin degradation.

There is a significant increase in adsorption during the first hour, after which the variation is insignificant, tending to a level reached after three hours for all samples. The absorption capacity increased in the following order: KTA < hYTA < YTA < MTA.

The higher degradation rate of antibiotics obtained for supported TiO$_2$ on zeolites and mesoporous silica was attributed to the higher adsorption capacity of the supports [55,56]. The dominant adsorption mechanism was attributed to the electrostatic interaction and H-bonding between ionized and unionized molecules of AMX and active groups from the support surface in acidic ($\equiv$Si-OH^{2+}) or basic conditions ($\equiv$Si-O$^-$) [57]. Additionally, the Brönsted acid sites due to the $\equiv$Si-OH-Al$\equiv$ group in the zeolite framework can be easily deprotonated to form $\equiv$Si-O$^-$ [58]. In AMX solution (pH = 7.2), the surface of supports is negatively charged and adsorption is the result of interaction between this deprotonated oxygen and hydrogen atoms from AMX antibiotic, weakly bound to O or N, as a result of intramolecular electronic effects (Figure S7). Therefore, when the photocatalyst is added to the AMX solution, the pH is increased to 9.4. Over time, there is a slight decrease in pH in both cases of adsorption and photocatalytic reaction. It was observed that after 5 h, the pH value was around 8.

Therefore, TiO$_2$ supporting on different materials determines both the number of centers and their strength. A different number of oxygen vacancies could be created by immobilization of TiO$_2$ on zeolite Y, as a result of interaction between TiO$_2$ and support. The electron pair remained in the vacancy after oxygen defect formation at the interface of TiO$_2$ and SiO$_2$ moves toward the neighboring Ti atoms [59].

In degradation of organic compounds from wastewater, adsorption and photocatalytic processes act simultaneously and the results showed a synergistic effect of these [60]. The processes that take place in degradation of amoxicillin (AMX) from water are presented in Figure S8 for the YTA sample. It can be seen that in the first 3 h, the variation in AMX concentration is strongly determined by adsorption. After 3 h, the significant decrease in concentration can be considered only as a result of the photocatalytic process. In photocatalytic degradation of AMX, adsorption can be considered as a rate-determining step. It has been shown that, depending on the photocatalytic process, its relationship with adsorption can be proportional or contradictory [61]. Because amoxicillin adsorption could be the rate-determining step for the photocatalytic reaction, we irradiated the samples after a 30 min stirring step in dark conditions. Degradation of AMX using the obtained photocatalysts showed the effect of a porous support type on the process efficiency. The synthesized photocatalysts were used in degradation of AMX under UV and visible light irradiation. The obtained results are shown in Figure 9A,B.

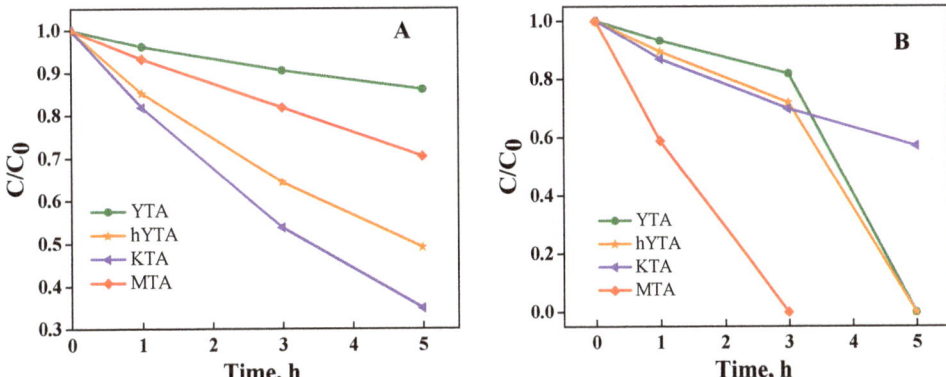

Figure 9. Photocatalytic degradation of amoxicillin under UV irradiation (**A**) and visible light (**B**) irradiation.

Under UV light irradiation, the best results in AMX degradation were obtained for the KTA sample (Figure 9A), with the highest titania dispersion (wide angle XRD- Figure 1A and Raman results—Figure 5) and highest percent of Au metalic nanoparticles on the surface (XPS results—Table 1). A better photoactivity obtained for the sample with smaller size of active species is in accordance with the other reported results [10]. This sample also presents larger ordered mesopores (Figure 1B, Table S1). The lowest photocatalytic degradation of AMX was obtained using the YTA sample, with small pore size. These results suggest an important effect of support textural properties, such as pore size, on obtaining active photocatalysts.

Under visible light irradiation, the best results were obtained for the MTA photocatalyst (Figure 9B). It is the only sample that contains anatase and rutile as titania crystalline species, according to the XRD and Raman results. Further, on the lower surface area of the MTA sample, the highest percent of metallic gold clusters (XPS results—Table 1) was evidenced. The rate of the photocatalytic reaction for this sample is significantly higher (3 h) and, unlike the others, less influenced by the adsorption process. Under these conditions, gold species absorbed the visible light and the photons induced an efficient separation of electron holes. The plasmon-induced electrons in Au NP clusters were transported to CB of TiO_2 and reacted with adsorbed oxygen, forming $\cdot O^{2-}$ active species.

After 5 h of visible light irradiation (Figure 9B), the lowest AMX degradation efficiency was obtained for KTA samples with the lowest titania content and highest percent of Au metallic nanoparticles on the surface. In this case, the activity is determined by the nature of the titanium species. In these samples, the majority are Ti-O-Si species, more active in photocatalytic reactions performed under UV light (KTA sample—Figure 9A).

The photocatalytic performances of the synthesized materials were also evaluated through the AMX degradation kinetics studies. The kinetics results obtained under UV and visible light irradiation are presented in Figure 10A,B. The AMX degradation in the presence of Ti-Au photocatalysts follows a pseudo first-order mechanism, expressed as $-\ln(C/C_0) = k_{app}t$. The high values of linear regression coefficients (R^2) of the kinetic plots confirm that photocatalytic degradation process of AMX using all the synthesized materials follows the pseudo first-order reaction, as reported by previous studies for photocatalytic degradation of AMX [62]. The values of apparent rate constant (k_{app}) of the studied systems are presented in Figure 10A,B.

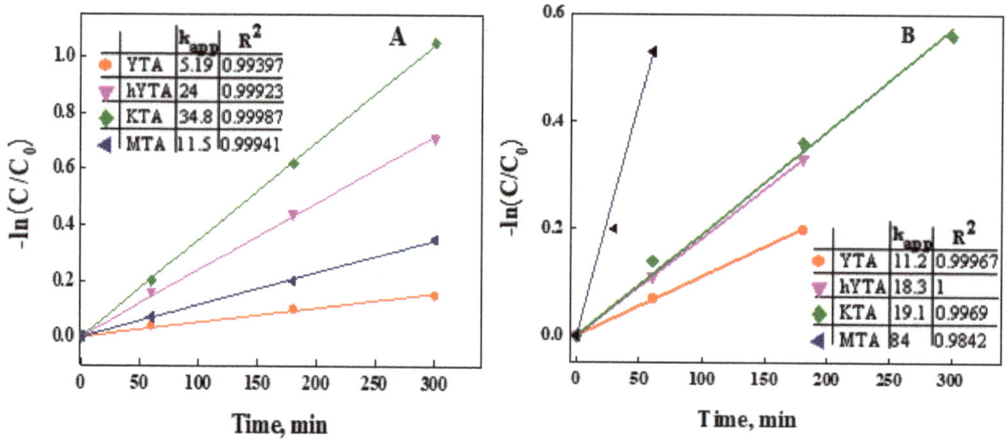

Figure 10. Kinetic and apparent rate constant k_{app} values ($\times 10^{-4}$ min^{-1}) of AMX photocatalytic degradation under UV (**A**) and visible light (**B**) irradiation.

Comparing the apparent rate constants obtained under UV light irradiation, it can be seen (Figure 10A) that KTA is 1.5-, 3- and 6-times higher more photoactive than hYTA, MTA and YTA, respectively. In the case of visible light irradiation, the photocatalytic process is promoted via plasmon activation of the Au NPs [63], followed by injection of photogenerated electrons in the conduction band of TiO$_2$. In these conditions, the best photocatalytic activity was obtained for the MTA sample with the highest value of k_{app} (88.3 × 10^{-4} min^{-1}), due to the presence of TiO$_2$ as a mixture of anatase and rutile phase, as Raman spectra showed (Figure 5). This behavior indicates that the presence of a mixed anatase/rutile TiO$_2$ near Au NP is a crucial factor, leading to a synergistic electron transfer, which enhances the photocatalytic performance. There is a consecutive electron transfer from photoactivated Au NPs to rutile and then to adjacent anatase TiO$_2$ phase [63]. It was noticed that the mixed anatase/rutile TiO$_2$ phases have a high photocatalytic activity comparatively with anatase or rutile single phases [64]. The explanation proposed and widely accepted in the literature is based on a favorable contact between anatase and rutile phases, which ensures the electron transfer, leading to a better separation of photogenerated charge [64].

The kinetics results evidenced that hierarchical structure of zeolite Y leads to an increase in the apparent rate constant k_{app}, probably due to the improvement in accessibility degree of AMX molecules to the active sites.

Gold immobilization on supported TiO$_2$ nanoparticles extends the sensitivity of the photocatalysts into the visible domain and increases their photoactivity through electron injection from activated Au species into the conduction band of the semiconductor. The coupling of electromagnetic field of the incident visible light with oscillations in the electrons in Au NPs leads to an enhancement in electromagnetic fields near the Au NP surface [65]. In the case of UV irradiation, Au NPs act as an electron trap for photogenerated electrons from TiO$_2$, avoiding electron-hole recombination. Generally, noble metals are used in photocatalyst synthesis to facilitate electron capture by forming a Schottky barrier on the metal–semiconductor junction [66]. Au produces the highest Schottky barrier among noble metals [65].

Although high values of amoxicillin degradation were obtained for catalysts as KTA, MTA and hYTA, total mineralization was not achieved. Therefore, the lowest value of total organic carbon elimination (TOC/TOC$_0$) was 0.6 for the hYTA sample after 5 h under UV light irradiation. The low degree of mineralization was also indicated in other studies [55] and has been assigned to certain products more resistant to further oxidation. However, these products may be more biodegradable and bacteria or fungi can mineralize them [67]. The employed HPLC analytical method allowed for the detection of 6 products after AMX degradation under UV irradiation and 11 after its degradation under visible light. Most of them had smaller peak areas. We compared these results with those reported in the literature about amoxicillin degradation products [68–70], which indicate great diversity in intermediates resulting from AMX degradation. The proposed possible reaction pathway showed hydroxylation as the first step of degradation. The HO· radical plays a major role in the attack of benzoic ring or nitrogen moiety of AMX, which possess electrophilic characteristics [69]. Opening of the β-lactam ring of AMX by destruction of lactamic bond can be another pathway for AMX degradation. In the slightly basic conditions of the photocatalytic reactions (pH~8), the hydrolysis of AMX is activated. Furthermore, basic properties of the photocatalyst surface can favor the adsorption of C=O from β-lactam ring of AMX with more electrophilic properties. Nevertheless, the study of [55] showed that hydrolysis may contribute significantly to overall degradation of AMX under chosen irradiation conditions.

2.3. Proposed Mechanism for the Photocatalytic Degradation of AMX

It was previously reported by our group [16] that the photodegradation process of organic molecules as AMX was carried out through highly oxidative species as $\cdot O_2^-$ and HO· radicals, obtained due to the presence of photogenerated charge carriers. Other studies highlighted that AMX photodegradation is mainly due to the reactive species produced during irradiation (e.g., hydroxyl radical, hole and superoxide ion) [68]. Figure 11 shows a schematic representation of the proposed mechanism for photocatalytic degradation of AMX using materials synthesized in this study.

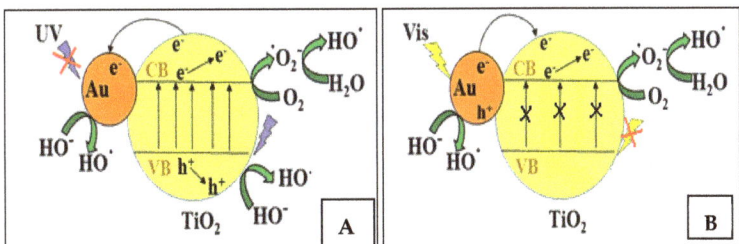

Figure 11. Schematic representation of the proposed mechanism for photocatalytic degradation of AMX under UV (**A**) and visible light irradiation (**B**).

Depending on light irradiation domain, the photocatalytic process could be explained by different roles of Au NPs, which act as electron traps in the case of UV irradiation or as an electron generator under visible light irradiation. When a photocatalytic experiment is carried out under UV irradiation (Figure 11A), TiO_2 nanoparticles are activated and there is an electron transfer from the valence band to the conduction band, leading to the formation of photogenerated charge carriers. Au NPs act as an electron trap, improving the separation of e^-/h^+ pairs. Therefore, unpaired electrons are used to obtain superoxide species ($\cdot O_2^-$) through reduction of O_2 molecules adsorbed on the photocatalyst surface, while unpaired holes ensure the oxidation process of hydroxil ions from aqueous medium to HO· radicals, responsible for AMX degradation. On the other hand, in the case of visible irradiation (Figure 11B), the activated species are Au NPs through the SPR effect. The photogenerated hot electrons from gold nanoparticles are injected into the conduction band of TiO_2 and used to obtain highly reactive oxygen species as $\cdot O_2^-$ and HO· after reduction of adsorbed oxygen molecules. Further, h^+ species remained in Au NPs after electron transfer is used to oxidize hydroxyl ions to HO· radicals.

Degradation efficiency is influenced among other important physico-chemical parameters of photocatalysts by the reactive oxygen species production under light irradiation, their number and the photocatalytic process time. Amoxicillin degradation occurs in several steps, the possible reaction products being AMX diketopiperazine, AMX penicilloic acid, phenol hydroxypyrazine, AMX S-oxide and AMX penilloic acid [69,70]. The variation in inhibition percentage with the irradiation conditions (UV and visible light) may be explained by different contributions of $\cdot O_2^-$ and HO· reactive species. The previous studies on the photocatalytic activity of supported titanium oxide [13] evidenced the main contribution of HO· reactive species to degradation of organic pollutants. In conditions of UV light and slightly alkaline pH, these radicals favor the hydrolysis reactions for the adsorbed amoxicillin molecules. In this way, AMX penicilloic acid, phenol hydroxypyrazine, AMX penilloic acid and AMX diketopiperazine can be obtained. These products, without β-lactamic ring, do not have an inhibitory effect against bacteria. Under visble light irradiation, the presence of gold, which induces the plasmonic effect, favors the formation of $\cdot O_2^-$ species that can lead to oxidation compounds containing the β-lactam ring as AMX S-oxide with an inhibitory effect.

3. Materials and Methods

3.1. Materials

The chemicals used in the photocatalyst synthesis are: sodium silicate solution ($Na_2O(SiO_2)_x \cdot xH_2O$, reagent grade), sodium aluminate ($NaAlO_2$), tetraethyl orthosilicate—TEOS ($Si(OC_2H_5)_4$, reagent grade, 98%), sodium hydroxide (NaOH, ≥98%), ethanol (CH_3CH_2OH, >99.8%), 1-butanol ($CH_3(CH_2)_3OH$, anhydrous, 99.8%), tetradecyltri- methylammonium bromide—TTAB ($CH_3(CH_2)_{13}N(Br)(CH_3)_3$, for synthesis), cetyltri- methylammonium bromide—CTAB ($CH_3(CH_2)_{15}N(Br)(CH_3)_3$, ≥98%) and triblock copolymer P123 (PEG-PPG-PEG, average M_n ~ 5800); all were purchased from Sigma Aldrich (Burlington, MA, USA). Titanium (IV) n-butoxide ($Ti(OCH_2CH_2CH_2CH_3)_4$, reagent grade, 97%) from ACROS Organics (Geel, Belgium) and chloroauric acid ($HAuCl_4 \cdot 3H_2O$, ≥99.9% trace metals basis) from Fluka (NC, USA). The amoxicillin ($C_{16}H_{19}N_3O_5S$, 95.0–102.0% anhydrous basis) for photocatalytic reactions was a Sigma Aldrich product.

3.2. Synthesis of Materials

All porous supports used in this study were synthesized as previously reported by our group [7,13,16]. All the supports were obtained by hydrothermal treatment. For synthesis of MCM-48 mesoporous silica, a cationic surfactant (CTAB), dispersed in a basic medium, was used as template while KIT-6 was obtained in acidic condition, using a non-ionic surfactant (P123). In both cases, alcohol was used with a double role: solvent and co-surfactant (ethanol—EtOH for MCM-48 and butanol—BuOH for KIT-6). Hierarchical zeolite Y was obtained by a seed assisted method, in the presence of a cationic surfactant (TTAB). Zeolite Y was obtained similarly, but in the absence of the surfactant. Cubic mesoporous silica MCM-48 and KIT-6 was suggestively noted M and K, while zeolite Y materials with microporous and hierarchical structure were denominated Y and hY, respectively.

Titanium and gold active sites were incorporated in the obtained supports by post-synthesis method (impregnation). The cubic silica and zeolite Y supports were impregnated with titanium species from alcoholic solution of titanium (IV) n-butoxide obtaining samples with 10% TiO_2 loading. After impregnation, materials were kept for 24 h at room temperature, dried at 80 °C and calcinated in air at 600°. The obtained samples were named YT, hYT, MT and KT and were used as supports for gold immobilization. Au was immobilized by a double-impregnation method using $HAuCl_4$ solution. The amount of gold precursor was equivalent to 1 wt.% of Au in each sample. In the second step the chlorine was removed using a Na_2CO_3 solution [35]. The obtained samples were washed with sodium carbonate solution and deionized water, centrifuged, dried overnight at room temperature and then at 120 °C for 6 h. This method was chosen due to the advantage of preparation of highly active materials by using a simple method and a suitable precursor for gold as chloroauric acid [26,35]. The obtained materials were named YTA, hYTA, MTA and KTA.

3.3. Characterization of Materials

The crystalline and ordered mesoporous structures of materials were examined at wide and low angles using powder X-ray diffraction (Rigaku Ultima IV diffractometer-Rigaku Corp., Tokyo, Japan-with Cu Kα, λ = 0.15406 nm). Crystallite size of supported TiO_2 was calculated using Scherrer's formula along (101) direction: D = k·λ/(FWHM)·cos(θ), where λ is the wavelength of the Cu Kα radiation (1.54056 Å), FWHM is full width at half maximum of the intensity vs. 2θ profile, θ is Bragg's diffraction angle and k is the shape factor (0.9). Elemental analysis of the obtained samples was determined using Rigaku ZSX Primus II spectrometer and the textural properties were characterized by Micromeritics ASAP 2020. Morphology, sample composition, ordered porous structure, zeolite crystals and dispersion of gold nanoparticles were evidenced by scanning electron microscopy (ZEISS EVO LS10 SEM, Germany) and transmission electron microscopy (TECNAI 10 G2-F30 and F20 G2 TWIN Cryo-TEM -FEI with EDX, OR, USA).

Surface analysis was performed by X-ray photoelectron spectroscopy (XPS—PHI Quantera equipment, ON, Canada). The X-ray source was monochromatized Al Kα radiation (1486.6 eV) and overall energy resolution was estimated at 0.6 eV by the full width at half-maximum (FWHM) of the Au4f7/2 photoelectron line (84 eV).

The interaction with light of photocatalysts was examined using Raman, UV-Vis diffuse reflectance (JASCO V570 spectrophotometer, Tokyo, Japan) and photoluminescence spectroscopy. Micro-Raman spectra were recorded by means of a RM-1000 Renishaw Raman (Renishaw, New Mills, Wotton-Under-Edge, UK) and a LabRam HR800 spectrometer (HORIBA FRANCE SAS, Palaiseau, France) equipped with CCD detectors and gratings of 2400 gr/mm. Two lasers, Ar$^+$-ion and He-Cd operating at 514.5 nm and 325 nm, respectively, were used as exciting radiations through the 50× and 40× NUV microscope objectives manufactured by Leica (Leica, Wetzlar, Germany) and Olympus (Olympus Corporation, Tokyo, Japan). Owing to two BragGrate filters, visible Raman spectra were recorded from 5 cm^{-1} in order to access pore and nanoparticle size information. The UV-Vis diffuse reflectance spectra were recorded on a JASCO V570 spectrophotometer. The fluorescence spectra were recorded with an FLSP 920 spectrofluorimeter (Edinburgh Instruments, Livingston, UK). The excitation source was an Xe lamp, excitation wavelength 550 nm and spectra were recorded between 570 and 800 nm. The excitation and emission slits were of 10 nm for all measurements.

3.4. Adsorption and Photocatalytic Experiments

The adsorption and photocatalytic degradation of amoxicillin reactions were carried out in a dark room thermostated at 30 °C. Next, 20 mg of the photocatalyst was added in quartz microreactors with 10 mL aqueous solution of amoxicillin—AMX (30 mg/L)—under stirring. Before reaction, mixture was first stirred in darkness for 30 min in order to ensure adsorption of the AMX on photocatalyst surface. Then, the reaction mixture was irradiated under UV light (λ = 254 nm), using 2 × 60 W halogen lamps, and laser-visible light (532 nm). At given irradiated time intervals (1, 3 and 5 h, respectively), 3 mL of suspension was taken out and the catalyst was separated from the suspension using Millipore syringe filter of 0.45 µm. The absorbance (A) of solution was measured by JASCO V570 UV-Vis spectrophotometer and degradation process evaluated by the variation in the maximum absorbance of AMX (λ = 230 nm). To evaluate AMX degradation, the solution phase concentration C (mg/L) at moment t and initial concentration C_0 (mg/L) at t = 0 were used.

Considering the AMX degradation as a first-order reaction, the apparent rate constant (k_{app}) was calculated using the relation: $-\ln(C/C_0)$ = kappt, where kapp is the apparent rate constant of the photodegradation process, C_0 and C represent AMX concentrations (mg/L) at the initial time and a given time t (min), respectively.

For AMX degradation monitoring RigolL300 HPLC equipped with an UV-Vis diode-array detector (λ: 210; 240 and 274 nm) and a Zorbax SB-C18 column (Agilent, Santa Clara, CA, USA) was employed. The mobile phase was a mixture of phosphate-buffer solution (20 mM)/acetonitrile, with a ratio of 80/20, flowing with 1 mL/min. The mineralization degree was monitored by total organic carbon (TOC) analysis. The TOC value was calculated by subtracting the value of the total inorganic carbon (TIC) from the value of the total carbon (TC). TC and TIC measurements were performed using an HiPer TOC Thermo Electron analyzer based on carbon dioxide infrared absorption. For TC, the UV-Persulfate oxidation method was employed, while TIC method involves the sample acidification and quantification of the released CO_2. Prior to analysis, all the samples collected during the runs were filtered through 0.20 μm regenerate cellulose filters. All chromatographic, TOC analyses and antibacterial activity were performed in duplicate.

4. Conclusions

High-performance photocatalysts were obtained by immobilization of Ti and Au NPs on different supports with specific structural, textural and morphological properties. For dispersion of Ti species and, implicitly, of Au NPs on the support, interaction between them and interaction of each one with the support are considered the key factors that have a synergistic effect on photocatalytic performances of synthesized materials. The pore size, surface and adsorption properties influence the photocatalytic process, both by providing active species available for AMX degradation, as well as by ensuring the reagent mass transfer. The best results (100% degradation efficiency) were obtained under visible light for samples with Au metallic clusters on the surface (XPS results). Between these samples, the highest apparent rate constant was obtained for that obtained by dispersion of Ti and Au on MCM-48 support, the only one containing both anatase and rutile as titania species. A high adsorption capacity was also evidenced for this sample. The reaction mechanism, in which Au is e- trap in UV and e- generator in visible light, explained the synergistic action of support, as Au and TiO_2 species in photocatalytic reactions.

Although high values of amoxicillin degradation were obtained, total mineralization was not achieved. The low degree of mineralization was assigned to certain products more resistant to further oxidation. A different number of AMX by-products was obtained under UV and visible light irradiation. These results were explained by the difference between reaction mechanisms under UV and visible light.

Supplementary Materials: The following supporting information can be downloaded at: https://www.mdpi.com/article/10.3390/catal12101129/s1, Figure S1: EDX results for KTA (A), hYTA (B), YTA (C) and MTA (D) samples and the TEM image of the evaluated area (Cu and C are components of TEM grid support for the powder of sample); Figure S2: TEM images of samples MTA (A,B) and hYTA (C,D), Figure S3: (a–d) Au 4f (7/2, 5/2) high resolution, deconvoluted XPS spectra for the synthesized samples, Figure S4: Ti2p (a), Si2p (b) and Al2p (c) high resolution XPS spectra, Figure S5: The deconvoluted O1s spectra for YTA (a), hYTA (b), MTA (c) and KTA (d), Figure S6: Raman spectra of the cubic mesoporous supported (M/K)TA photocatalysts (a), zeolite supported (hY/Y)TA photocatalysts (b) and Y, (Y/K)TA photocatalysts (c). $λ_L$ is the laser excitation line and A is anatase, Figure S7: Chemical structure of amoxicillin (AMX), Figure S8: The processes that take place in the removal of amoxicillin (AMX) from water (YTA photocatalyst); Table S1: Texture properties of samples and TiO_2 crystallite size of the samples after each impregnation step.

Author Contributions: Conceptualization, V.P.; methodology, V.P. and G.P.; validation, V.P. and E.M.A.; formal analysis, G.P., I.A. and M.C.; investigation, G.P., E.M.A., E.B., S.S. and C.B.; data curation, G.P., E.M.A., I.A., E.B., A.B., D.C.C. and B.T.; writing—original draft preparation, G.P., E.M.A. and S.S.; writing—review and editing, V.P.; visualization, V.P.; supervision, V.P. All authors have read and agreed to the published version of the manuscript.

Funding: This research received no external funding.

Data Availability Statement: The data presented in this study are available on request from the corresponding author.

Acknowledgments: The authors are thankful for the financial support from EU (ERDF) INFRANANOCHEM–No. 19/01.03.2009 project. The research infrastructure developed through this project and POS-CCE "AGRI-FLUX" project, no. 645/18.03.2014, SMIS-CSNR 48695 was partially used for sample characterization. Further, the authors would like to thank Alin Enache (Apel Laser) for the visible light laser (DPSS-532-100, λ = 532 nm) needed for the photocatalytic experiments.

Conflicts of Interest: The authors declare no conflict of interest.

References

1. Yoshiiri, K.; Wang, K.; Kowalska, E. TiO_2/Au/TiO_2 Plasmonic Photocatalysts: The Influence of Titania Matrix and Gold Properties. *Inventions* **2022**, *7*, 54. [CrossRef]
2. Do, T.C.M.V.; Nguyen, D.Q.; Nguyen, K.T.; Le, P.H. TiO_2 and Au-TiO_2 Nanomaterials for Rapid Photocatalytic Degradation of Antibiotic Residues in Aquaculture Wastewater. *Materials* **2019**, *12*, 2434. [CrossRef] [PubMed]
3. Yang, C.; Kalwei, M.; Schüth, F.; Chao, K. Gold nanoparticles in SBA-15 showing catalytic activity in CO oxidation. *Appl. Catal. A* **2003**, *254*, 289–296. [CrossRef]
4. Moragues, A.; Puértolas, B.; Mayoral, Á.; Arenal, R.; Hungría, A.B.; Murcia-Mascarós, S.; Taylor, S.H.; Solsona, B.; García, T.; Amorós, P. Understanding the role of Ti-rich domains in the stabilization of gold nanoparticles on mesoporous silica-based catalysts. *J. Catal.* **2018**, *360*, 187–200. [CrossRef]
5. Luna, M.; Gatica, J.M.; Vidal, H.; Mosquera, M.J. Au-TiO_2/SiO_2 photocatalysts with NOx depolluting activity: Influence of gold particle size and loading. *Chem. Eng. J.* **2019**, *368*, 417–427. [CrossRef]
6. Qi, F.; Wang, C.; Cheng, N.; Liu, P.; Xiao, Y.; Li, F.; Sun, X.; Liu, W.; Guo, S.; Zhao, X.-Z. Improving the performance through SPR effect by employing Au@SiO_2 core-shell nanoparticles incorporated TiO_2 scaffold in efficient hole transport material free perovskite solar cells. *Electrochim. Acta* **2018**, *282*, 10–15. [CrossRef]
7. Filip, M.; Todorova, S.; Shopska, M.; Ciobanu, M.; Papa, F.; Somacescu, S.; Munteanu, C.; Parvulescu, V. Effects of Ti loading on activity and redox behavior of metals in PtCeTi/KIT-6 catalysts for CH_4 and CO oxidation. *Catal. Today* **2018**, *306*, 138–144. [CrossRef]
8. Rizzi, F.; Castaldo, R.; Latronico, T.; Lasala, P.; Gentile, G.; Lavorgna, M.; Striccoli, M.; Agostiano, A.; Comparelli, R.; Depalo, N.; et al. High Surface Area Mesoporous Silica Nanoparticles with Tunable Size in the Sub-Micrometer Regime: Insights on the Size and Porosity Control Mechanisms. *Molecules* **2021**, *26*, 4247. [CrossRef]
9. Lee, Y.Y.; Jung, H.S.; Kim, J.M.; Kang, Y.T. Photocatalytic CO_2 conversion on highly ordered mesoporous materials: Comparisons of metal oxides and compound semiconductors. *Appl. Catal. B Environ.* **2018**, *224*, 594–601. [CrossRef]
10. Sun, Z.; Bai, C.; Zheng, S.; Yang, X.; Frost, R.L. A comparative study of different porous amorphous silica mineral ssupported TiO_2 catalysts. *Appl. Catal. A Gen.* **2013**, *458*, 103–110. [CrossRef]
11. Vodyankin, A.A.; Vodyankin, O.V. The Effect of Support on the Surface Properties and Photocatalytic Activity of Supported TiO_2 Catalysts. *Key Eng. Mater.* **2016**, *670*, 224–231. [CrossRef]
12. Yadav, R.; Amoli, V.; Singh, J.; Tripathi, M.K.; Bhanja, P.; Bhaumik, A.; Sinha, A.K. Plasmonic gold deposited on mesoporous $Ti_xSi_{1-x}O_2$ with isolated silica in lattice: An excellent photocatalyst for photocatalytic conversion of CO_2 into methanol under visible light irradiation. *J. CO2 Util.* **2018**, *27*, 11–21. [CrossRef]
13. Mureseanu, M.; Filip, M.; Somacescu, S.; Baran, A.; Carja, G.; Parvulescu, V. Ce, Ti modified MCM-48 mesoporous photocatalysts: Effect of the synthesis route on support and metal ion properties. *Appl. Surf. Sci.* **2018**, *444*, 235–342. [CrossRef]
14. Peng, R.; Zhao, D.; Dimitrijevic, N.M.; Rajh, T.; Koodali, R.T. Room Temperature Synthesis of Ti–MCM-48 and Ti–MCM-41 Mesoporous Materials and Their Performance on Photocatalytic Splitting of Water. *J. Phys. Chem. C* **2012**, *116*, 1605–1613. [CrossRef]
15. Jianga, C.; Lee, K.Y.; Parlett, C.M.A.; Bayazit, M.K.; Lau, C.C.; Ruan, Q.; Moniz, S.J.A.; Lee, A.F.; Tang, J. Size-controlled TiO_2 nanoparticles on porous hosts for enhanced photocatalytic hydrogen production. *Appl. Catal. A Gen.* **2016**, *521*, 133–139. [CrossRef]
16. Petcu, G.; Anghel, E.M.; Somacescu, S.; Preda, S.; Culita, D.; Mocanu, S.; Ciobanu, M.; Parvulescu, V. Hierarchical Zeolite Y Containing Ti and Fe Oxides as Photocatalysts for Degradation of Amoxicillin. *J. Nanosci. Nanotechnol.* **2020**, *20*, 1158–1169. [CrossRef] [PubMed]
17. Kim, M.-R.; Kim, S. Enhanced Catalytic Oxidation of Toluene over Hierarchical Pt/Y Zeolite. *Catalysts* **2022**, *12*, 622. [CrossRef]
18. Ayati, A.; Ahmadpour, A.; Bamoharram, F.F.; Tanhaei, B.; Manttari, M.; Sillanpaa, M. A review on catalytic applications of Au/TiO_2 nanoparticles in the removal of water pollutant. *Chemosphere* **2014**, *107*, 163–174. [CrossRef]
19. Zhu, H.; Chen, X.; Zheng, Z.; Ke, X.; Jaatinen, E.; Zhao, J.; Guo, C.; Xied, T.; Wangd, D. Mechanism of supported gold nanoparticles as photocatalysts under ultraviolet and visible light irradiation. *Chem. Commun.* **2009**, *7524*, 7524–7526. [CrossRef]
20. Tang, K.Y.; Chen, J.X.; Legaspi, E.D.R.; Owh, C.; Lin, M.; Tee, I.S.Y.; Kai, D.; Loh, X.J.; Li, Z.; Regulacio, M.D.; et al. Gold-decorated TiO_2 nanofibrous hybrid for improved solar-driven photocatalytic pollutant degradation. *Chemosphere* **2021**, *265*, 129114. [CrossRef]
21. Sacaliuc, E.; Beale, A.M.; Weckhuysen, B.M.; Nijhuis, T.A. Propene epoxidation over Au/Ti-SBA-15 catalysts. *J. Catal.* **2007**, *248*, 235–248. [CrossRef]

22. Gutiérrez, L.F.; Hamoudi, S.; Belkacemi, K. Synthesis of Gold Catalysts Supported on Mesoporous Silica Materials: Recent Developments. *Catalysts* **2011**, *1*, 97–154. [CrossRef]
23. Zhou, J.; Yang, X.; Wang, Y.; Chen, W. An efficient oxidation of cyclohexane over Au@TiO$_2$/MCM-41 catalyst prepared by photocatalytic reduction method using molecular oxygen as oxidant. *Catal. Commun.* **2014**, *46*, 228–233. [CrossRef]
24. Xu, J.; Zhang, Z.; Wang, G.; Duan, X.; Qian, G.; Zhou, X. Zeolite crystal size effects of Au/uncalcined TS-1 bifunctional catalysts on direct propylene epoxidation with H$_2$ and O$_2$. *Chem. Eng. Sci.* **2020**, *227*, 115907. [CrossRef]
25. Liu, Z.; Shi, C.; Wu, D.; He, S.; Ren, B. A Simple Method of Preparation of High Silica Zeolite Y and Its Performance in the Catalytic Cracking of Cumene. *J. Nanotehnol.* **2016**, *2016*, 1486107. [CrossRef]
26. Basha, S.; Barr, C.; Keane, D.; Nolan, K.; Morrissey, A.; Oelgemoller, M.; Tobin, J.M. On the adsorption/photodegradation of amoxicillin in aqueous solutions by an integrated photocatalytic adsorbent (IPCA): Experimental studies and kinetics analysis. *Photochem. Photobiol. Sci.* **2011**, *10*, 1014–1022. [CrossRef]
27. Kozlova, A.; Kirik, S.D. Post-synthetic activation of silanol covering in the mesostructured silicate, materials MCM-41 and SBA-15. *Microporous Mesoporous Mater.* **2010**, *133*, 124–133. [CrossRef]
28. Kumar, D.; Schumacher, K.; du Fresne von Hohenesche, C.; Grun, M.; Unger, K.K. MCM-41, MCM-48 and related mesoporous adsorbents: Their synthesis and characterization. *Colloids Surf. A Physicochem. Eng. Asp.* **2001**, *187–188*, 109–116. [CrossRef]
29. Kishor, R.; Singh, S.B.; Ghoshal, A.K. Role of metal type on mesoporous KIT-6 for hydrogen storage. *Int. J. Hydrogen Energy* **2018**, *43*, 10376–10385. [CrossRef]
30. Duan, Y.; Zhai, D.; Zhang, X.; Zheng, J.; Li, C. Synthesis of CuO/Ti-MCM-48 Photocatalyst for the Degradation of Organic Pollutions Under Solar-Simulated Irradiation. *Catal. Lett.* **2017**, *148*, 51–61. [CrossRef]
31. Purushothaman, R.; Palanichamy, M.; Bilal, I.M. Functionalized KIT-6/Terpolyimide Composites with Ultra-Low Dielectric Constant. *J. Appl. Polym. Sci.* **2014**, *131*, 40508. [CrossRef]
32. Naumkin, A.V.; Kraut-Vass, A.; Gaarenstroom, S.W.; Poell, C.J. *NIST X-ray Photoelectron Spectroscopy Database*; Version 4.1; NIST Standard Reference Database NIST SRD 20, National Institute of Standards and Technology: Gaithersburg, MD, USA, 2012.
33. Moulder, F.; Stickle, W.F.; Sobol, P.E.; Bomben, K.D. *Handbook of X-Ray Photoelectron Spectroscopy*, ULVAC-PHI, Inc, 370 Enzo, Chigasaki 253-8522, Japan; Perkin-Elmer Corporation: Eden Prairie, MN, USA, 1995.
34. Zhang, Y.; Liu, J.X.; Qian, K.; Jia, A.; Li, D.; Shi, L.; Hu, J.; Zhu, J.; Huang, X. Structure Sensitivity of Au-TiO$_2$ Strong Metal–Support Interactions. *Angew. Chem.* **2021**, *60*, 12074–12081. [CrossRef] [PubMed]
35. Perera, A.S.; Trogadas, P.; Nigra, M.M.; Yu, H.; Coppens, M.O. Optimization of mesoporous titanosilicate catalysts for cyclohexene epoxidation via statistically guided synthesis. *J. Mater. Sci.* **2018**, *53*, 7279–7293. [CrossRef] [PubMed]
36. Zhang, H.; Tang, C.; Lv, Y.; Gao, F.; Dong, L. Direct synthesis of Ti-SBA-15 in the self-generated acidic environment and its photodegradation of Rhodamine. *J. Porous Mater.* **2014**, *21*, 63–70. [CrossRef]
37. Montagna, M. Characterization of Sol-Gel Materials by Raman and Brillouin Spectroscopies. In *Handbook of Sol-Gel Science and Technology*; Klein, L., Aparicio, M., Jitianu, A., Eds.; Springer International Publishing: Cham, Switzerland, 2016; pp. 1–32.
38. Sauvajol, J.; Pelous, J.; Woignier, T.; Vacher, R. Low frequency Raman study of the harmonic vibrational modes in silica aerogels. *J. Phys. Colloq.* **1989**, *50*, 167–169. [CrossRef]
39. Machon, D.; Bois, L.; Fandio, D.J.J.; Martinet, Q.; Forestier, A.; LeFloch, S.; Margueritat, J.; Pischedda, V.; Morris, D.; Saviot, L. Revisiting Pressure-induced Transitions in Mesoporous Anatase TiO$_2$. *J. Phys. Chem. C* **2019**, *123*, 23488–23496. [CrossRef]
40. Ivanda, M.; Musić, S.; Gotić, M.; Turković, A.; Tonejc, A.M.; Gamulin, O. XRD, Raman and FT-IR spectroscopic observations of nanosized TiO$_2$ synthesized by the sol–gel method based on an esterification reaction. *J. Mol. Struct.* **1999**, *480*, 645–649. [CrossRef]
41. Zhu, K.-R.; Zhang, M.S.; Chen, Q.; Yin, Z. Size and phonon-confinement effects on low-frequency Raman mode of anatase TiO$_2$ nanocrystal. *Phys. Lett. A* **2005**, *340*, 220–227. [CrossRef]
42. Yang, X.; Wang, Y.; Zhang, L.; Fu, H.; He, P.; Han, D.; Lawson, T.; An, X. The Use of Tunable Optical Absorption Plasmonic Au and Ag Decorated TiO$_2$ Structures as Efficient Visible Light Photocatalysts. *Catalysts* **2020**, *10*, 139. [CrossRef]
43. Georgescu, D.; Baia, L.; Ersen, O.; Baia, M.; Simon, S. Experimental assessment of the phonon confinement in TiO$_2$ anatase nanocrystallites by Raman spectroscopy. *J. Raman Spectrosc.* **2012**, *43*, 876–883. [CrossRef]
44. Wang, T.; Luo, S.; Tompsett, G.A.; Timko, M.T.; Fan, W.; Auerbach, S.M. Critical Role of Tricyclic Bridges Including Neighboring Rings for Understanding Raman Spectra of Zeolites. *J. Am. Chem. Soc.* **2019**, *141*, 20318–20324. [CrossRef] [PubMed]
45. Luan, Z.; Maes, E.M.; van der Heide, P.A.W.; Zhao, D.; Czernuszewicz, R.S.; Kevan, L. Incorporation of Titanium into Mesoporous Silica Molecular Sieve SBA-15. *Chem. Mater.* **1999**, *11*, 3680–3686. [CrossRef]
46. Aguiar, H.; Serra, J.; González, P.; León, B. Structural study of sol–gel silicate glasses by IR and Raman spectroscopies. *J. Non-Cryst. Solids* **2009**, *355*, 475–480. [CrossRef]
47. Liu, Q.; Li, J.; Zhao, Z.; Gao, M.; Kong, L.; Liu, J.; Wei, Y. Design, synthesis and catalytic performance of vanadium-incorporated mesoporous silica KIT-6 catalysts for the oxidative dehydrogenation of propane to propylene. *Catal. Sci. Technol.* **2016**, *6*, 5927–5941. [CrossRef]
48. Fan, F.; Feng, Z.; Li, C. UV Raman Spectroscopic Studies on Active Sites and Synthesis Mechanisms of Transition Metal-Containing Microporous and Mesoporous Materials. *Acc. Chem. Res.* **2010**, *43*, 378–387. [CrossRef]
49. Ricchiardi, G.; Damin, A.; Bordiga, S.; Lamberti, C.; Spano, G.; Rivetti, F.; Zecchina, A. Vibrational Structure of Titanium Silicate Catalysts. A Spectroscopic and Theoretical Study. *J. Am. Chem. Soc.* **2001**, *123*, 11409–11419. [CrossRef]

50. Parvulescu, V.; Ciobanu, M.; Petcu, G. Immobilization of semiconductor photocatalysts. In *Handbook of Smart Photocatalytic Materials Fundamentals, Fabrications, and Water Resources Applications*; Hussain, C.M., Mishra, A.K., Eds.; Elsevier: Amsterdam, The Netherlands, 2020.
51. Zhu, Q.; Lu, J.; Wang, Y.; Qin, F.; Shi, Z.; Xu, C. Burstein-Moss Effect Behind Au Surface Plasmon Enhanced Intrinsic Emission of ZnO Microdisks. *Sci. Rep.* **2016**, *6*, 36194. [CrossRef]
52. Wojcieszak, D.; Kaczmarek, D.; Domaradzki, J.; Mazur, M. Correlation of Photocatalysis and Photoluminescence Effect in Relation to the Surface Properties of TiO_2:Tb Thin Films. *Int. J. Photoenergy* **2013**, *2013*, 526140. [CrossRef]
53. Liqiang, J.; Yichun, Q.; Baiqi, W.; Shudan, L.; Baojiang, J.; Libin, Y.; Wei, F.; Honggang, F.; Jiazhong, S. Review of photoluminescence performance of nano-sized semiconductor materials and its relationships with photocatalytic activity. *Sol. Energy Mater. Sol. Cells* **2006**, *90*, 1773–1787. [CrossRef]
54. De Lourdes Ruiz Peralta, M.; Pal, U.; Sanchez Zeferino, R. Photoluminescence (PL) Quenching and Enhanced Photocatalytic Activity of Au-Decorated ZnO Nanorods Fabricated through Microwave-Assisted Chemical Synthesis. *ACS Appl. Mater. Interfaces* **2012**, *4*, 4807–4816. [CrossRef]
55. Kanakaraju, D.; Kockler, J.; Motti, C.A.; Glass, B.D.; Oelgemoller, M. Titanium dioxide/zeolite integrated photocatalytic adsorbents for the degradation of amoxicillin. *Appl. Catal. B Environ.* **2014**, *166–167*, 45–55. [CrossRef]
56. Nairi, V.; Medda, L.; Monduzzi, M.; Salis, A. Adsorption and release of ampicillin antibiotic from ordered mesoporous silica. *J. Colloid Interface Sci.* **2017**, *497*, 217–225. [CrossRef] [PubMed]
57. de Sousa, D.N.R.; Insa, S.; Mozeto, A.A.; Petrovic, M.; Chaves, T.F.; Fadini, P.S. Equilibrium and kinetic studies of the adsorption of antibiotics from aqueous solutions onto powdered zeolites. *Chemosphere* **2018**, *205*, 137–146. [CrossRef] [PubMed]
58. Kuzniatsova, T.; Kim, Y.; Shqau, K.; Dutta, P.K.; Verweij, H. Zeta potential measurements of zeolite Y: Application in homogeneous deposition of particle coatings. *Micropor. Mesopor. Mat.* **2007**, *103*, 102–107. [CrossRef]
59. Chinh, V.D.; Broggi, A.; Di Palma, L.; Scarsella, M.; Speranza, G.; Vilardi, G.; Thang, P.N. XPS Spectra Analysis of Ti^{2+}, Ti^{3+} Ions and Dye Photodegradation Evaluation of Titania-Silica Mixed Oxide Nanoparticles. *J. Electron. Mater.* **2018**, *47*, 2215–2224. [CrossRef]
60. Liu, W.; He, T.; Wang, Y.; Ning, G.; Xu, Z.; Chen, X.; Hu, X.; Wu, Y.; Zhao, Y. Synergistic adsorption-photocatalytic degradation efect and norfoxacin mechanism of ZnO/ZnS@BC under UV-light irradiation. *Sci. Rep.* **2020**, *10*, 11903. [CrossRef]
61. Emara, M.M.; Ali, S.H.; Hassan, A.A.; Kassem, T.S.E.; Van Patten, P.G. How does photocatalytic activity depend on adsorption, composition, and other key factors in mixed metal oxide nanocomposites. *Colloid Interfac. Sci.* **2021**, *40*, 100341. [CrossRef]
62. Li, D.; Zhu, Q.; Han, C.; Yang, Y.; Jiang, W.; Zhang, Z. Photocatalytic degradation of recalcitrant organic pollutants in water using a novel cylindrical multi-column photoreactor packed with TiO_2-coated silica gel beads. *J. Hazard. Mater.* **2015**, *285*, 398–408. [CrossRef]
63. Tsukamoto, D.; Shiraishi, Y.; Sugano, Y.; Ichikawa, S.; Tanaka, S.; Hirai, T. Gold Nanoparticles Located at the Interface of Anatase/Rutile TiO_2 Particles as Active Plasmonic Photocatalysts for Aerobic Oxidation. *J. Am. Chem. Soc.* **2012**, *134*, 6309–6315. [CrossRef]
64. Siah, W.R.; Lintang, H.O.; Shamsuddin, M.; Yuliati, L. High photocatalytic activity of mixed anatase-rutile phases on commercial TiO_2 nanoparticles. *IOP Conf. Ser. Mater. Sci. Eng.* **2016**, *107*, 012005. [CrossRef]
65. Tseng, Y.H.; Chang, I.G.; Tai, Y.; Wu, K.W. Effect of surface plasmon resonance on the photocatalytic activity of Au/TiO_2 under UV/visible illumination. *J. Nanosci. Nanotechnol.* **2012**, *12*, 416–422. [CrossRef] [PubMed]
66. Chen, J.; Zhang, J.; Ye, M.; Rao, Z.; Tian, T.; Shu, L.; Lin, P.; Zeng, X.; Ke, S. Flexible TiO_2/Au thin films with greatly enhanced photocurrents for photoelectrochemical water splitting. *J. Alloys Compd.* **2020**, *815*, 152471. [CrossRef]
67. Wang, Y.; Chen, C.; Zhou, D.; Xiong, H.; Zhou, Y.; Dong, S.; Rittmann, B.E. Eliminating partial-transformation products and mitigating residual toxicity of amoxicillin through intimately coupled photocatalysis and biodegradation. *Chemosphere* **2019**, *237*, 124491. [CrossRef] [PubMed]
68. Elmolla, E.S.; Chaudhuri, M. Photocatalytic degradation of amoxicillin, ampicillin and cloxacillin antibiotics in aqueous solution using UV/TiO_2 and $UV/H_2O_2/TiO_2$ photocatalysis. *Desalination* **2010**, *252*, 46–52. [CrossRef]
69. Gozlan, I.; Rotstein, A.; Avisar, D. Amoxicillin-degradation products formed under controlled environmental conditions: Identification and determination in the aquatic environment. *Chemosphere* **2013**, *91*, 985–992. [CrossRef]
70. Mirzaei, A.; Chen, Z.; Haghighat, F.; Yerushalmi, L. Magnetic fluorinated mesoporous g-C_3N_4 for photocatalytic degradation of amoxicillin: Transformation mechanism and toxicity assessment. *Appl. Catal. B Environ.* **2019**, *242*, 337–348. [CrossRef]

Review

Insights into the Redox and Structural Properties of CoOx and MnOx: Fundamental Factors Affecting the Catalytic Performance in the Oxidation Process of VOCs

Veronica Bratan, Anca Vasile *, Paul Chesler * and Cristian Hornoiu

"Ilie Murgulescu" Institute of Physical-Chemistry of the Romanian Academy, 202 Spl Independentei, 060021 Bucharest, Romania
* Correspondence: avasile@icf.ro (A.V.); pchesler@icf.ro (P.C.); Tel.: +40-21-318-85-95

Abstract: Volatile organic compound (VOC) abatement has become imperative nowadays due to their harmful effect on human health and on the environment. Catalytic oxidation has appeared as an innovative and promising approach, as the pollutants can be totally oxidized at moderate operating temperatures under 500 °C. The most active single oxides in the total oxidation of hydrocarbons have been shown to be manganese and cobalt oxides. The main factors affecting the catalytic performances of several metal-oxide catalysts, including CoO_x and MnO_x, in relation to the total oxidation of hydrocarbons have been reviewed. The influence of these factors is directly related to the Mars–van Krevelen mechanism, which is known to be applied in the case of the oxidation of VOCs in general and hydrocarbons in particular, using transitional metal oxides as catalysts. The catalytic behaviors of the studied oxides could be closely related to their redox properties, their nonstoichiometric, defective structure, and their lattice oxygen mobility. The control of the structural and textural properties of the studied metal oxides, such as specific surface area and specific morphology, plays an important role in catalytic applications. A fundamental challenge in the development of efficient and low-cost catalysts is to choose the criteria for selecting them. Therefore, this research could be useful for tailoring advanced and high-performance catalysts for the total oxidation of VOCs.

Keywords: CoO_x; MnO_x; hydrocarbon total oxidation; redox properties; catalytic performances; lattice oxygen mobility

1. Introduction

1.1. General Features

According to the European Directive no. 1999/13/EC, a volatile organic compound (VOC) is defined as "any organic compound having at 293.15 K a vapor pressure of 0.01 kPa or more, or having a corresponding volatility under the particular conditions of use". Due to their high volatility, VOCs are considered the main pollutants in the air. They are emitted from oil and gas fields, in diesel exhaust, and they are also encountered in many home activities, such as painting, cooking, grass cutting, etc. These compounds are generally toxic and can cause eye irritation, respiratory problems, and even cancer [1–4].

In addition to the direct harmful effects of these compounds on human health, VOCs also have an indirect effect: they contribute to atmospheric pollution, being the precursors of ground-level ozone and photochemical smog, in the presence of NO_x and solar radiation [1,5–7].

Photochemical smog is a mixture of nitrogen oxides and volatile organic compounds which are able to react under the action of sunlight:

$$NO_2 + UV\ radiation = NO + O \tag{1}$$

The resulting oxygen radicals further react with oxygen in the air to form ozone.

$$O + O_2 = O_3 \tag{2}$$

Atmospheric ozone can be dangerous for people (it is irritating and causes respiratory problems, including asthma) and can also attack certain materials, such as textiles, paints, works of art, books, etc. However, if NO_x particles were the only pollutants in the atmosphere, ozone, resulting as shown in Equation (2), would be consumed in the reaction with NO, according to reaction (1):

$$O_3 + NO = NO_2 + O_2 \tag{3}$$

In this case, the ozone concentration in air will reach an equilibrium value and no longer be dangerous. However, if hydrocarbons (HCs) are also present in the atmosphere, part of the ozone reacts with them and produces toxic products, such as peroxyacetyl nitrate (PAN). PAN, in addition to having harmful effects on human health, reduces and may even stop the growth of plants, and in polluted regions can act as a source of NO_x [8].

Depending on the nature of the functional group, the main volatile organic compounds emitted are classified into: aliphatic hydrocarbons (alkanes, alkenes, and alkynes), aromatic hydrocarbons, oxygenated organic compounds (alcohols, ketones, and esters), and halogenated organic compounds. Their impact on the environment and, implicitly, on human health, depends both on the nature of the VOCs and on their concentration in air. Unsaturated and aromatic hydrocarbons are the most polluting, especially due to their large contribution to the formation of "photochemical ozone". Thus, according to reference [5], in 2010, in China, alkenes/alkynes and aromatic hydrocarbons accounted for 28% and 54%, respectively, of the total VOCs with ozone-forming potential, although their effective contributions to the total emissions of organic compounds was only 8.5% and 34%, respectively.

1.2. VOC Removal: General Methods and Materials

Due to their harmful effects on human health and the environment, VOC abatement has become imperative. Nowadays, there are several methods generally used to eliminate VOCs from the air stream, such as adsorption onto activated carbons or zeolites, thermal combustion, and photocatalytic and catalytic oxidation [6,9–13]. The emission sources for volatile organic compounds, as well as the variety of compounds emitted by each source, are so diverse that all disposal methods have practical limitations. Adsorption is restricted to highly diluted VOC emissions and is limited by the difficult recovery of the adsorbent [12,14]. Thermal combustion needs high temperatures in order to achieve full oxidation of concentrated VOC streams (higher than 800 °C), making it not very economical. Furthermore, incomplete thermal oxidation of VOCs can produce numerous harmful byproducts (dioxins, nitrogen oxides, etc.) [2,14]. Photocatalytic oxidation works at low temperatures, but it has lower efficiency and can also produce harmful byproducts [1].

Thus, catalytic oxidation remains a better alternative, as the pollutants can be totally oxidized at lower operating temperatures (200–500 °C). Recently, high-performance catalysts have been developed for the removal of pollutants at low temperatures. In general, by the deposition of noble metals on various oxide supports, superior performances in VOC catalytic oxidation have been achieved [15–20]. Some problems with using noble-metal-based catalysts are their high cost and their poor resistance to poisoning, hence the need to replace these materials with abundant ones which are cheaper, less susceptible to supply fluctuations, and more environmentally friendly.

Transition metal oxides are industrially important materials, and many experimental studies have been conducted on the oxidation of VOCs over them. The most commonly used metal-oxide catalysts include Cr, Mn, Co, Ni, Fe, Cu, and V, used pristine or mixed, with or without supports [1,13,14,20–27]. Their behaviors in oxidation reactions of hydrocarbons have been studied for many years. Blazowski et al. [28] found that Co_3O_4, Cr_2O_3,

CuO, NiO, MnO$_2$, and V$_2$O$_5$ presented good performances in catalyzing the oxidation of hydrocarbons. Dixon et al. [29] stated that, among the metal oxides corresponding to the fourth-period transition-metal series, higher activity was presented by Cr$_2$O$_3$ and Co$_3$O$_4$. Moro-oka et al. [30] also reported that the highest rate of reaction in the oxidation of several hydrocarbons (acetylene, ethylene, propane, and isobutene) over a series of metal oxides (Co$_3$O$_4$, NiO, MnO$_2$, Fe$_2$O$_3$, Cr$_2$O$_3$, and CeO$_2$) was found for Co$_3$O$_4$ and MnO$_2$. However, the best catalytic performances in the latter study were obtained for Pt and Pd catalysts.

Although they are generally less active than noble-metal catalysts, they have the advantages of much lower costs, higher thermal stabilities, and resistance to poisoning, which are ultimately reflected in a decrease in the total cost of the depollution technology. Recent studies have confirmed that the most active single oxides in the total oxidation of hydrocarbons are manganese and cobalt oxides [31–38]. Hence, Lahousse et al. showed that the γ-MnO$_2$ catalyst outperformed the 0.3 wt% Pt/TiO$_2$ catalyst in the oxidation of n-hexane [39], and Lin et al. [40] synthesized a mesoporous α-MnO$_2$ microsphere with high toluene-combustion activity comparable to that of a Pd-based FeCrAl catalyst. Among Co oxides, the most active catalyst in the complete oxidation of hydrocarbons has been shown to be Co$_3$O$_4$. The studies performed in references [41,42] showed better performances in the propane oxidation reaction of Co$_3$O$_4$, even when it was compared with some commercial as well as self-made Pt, Pd deposited on different support catalysts (Figure 1).

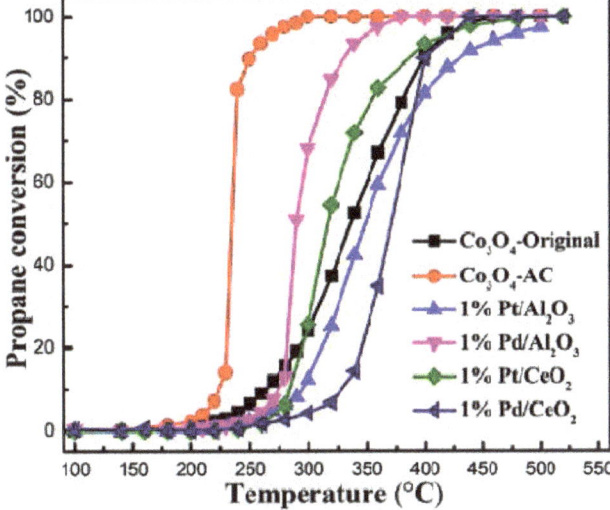

Figure 1. Catalytic combustion of propane (0.3% C$_3$H$_8$, 10% O$_2$, and N$_2$ balanced) as a function of temperature over Co$_3$O$_4$, commercial 1%Pt/Al$_2$O$_3$ and 1%Pd/Al$_2$O$_3$, and self-made 1%Pt/CeO$_2$ and 1%Pd/CeO$_2$, at WHSV of 240,000 mL g^{-1} h^{-1}. Reprinted and adapted with permission from Ref. [41]. Copyright 2018 Elsevier.

In this review, we mainly focus on the complete catalytic oxidation of hydrocarbons at moderate temperatures (below 500 °C) using transition metal oxides, namely, cobalt and manganese oxides. The review will be divided into three parts. First, a brief description of the reaction mechanisms is presented. The second and third parts deal with CoO$_x$ and MnO$_x$ oxides, respectively. In both cases, we examined these single-metal oxides that are widely studied as active components in the catalytic oxidation of VOCs and investigated some important influencing factors affecting their catalytic performances. As an important characteristic of the metal-oxide catalysts, the mobility of oxygen species and redox properties has been emphasized. Several case studies have been presented to illustrate the correlations between structure and morphology, textural properties (surface area and porosity), the nature of the exposed facets and crystal defects, redox properties, and catalytic activity. Finally, we present the conclusions and our perspectives on future developments in this field.

2. Reaction Mechanism Overview

A fundamental challenge in the development of efficient and low-cost catalysts for the removal of pollutants from the air is to first understand and then choose the main criteria for selecting them. This can be undertaken as a strategy for tailoring new catalysts with improved structures and properties. For this purpose, the most important step is to understand the reaction mechanism. Three kinetic models are usually used to describe the mechanism for the complete oxidation of hydrocarbons: the Langmuir–Hinshelwood (L–H), the Eley–Rideal (E–R), and the Mars–van Krevelen (MvK) models.

Langmuir was the first to describe how a bimolecular reaction takes place at the surface of a catalyst [43,44]. He identified two types of surface interactions: (i) interaction between molecules or atoms adsorbed at adjacent sites on the surface, which is known as the Langmuir–Hinshelwood (L–H) mechanism; and (ii) an interaction that takes place as a result of the collision of the gas molecules of one of the reactants and the adsorbed molecules of the other, which is known as the Eley–Rideal (E–R) mechanism (Figure 2).

Therefore, in the L–H model, both the VOC and oxygen molecules are adsorbed on the surface of the catalyst. The two reactant molecules could be adsorbed at different sites or they could compete for the same site. After the reaction between adsorbed reactants takes place, the products (CO_2 and H_2O) desorb from the surface of the catalyst. In the Eley–Rideal (ER) mechanism, oxygen is too weakly adsorbed, so that it reacts directly from the gas phase with adsorbed VOC, or the inverse [45].

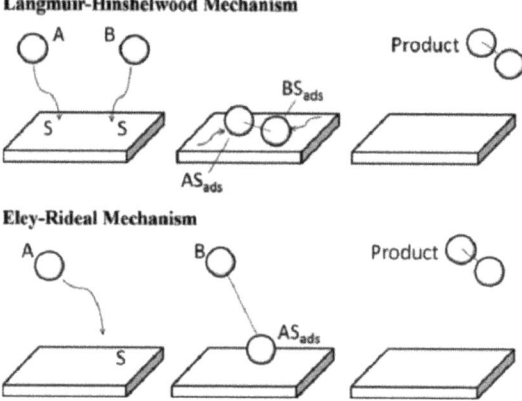

Figure 2. Top model: The Langmuir–Hinshelwood mechanism: two molecules adsorb onto the surface and diffuse and interact with each other until a product is formed and desorbs from the surface. Bottom model: The Eley–Rideal mechanism: a molecule adsorbs onto the surface and another molecule interacts with the adsorbed one until a product is formed and desorbs from the surface. Reprinted with permission from Ref. [46]. Copyright 2014 Springer Nature.

Almost at the same time as the formulation of these two mechanisms, another one emerged based on the idea that the lattice components of the catalyst appear in the reaction products [47–49]. In 1954, Mars and van Krevelen [50] proposed this new mechanism, and now it is universally accepted as being the one that describes the oxidation of hydrocarbons on metal-oxide catalysts at moderate temperatures [3,14,51]. According to this mechanism, denominated the Mars–van Krevelen or redox mechanism, the reaction involves two consecutive steps.

In the first step, the hydrocarbon adsorbed on the surface of the catalyst is oxidized by the oxygen atom from the oxide network. As a consequence, an oxygen vacancy (–□–) appears on the surface, leaving the catalyst's surface in a reduced state.

$$C_xH_{y(ads)} + -M^{n+}-O^{2-}-M^{n+}- \rightarrow xCO_{2(g)} + y/2 H_2O_{(g)} + -M^{(n-1)+}-\square-M^{(n-1)+}, \quad (4)$$

In the second step, re-oxidation of the catalyst's surface takes place. Re-oxidation actually means the filling of the oxygen vacancy generated according to Equation (4). This could happen either directly, through the oxygen in the gas phase [52], following the equation:

$$-M^{(n-1)+}-\square-M^{(n-1)+}- + 1/2 O_2 \rightarrow -M^{n+}-O^{2-}-M^{n+}, \quad (5)$$

or indirectly, through the diffusion of the oxygen atom from the bulk of the oxide to the reduced site [53]. Thus, the mobility of the oxygen in the catalyst has to be high enough that an oxygen ion from the lattice can diffuse toward the newly formed anionic vacancy.

In view of these facts, the catalytic performance of a metal oxide in the total oxidation reaction of hydrocarbons will be dictated by:

(1) Its ability to adsorb hydrocarbons;
(2) Its redox properties (the ease of the reduction and reoxidation of its surface);
(3) The mobility of the oxygen atoms in the oxide lattice.

The adsorption of a HC on the catalyst's surface depends both on the catalyst's properties and the HC's characteristics. For a metal oxide to be easily reducible, it has to possess the capacity to easily release an oxygen atom from the lattice. By releasing an oxygen atom, the metallic ion changes towards a lower valence state, leaving the oxide stable; this is characteristic mainly of transitional-metal oxides with metallic ions in high oxidation states. In addition, the reducibility of oxides could be improved if oxygen vacancies are present. The presence of an oxygen vacancy makes the adjacent lattice oxygen more easily released and/or transferred.

On the other hand, a good catalyst must also be readily re-oxidized, otherwise the reaction will not advance. In this respect, the presence of oxygen-vacancy defects is also an important factor because they favor the activation of gaseous oxygen (which is usually adsorbed on the anionic vacancies near reduced metallic ions, $M^{(n-1)+}$), according to the following reaction [54]:

$$O_{2(g)} \underset{-e}{\overset{+e}{\rightleftarrows}} O_{2(ads)}^- \underset{-e}{\overset{+e}{\rightleftarrows}} O_{2(ads)}^- \rightleftarrows 2O_{(ads)}^- \underset{-2e}{\overset{+2e}{\rightleftarrows}} 2O_{(ads)}^{2-} \rightleftarrows 2O_{(lattice)}^{2-}, \quad (6)$$

The ionic forms of oxygen which are stable on the metal-oxide surface, O_2^- and O^-, are strong electrophilic species and could be themselves directly involved in the oxidation reaction of hydrocarbons by causing the breakage of C-C bonds [55]. In the temperature range of 100–300 °C, complex surface oxidation/re-oxidation mechanisms are operative [56,57] because of the lower activation energies of processes that involve adsorbed oxygen instead of lattice oxygen. The predominant mechanism will be established by the oxygen-storage capacity of the oxide and the oxygen-lattice availability.

3. Cobalt Oxides: The Influence of Structural and Redox Properties on Catalytic Performance

Co_3O_4 (with the extended formula $Co^{2+} Co^{3+}_2 O_4$) has a spinel-type crystal structure in which Co^{2+} occupies 8 tetragonal sites, Co^{3+} takes 16 octahedral sites, and 32 sites are occupied by O ions (Figure 3). The lattice oxygen is cubically close-packed in a unit cell, with 1/8 of the tetrahedral sites occupied by Co^{2+} and half of the octahedral sites occupied by Co^{3+} [58].

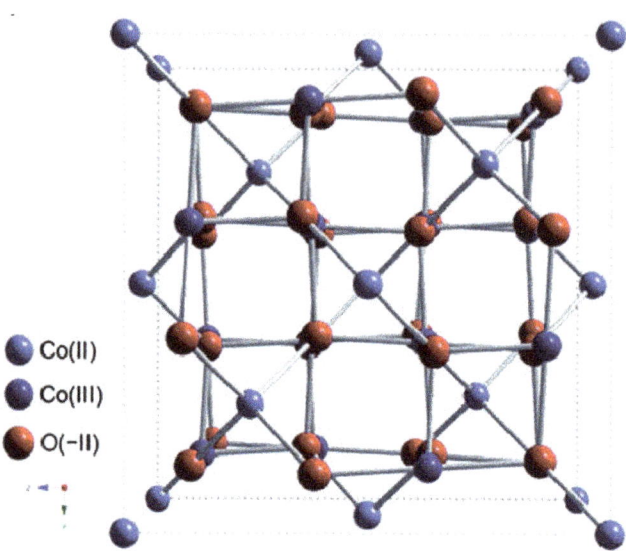

Figure 3. Scheme of the spinel structure of cobalt(II, III) oxide. Reprinted with permission from Ref. [58]. Copyright 2012 Royal Society of Chemistry.

Although cobalt oxides are the most efficient oxides used as catalysts in the total oxidation of pollutants in the atmosphere, at low temperatures a disadvantage in their use is their deactivation during the reactions. The deactivation occurs either at low temperatures (below 100 °C) under reaction conditions (since the reoxidation of the catalyst is assumed to take place slowly), or at temperatures above 500 °C, when sintering of the oxide particles occurs, producing a significant decrease in the specific surface area and, implicitly, a decrease in catalytic performance. Liu et al. [59] observed an accelerated deactivation after cycling of the reactor temperature between 210 and 500 °C. However, it is worth noting that in the second cycle no further deactivation occurred. Therefore, by working at intermediate temperatures, these shortcomings could be prevented.

Zhang et al. [60] prepared Co_3O_4 catalysts via a simple precipitation method using various precipitants or precipitant precursors: oxalic acid, sodium carbonate, sodium hydroxide, ammonium hydroxide, and urea. A commercial Co_3O_4 catalyst (Co-com) was also investigated for comparison purposes. They studied the influence of various redox and structural properties on the performances of the obtained samples in the catalytic oxidation of propane and toluene. The sample synthesized using carbonate had the highest catalytic activity, and this was related to the enhancement of the surface area, the number of lattice defects, surface Co^{2+} concentration, and the reducibility of the sample (Figure 4). In fact, although the catalyst Co-CO_3 exhibited a specific reaction rate for propane oxidation much higher than that of other catalysts, its performance was comparable to that of the Co-com catalyst, the latter having a larger specific surface area that overcompensated its lower intrinsic activity. However, Co-CO_3 proved an excellent ability to oxidize toluene and propane completely, while Co-com exhibited lower reducibility, which negatively affected the catalytic performance in toluene oxidation.

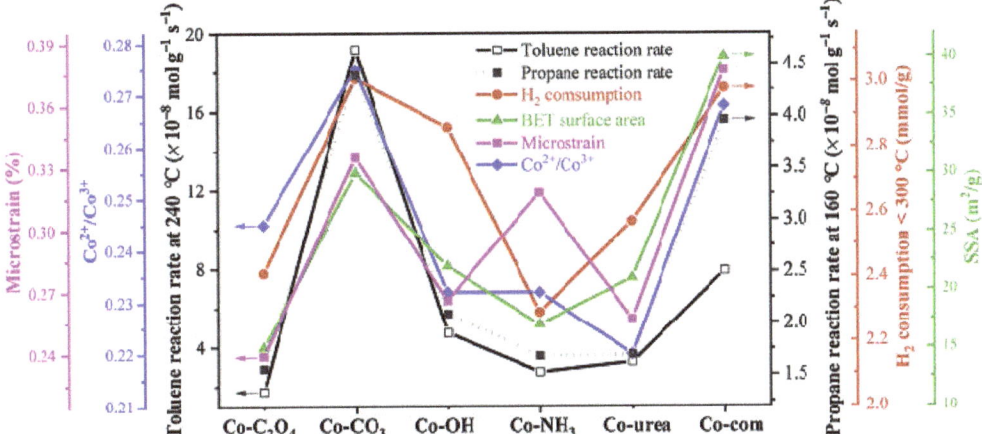

Figure 4. The relationship between the catalytic reaction rate and several physical and chemical parameters. Reprinted with permission from Ref. [60]. Copyright 2020 Elsevier.

The experimental studies focused on synthesizing cobalt oxides with small particle sizes, large specific surface areas, and different morphologies in order to obtain materials with large numbers of active sites, high concentrations of surface defects (especially oxygen vacancies), and better reducibility—factors which could improve these oxides' performances as oxidation catalysts.

Solsona et al. [61] successfully obtained Co_3O_4 nanoparticles and then tested them in propane total oxidation. The decrease in particle size resulted in an increase in specific surface area—a paramount factor in heterogeneous catalysis. Among the studied Co_3O_4 catalysts, those with the higher surface areas were the most reactive, total conversion being obtained at a reaction temperature lower than 250 °C. The authors found that the small size of the crystallite had a beneficial effect on catalytic performance: not only did it have a large number of active sites (the larger the exposed surface area, the greater the number of accessible active centers), it also produced a decrease in the energy of the Co-O bonds, which meant that oxygen could be easily released, increasing the oxide reducibility of cobalt oxide.

Other authors also found that the activity of Co catalysts in the total oxidation reaction of propane was mainly determined by the specific surface. Puertolas et al. [55], studying several Co_3O_4 oxides obtained by a hydrothermal method in the presence of various organic acids, observed a decrease in surface area compared to the reference sample (prepared without the addition of organic acids). Although the reasons for this decrease were not completely explained, it was found that the reference sample had better catalytic performances than the oxides obtained in the presence of organic acids. As the specific reaction rates (normalized on the surface) calculated for all of the studied cobalt-oxide catalysts had almost the same values, the authors assumed that the bulks and surfaces of the catalysts possessed similar redox characteristics. In conclusion, it was stated that the differences in catalytic performances were mainly determined by the different surface areas, establishing that there is a direct relationship between specific surface area and Co_3O_4 activity.

In this regard, mesoporous transition metal oxides could be prominent candidates for use as catalysts. They possess large specific surface areas but also ordered pore structures, which could facilitate the diffusion of reactant molecules towards active centers where they can be adsorbed, which is the first important step in the enhancement of catalytic performance. This is why numerous studies have been performed in order to obtain mesoporous materials with improved physicochemical properties [35,42,62,63]. Thus, it was possible to obtain Co_3O_4 oxides with specific surfaces of over 100 m²/g, sometimes even higher than 300 m²/g [35,63], and 3D ordered mesoporous structures, usually generated using mesoporous silica (KIT-6, SBA-15, or SBA-16) as a hard template. The as-prepared Co_3O_4 catalysts exhibited exceptional catalytic properties, reaching a T_{90} (the temperature at which the conversion reaches a value of 90%, used for measuring catalytic performances) lower than 200 °C for the total oxidation of toluene, for example [63].

Another important factor that influences catalytic activity is the oxidation state of metallic ions. In reference [64], the authors followed the effect of the oxidation states of metallic ions by studying the oxidation reaction of xylene over three mesoporous cobalt-oxide catalysts with various Co^{3+}/Co^{2+} ratios. It was found that the conversion measured at 240 °C increased in the order: Co_3O_4-CoO mixture > Co_3O_4 > CoO. The authors stated that, to obtain a high conversion, the presence of ions in a high oxidation state is necessary, but that better performances also require surface Co^{2+} species, which could be slightly oxidized to Co^{3+} species by oxygen gas (Equation (5)). The first step in the activation of gaseous oxygen on the oxygen vacancies near Co^{2+} is the chemisorption of oxygen (Equation (6)), resulting in electrophilic adsorbed oxygen (O_2^-)—a very active species involved in the oxidation of VOCs. Liu et al. [65] also found that the extra CoO phase is beneficial for catalytic activity, decreasing the crystalline size and increasing the specific surface area and the Co^{2+}/Co^{3+} and O_{ads}/O_{latt} ratios (where O_{ads} is adsorbed oxygen and O_{latt} refers to the oxygen atoms from the oxide lattice).

In reference [66], the authors investigated the effects of the coexistence of cobalt defects and oxygen vacancies on the performances and mechanisms of a $Co_{3-x}O_{4-y}$ catalyst for the toluene-degradation reaction. Their studies proved that cobalt defects are favorable for the formation of oxygen vacancies that can accelerate the conversion of intermediates, thus obtaining over 90% toluene conversion at a temperature lower than 190 °C (300 ppm VOC, GHSV = 72,000 mL g^{-1} h^{-1}).

Traditionally, reducing the size of particles is the main goal, but in recent years oxide catalysts with desired structural and physicochemical parameters have been obtained by controlling their morphologies by synthesis. For example, obtaining Co_3O_4 in the form of halospheres is a method of increasing the specific surface area. Such structures have the advantages of a very high surface area–volume ratio, low density, and high permeability for reactants [58,67]. At the same time, a particular morphology allows the predominant exposure of a certain facet with higher reactivity, which has a significant impact on catalytic performance. Thus, different morphologies of Co_3O_4, such as octahedra [68], cubes [68–70], rods, sheets [69,70], plates [69], etc., have been obtained and studied in the total oxidation of hydrocarbons. For this purpose, the conditions of the synthesis reaction, the precursor, the heat treatment, the surfactant, etc., could be varied (Figure 5).

Generally, the experimental studies on the oxidation of hydrocarbons (except methane) over Co_3O_4 have shown that the facet {110} is more active than other facets [70–72]. This result was explained by the higher mobility of oxygen vacancies on these planes, which permit two vacancies easily meeting together to react with O_2 [73]. Meanwhile, the {110} facet consists mainly of Co^{3+} cations [69,70].

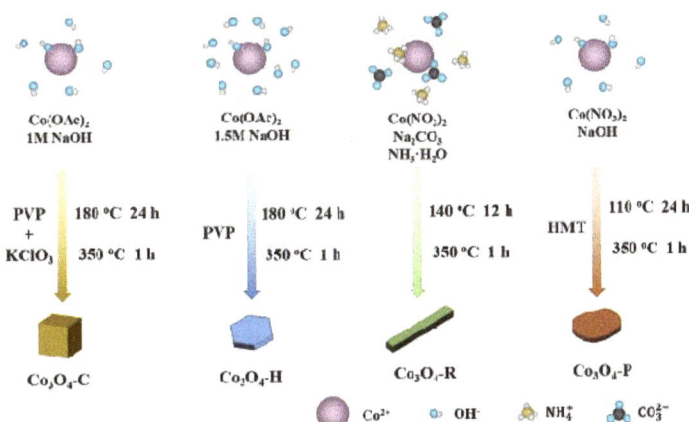

Figure 5. Schematic illustration of the synthetic routes for the investigated Co_3O_4 catalysts. Reprinted with permission from Ref. [69]. Copyright 2019 John Wiley and Sons.

Thus, in reference [70], Co_3O_4 rods, sheets, and cubes were synthesized with the exposed facets {110}, {111}, and {100}, respectively (Figure 6). Using various experimental techniques and theoretical calculations (by DFT), the authors comprehensively studied the behavior of the prepared catalysts in propane combustion, and they related the improved catalytic activity of the rod-type Co_3O_4 (Figure 6) to the higher reactivity of the predominantly exposed facets {110}. The facet {110}, mainly exposed on Co_3O_4 rods, had the highest calculated area and a higher density of low-coordination Co atoms, which can promote the generation of active oxygen. From DFT studies, the authors determined that this facet also had the lowest energy for oxygen-vacancy formation, which is beneficial for the adsorption and activation of O_2. This was experimentally confirmed by O_2-TPD and TPR, from which results it was confirmed that Co_3O_4-R had the most active electrophilic oxygen species (O^-) and was the more reducible sample.

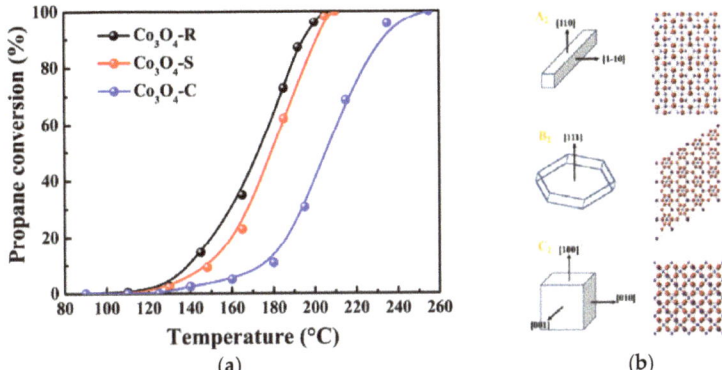

Figure 6. (a) Ignition curves of different catalysts for propane oxidation. (b) Corresponding geometric models of Co_3O_4 (A2, B2, and C2) and surface atomic structure models of (110), (111), and (100) facets. Reprinted and adapted with permission from Ref. [70]. Copyright 2021 Elsevier.

Electrophilic oxygen species, such as O^- and/or $O_2{}^-$, are thought to be easily formed on the surface of Co_3O_4 and could be incorporated into the oxide lattice to replace the oxygen atoms which oxidize the hydrocarbons. Additionally, such reactive oxygen species are expected to directly act through an electrophilic attack on the C=C or C-H bonds, playing a decisive role in the catalytic degradation of hydrocarbons [74]. Several in situ studies have shown that propene/propane [75] and toluene [76] oxidations over Co_3O_4 follow a typical Mars–van Krevelen type mechanism. However, in the latter study, Zhong et al. [76] investigated the oxidation of toluene over Co_3O_4 through in situ DRIFTS combined with quasi-in situ XPS and UV–Vis diffuse reflectance spectroscopy. They proposed a redox mechanism in which toluene and oxygen molecules are firstly adsorbed on the oxide surface and oxygen is activated via oxygen vacancies. Afterward, the activated toluene molecules react with both the lattice and chemisorbed oxygen species, forming oxygenated intermediates which finally desorb as CO_2 and H_2O. Although the study revealed the decisive role of surface lattice oxygen, the authors also identified the importance of the presence of gaseous oxygen in the mineralization of the intermediate products and claimed surface adsorbed oxygen as active oxygen.

Liu et al. [77] also stated that the catalytic performances were related to the reactivity of the preferentially exposed crystallographic planes. Three Co_3O_4 samples were prepared using urea and three different cobalt precursors. The as-synthesized samples possessed different morphologies: hydrangea-like (H), spiky (S), and pompon-like (P) (Figure 7). The best catalyst, sample H, performed a 90% toluene conversion at about 248 °C, which was 60 °C lower than that of sample P (Figure 8). Its higher catalytic activity was mainly explained by the more active {110} exposed planes, in contrast to the P sample which had exposed {111} facets. Compared with the S sample, which possessed the same primarily exposed facets, the enhanced activity of H could be attributed to the larger number of active sites provided by its larger surface area (Table 1). Catalytic activity varies in line with the O_{ads}/O_{latt} ratio (obtained via XPS) and with the number of oxygen vacancies (determined by Raman analysis). Meanwhile, the H sample contained a higher concentration of Co^{3+}.

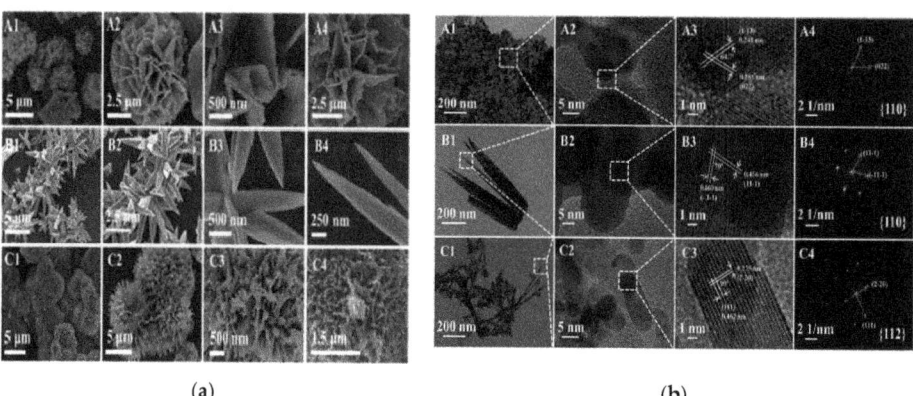

Figure 7. (a) SEM images of (A1–A3) H, (B1–B3) S, and (C1–C3) P before calcination and (A4), (B4), and (C4) after calcination at 350 °C. (b) TEM, HRTEM, and fast-Fourier-transform (FFT) images of the Co_3O_4 samples: (A1–A4) H, (B1–B4) S, and (C1–C4) P. Reprinted with permission from Ref. [77]. Copyright 2020 Elsevier.

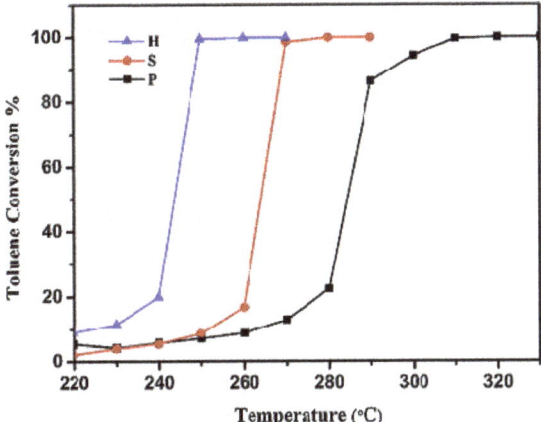

Figure 8. Catalytic performance of the synthesized Co_3O_4 as a function of reaction temperature for toluene oxidation under the following conditions: toluene 500 ppm, O_2/N_2 20 vol%, total flow rate = 100 mL/min, WHSV = 60,000 mL/(g h). Reprinted with permission from Ref. [77]. Copyright 2020 Elsevier.

In another study [78], a Co_3O_4 mesoporous catalyst was synthesized and shown to be very active in removing traces of ethylene. This catalyst presented 30% ethylene conversion at 0 °C, compared with a nanosheet-type Co_3O_4 sample which did not present catalytic activity, even at 20 °C. The authors proposed that the outstanding catalytic activity of the first sample was due to the combination of the higher reactivity of the exposed facets (the {110} planes being more active than the {112} planes, mainly exposed by Co_3O_4 nanosheets) and its porous structure, which permitted the reactants to pass and be adsorbed into the pores, where they were chemically activated.

However, there have also been some studies that have found another sequence for the reactivity of the different exposed facets. For example, Yao et al. [69] obtained Co_3O_4 nanocatalysts with different morphologies and different exposed facets through hydrothermal routes: nanocubes with exposed {100} facets, hexagonal nanoplates with {111} facets, nanorods with {110} facets, and nanoplates with dominantly {112} exposed facets (Figure 9). In this research, the catalytic activities of the different facets were found to vary in the order {111} > {100} > {110} > {112}, and the superior activity of the {111} facet was attributed to the easier adsorption of propane. Moreover, the lattice oxygens in the Co_3O_4 nanoplates were more active than those of the other samples (as confirmed by H_2-TPR and C_3H_8-TPSR), and the activation of gas-phase oxygen on the Co_3O_4 hexagonal nanoplates surface was more facile, which is also beneficial for catalytic activity. However, the differences in reactivity, expressed as T_{90} values, between the {111}, {100}, and {110} facets were, in this case, very small.

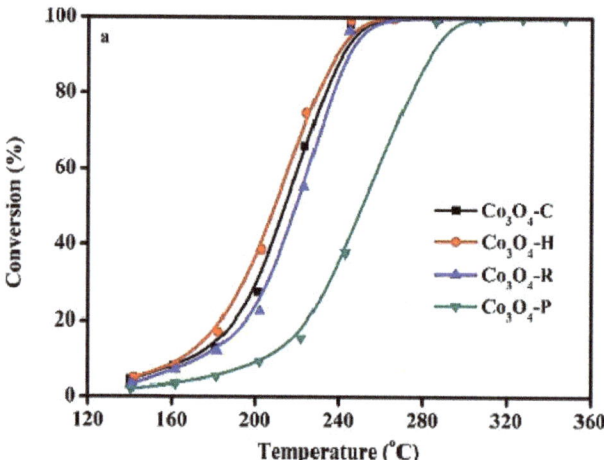

Figure 9. C_3H_8 combustion activities vs. temperature profiles of investigated catalysts. Reprinted and adapted with permission from Ref. [69]. Copyright 2019 John Wiley and Sons.

In other studies, it was found that the reactivity of different Co_3O_4 planes in propane oxidation: sometimes it was higher for {111} facets [79] and other times, for the preferable exposed {112} planes [80].

The contradictory results obtained in the literature are an indication that the problem of studying the factors that influence reactivity remains and that catalytic activity is determined by a combination of different parameters.

In Table 1, some of the results reported in the literature for the oxidation of propane and toluene, respectively, over cobalt oxides are summarized [35,36,55,59,61,63,65,69,70, 72,77,79,81–84]. The results include the specific reaction rate, measured at 200 °C and the temperature at which the conversion reaches a value of 90% (for measuring catalytic performance), the conditions of the reaction, and the main factors which were expected to determine the catalytic behavior of the oxidic materials: the specific surface area (BET), Co^{3+}/Co^{2+} ratio, O_{ads}/O_{latt} (both determined by XPS measurement), and the reduction temperature (determined by TPR measurements), as indicators of sample reducibility.

Table 1. Propane and toluene total oxidation over Co_3O_4.

CoO_x Sample	Reaction Conditions	BET (m²/g)	Co^{3+}/Co^{2+} Ratio /O_{ads}/O_{latt} Ratio	TPR Data (T_{max})	T_{90} (°C)	Reaction Rate (µmol·m⁻²·s⁻¹)	Ref.
Propane							
Co_3O_4 nanoparticles	8000 ppm C_3H_8 in air 50 mL/min	47	0.47/0.52	352	235	0.14×10^{-5}	[55]
Co_3O_4 nanoparticles	1000 ppm C_3H_8 in air 100 mL/min	48	1.28/0.52	107; 270; 366	260	0.017	[81]
Co_3O_4 rods {110}	2500 ppm C_3H_8 in air 30,000 mL/(g_{cat}·h)	48.5	0.46/0.57	113; 316	195	0.57×10^{-5}	[70]
Co_3O_4 nanoparticles	C_3H_8:O_2:N_2 1:10:89 100 mL/min	120	/0.78	~300; 350	240 [a]	0.03	[59]
CoO-Co_3O_4	0.3% C_3H_8 in air 30.000 mL/(g_{cat}·h)	62.8	0.65/1.47	333; 380	235	-	[65]

Table 1. Cont.

CoO_x Sample	Reaction Conditions	BET (m²/g)	Co^{3+}/Co^{2+} Ratio $/O_{ads}/O_{latt}$ Ratio	TPR Data (T_{max})	T_{90} (°C)	Reaction Rate (µmol·m⁻²·s⁻¹)	Ref.
Co_3O_4	0.4% C_3H_8:20% O_2 in Ar 98 mL/min	114	-	-	~210	0.015	[36]
Co_3O_4 nanoparticles	8000 vppm C_3H_8 in air 50 mL/min	99	-	268; 313; 364	~250 [a]	-	[61]
Co_3O_4 nanoparticles	1000 ppm C_3H_8 in air 100 mL/min 120.000 mL/(g_{cat}·h)	82	1.16/1.00	104; 280; 370	225	0.01	[82]
Co_3O_4 nanoparticles	1000 ppm C_3H_8 in 21% O_2/N_2 100 mL/min 40.000 mL/(g_{cat}·h)	51	-	270 [b]	224	32.9 × 10⁻⁵	[83]
Nanocubes {100}	1% C_7H_8; 10% O_2/N_2 33.3 mL/min 10.000 mL/(g_{cat}·h)	50.2	1.20/0.78	317	242		[69]
Nanosheets {111}		41.6	1.14/0.85	303	239		
Nanorods {110}		43.9	1.39/0.69	310	245		
Nanoplates {112}		23.6	1.25/044	344	283		
Toluene							
Mesoporous (Kit6)	1000 ppm C_7H_8 in O_2 (1/20 molar ratio) in N_2 20.000 mL/(g_{cat}·h)	121	1.46/0.95	283; 363	180	-	[63]
Mesoporous (SBA16)		118	1.54/0.98	299; 382	188		
Mesoporous (SBA16)	1000 ppm C_7H_8 in O_2 (1/200 molar ratio) in N_2 33 mL/min	313	-	273; 348	240 [a]	-	[35]
Co_3O_4 nanoparticles	1000 ppm C_7H_8 in 21% O_2/N_2 100 mL/min	51	-	270 [b]	242	150.8 × 10⁻⁵	[83]
ZIF-67, dodecahedral	1000 ppm C_7H_8 in air 66.7 mL/min 20.000 mL/(g_{cat}·h)	31.4	0.36/0.65	328; 395	254	-	[84]
MOF-74, rods		32.2	0.45/0.83	300; 397	248		
ZSA-1, octahedral		63.4	0.57/1.17	294; 380	239		
Hydrangea {110}	500 ppm C_7H_8, 20% O_2/N_2 100 mL/min 60.000 mL/(g_{cat}·h)	38.7	0.55/0.78	324; 396	248	-	[77]
Spiky {100}		29.4	0.53/0.75	349; 455	269		
Pompon {111}		96.6	0.41/0.59	393; 482	298		
Flower {110}	1000 ppm C_7H_8 in 20% O_2/N_2 80 mL/min 20.000 mL/(g_{cat}·h)	-	1.20/1.13	274; 344	228	-	[72]
Rods {112}		-	1.02/0.69	287; 362	249		
3D flower {111}	1000 ppm C_7H_8, 20% O_2/N_2 48.000 mL/(g_{cat}·h)	84.6	1.73/0.94	227; 301	238	-	[79]
2D plate {112}		24.9	1.34/0.62	294; 385	249		
1D needle {110}		24.9	1.26/0.51	308; 407	257		

[a] T_{100}. [b] Lowest temperature peak.

4. Manganese Oxides as Catalysts in the Combustion Reaction

Manganese is one of the most abundant metals on the earth's surface. Its compounds are not expensive and are non-toxic, so they can be considered "environmentally friendly" materials.

Manganese is a transitional metal with five unpaired electrons in the d level (electronic configuration: [Ar] $3d^54s^2$), which allows it to adopt multiple oxidation states, hence its special properties. [85]. Manganese has the ability to form numerous stable stoichiometric oxides, MnO, Mn_2O_3, Mn_3O_4, and MnO_2 (Figure 10), but also metastable (Mn_5O_8) or unstable ones (Mn_2O_7—a green explosive oil).

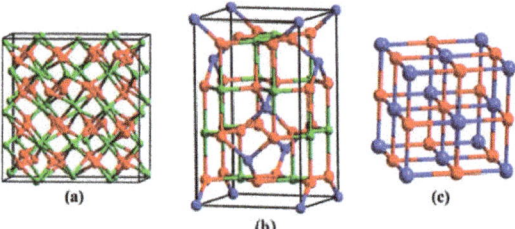

Figure 10. Crystal structures of mineral phases of (**a**) Mn_2O_3, (**b**) Mn_3O_4, and (**c**) MnO. All three manganese oxides are closely packed by lattice oxygen and differ only in the arrangement of octahedral and tetrahedral sites with various oxidation states of manganese. The green, blue, and red atoms represent Mn^{III}, Mn^{II}, and O^{2-}, respectively. Reprinted and adapted with permission from Ref. [86]. Copyright 2014 John Wiley and Sons.

The trivalent oxide, Mn_2O_3, occurs in two structural forms: α-Mn_2O_3 (bixbyte mineral), stabilized as the body-centered cubic crystal structure; and γ-Mn_2O_3 (with spinel structure), which is less stable and not present in nature.

Mn_3O_4 (hausmannite) is a stable phase of manganese oxide and occurs as a black mineral. Hausmannite possesses a normal spinel structure, $Mn^{2+}Mn^{3+}{}_2O_4$, in which Mn^{2+} and Mn^{3+} ions occupy the tetrahedral and octahedral sites, respectively.

Manganese dioxide (MnO_2) can be found in the Earth's crust in massive deposits and in important deep-sea minerals. MnO_2 polymorphic crystallographic forms (α-, β-, γ-, and δ-) occur as a consequence of divergence in the connectivity of corner- and/or edge-shared octahedral [MnO_6] building blocks. α-, β-, and γ-MnO_2 (also known as hollandite, pyrolusite, and nsutite, respectively) have tunnel structures, while δ-MnO_2 (birnessite) has a layered structure composed of two-dimensional sheets of edge-shared [MnO_6] octahedra (Figure 11) [85,87].

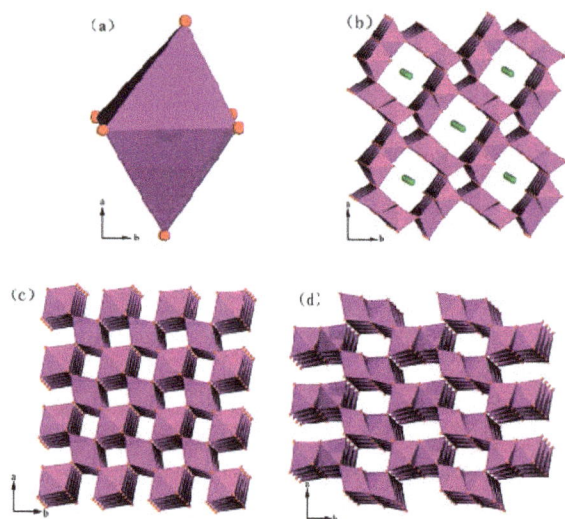

Figure 11. Polyhedral representations of the crystal structures of varieties of MnO_2: (**a**) [MnO_6] octahedron, (**b**) cryptomelane-type α-MnO_2, (**c**) pyrolusite-type β-MnO_2, and (**d**) nsutite-type γ-MnO_2. Reprinted and adapted with permission from Ref. [87]. Copyright 2014 Royal Society of Chemistry.

MnO$_x$ with different oxidation states can either coexist or gradually become interconvertible from one form to another during the oxidation process controlled by the diffusion of oxygen. The strong ability to switch oxidation states and the formation of structural defects are beneficial to high oxygen mobility and oxygen storage, and are responsible for the catalytic properties of MnO$_x$ [31,88]. Due to their low cost, low toxicity, simple ease of preparation, and high stability, manganese oxides can be promising candidates for use in depollution catalysis. Among earlier publications about the catalytic properties of manganese oxides in hydrocarbon oxidation, the studies of Busca, Finocchio, Baldi, and their group [32,89–92] and the studies of Lahousse et al. [39,93] deserve special mention. They followed the total oxidation of propane over manganese oxides in various oxidation states, both pure and mixed oxides (with TiO$_2$, Al$_2$O$_3$, or Fe$_2$O$_3$) [89–92], and found that single Mn-based oxides were very active materials in the total oxidation of propane, propene, and hexane—generally more active than mixed oxides.

Existing types of manganese oxides, such as MnO, MnO$_2$, Mn$_2$O$_3$, and Mn$_3$O$_4$, have different structural characteristics which have a direct influence on their use as active catalysts for the oxidation of hydrocarbons. Numerous studies have been conducted in order to determine which of these oxides has the best performance in the oxidation of hydrocarbons. In a review of gaseous heterogeneous reactions over Mn-based oxides [31], it was claimed that MnO$_2$ usually has a higher catalytic oxidation performance than Mn$_2$O$_3$ or MnO, as was expected based on the MvK mechanism. However, there are other studies that have proved that catalytic activity does not always increase with the increasing oxidation state of manganese; different morphologies and different reaction conditions lead to different results. Therefore, Kim et al. [94] found that the activity of manganese oxides in the catalytic combustion of benzene and toluene on pure Mn oxides varied in the order Mn$_3$O$_4$ > Mn$_2$O$_3$ > MnO$_2$ and was related to the higher oxygen mobility and larger surface area of Mn$_3$O$_4$. Almost the same results were reported by Piumetti et al. [95], the sequence for the catalytic activity in the total oxidation of VOCs (ethylene, propylene, toluene, and their mixture) being: Mn$_3$O$_4$ > Mn$_2$O$_3$ > Mn$_x$O$_y$ (where Mn$_x$O$_y$ was a mixture of Mn$_3$O$_4$ and MnO$_2$). In this case, the higher activity of Mn$_3$O$_4$ was explained on the basis of the large number of electrophilic oxygen species adsorbed on its surface, proving their beneficial role in VOC total oxidation.

The superior catalytic performance of a metal oxide is closely related to its redox properties and hence to its nonstoichiometric, defective structure. The existence of Mn^{3+} in a MnO$_2$ lattice usually generates oxygen vacancies and crystal defects, which increase its catalytic performance: the higher the oxygen vacancy density, the easier the activation adsorption of O$_2$. The formation of oxygen vacancies implies, also, the appearance of reduced Mn ions. The presence of both Mn^{3+} and Mn^{4+} has a beneficial effect because of the catalytic activity of MnO$_x$ generated by the electron transfer cycle between them (Equations (4) and (5)) [55,96]. Thus, in reference [96], various MnO$_x$ materials were synthesized using a coprecipitation method and then tested in the oxidation of toluene; in this case, the order of the catalytic performances of the obtained samples was also found to be directly related to the O$_{ads}$/O$_{latt}$ ratio values, which varied in relation to Mn^{3+}/Mn^{4+} ratios.

The ability of manganese oxides to crystallize into various forms could be exploited to obtain catalysts with improved performances. There are many studies in the literature that have attempted to establish which is the most active crystallographic form.

Reference [93] reported on n-hexane oxidation over three crystallographic forms of manganese dioxides, namely, pyrolusite (β), nsutite (γ), and ramsdellite. The nsutite MnO$_2$ was proved to be about five times more active than the other forms in the reaction of VOC removal in terms of conversion. This behavior was explained by the increase in the oxygen lability in this case as a result of the appearance of Mn vacancies. The complete oxidation of n-hexane is a reaction demanding many oxygen atoms. So, each Mn vacancy (if it is situated close to the surface) offers six O atoms that could be easily removed to take an active part in the catalytic process. In another study [97], two manganese oxides, γ- and β-MnO$_2$, were synthesized and the γ-MnO$_2$ structure was also proved to be more active;

its activity was related to the existence of the Mn^{3+}/Mn^{4+} couple and to the presence of lattice defects.

Huang et al. [98] synthesized α-, β-, and special biphase (denoted as α@β-) MnO_2 catalysts for toluene combustion (Figure 12a). α@β-MnO_2 presented a larger specific surface area and higher oxygen vacancy concentration, which made it a better catalyst. The best performances were obtained for α-MnO_2:β-MnO_2, with a mole ratio of 1:1 (Figure 12b).

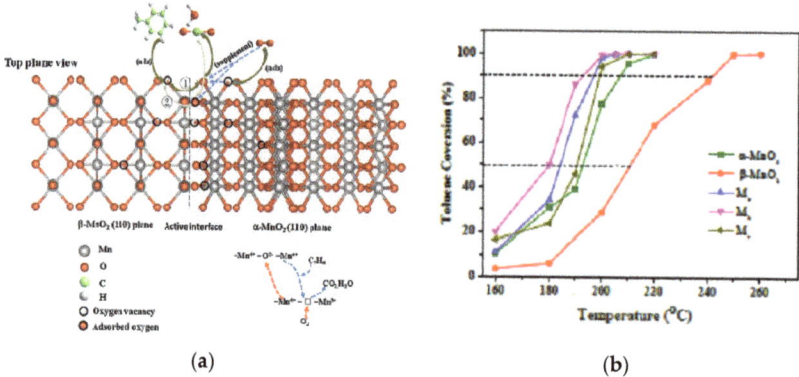

Figure 12. (a) Reaction scheme of toluene oxidation over α@β-MnO_2 catalysts. (b) Activity profiles of the five samples for toluene oxidation as a function of temperature. Reprinted and adapted with permission from Ref. [98]. Copyright 2018 Elsevier.

In reference [99], the authors presented a schematic diagram for toluene oxidation over some α-MnO_2 samples in order to highlight the role of the surface oxygen vacancies. The proposed mechanism was again MvK-type (Figure 13). The best catalytic performances were obtained over the catalyst with more surface oxygen vacancies and consequently more adsorbed oxygen, which supplemented the lattice oxygen consumed by the oxidation of toluene. On the other hand, more oxygen vacancies on the surface produced more Mn^{3+} ions, which elongated the bond length of Mn-O and contributed to the easier release of surface lattice oxygen, which finally also led to better catalytic activity.

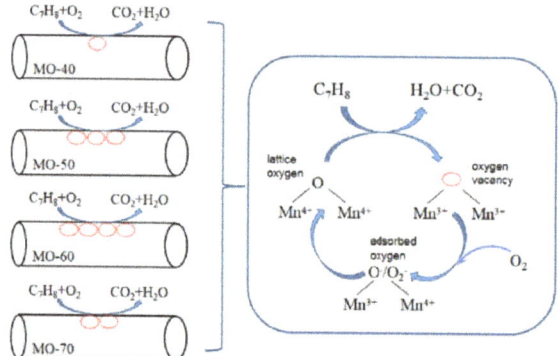

Figure 13. Mechanism of toluene oxidation on α-MnO_2. Reprinted with permission from Ref. [99]. Copyright 2021 Elsevier.

Xie et al. [100], by a hydrothermal method, synthesized MnO_2 nanoparticles with α-, β-, γ-, and δ-crystal phases. α-, β-, and γ-MnO_2 samples exhibited a 1D structure, and δ-MnO$_2$ was a 2D layered-structure material. The catalytic activities of the MnO$_2$ samples decreased in the order of α- ≈ γ- > β- > δ-MnO$_2$. In this case, surface area was not the determinant in relation to catalytic activity, while δ-MnO$_2$ presented a higher surface area than the others; furthermore, a direct relationship between the reducibility and the activity of the studied samples could not be established, the easily reducible sample (δ-MnO$_2$ again) having the worst catalytic performance. The authors concluded that, in this case, the main influencing factors that determined the catalytic activity were morphology and crystal structure.

Controlling morphology is of great importance in order to obtain materials with desired properties. For this purpose, Wang et al. [101] prepared nanosized rod-, wire-, and tube- like morphologies for α-MnO$_2$ and flower-like morphologies for α-Mn$_2$O$_3$ and compared their physicochemical properties and catalytic activities in toluene combustion (Figures 14 and 15). The rod-like α-MnO$_2$ possessed the best catalytic activity for toluene combustion, mainly due to the higher amount of adsorbed oxygen and to the increased low-temperature reducibility. Their results were presented in Table 2.

Figure 14. SEM images of (**a,b**) rod-like MnO$_2$, (**c,d**) wire-like MnO$_2$, (**e,f**) tube-like MnO$_2$, and (**g,h**) flower-like Mn$_2$O$_3$. Reprinted with permission from Ref. [101]. Copyright 2012 American Chemical Society.

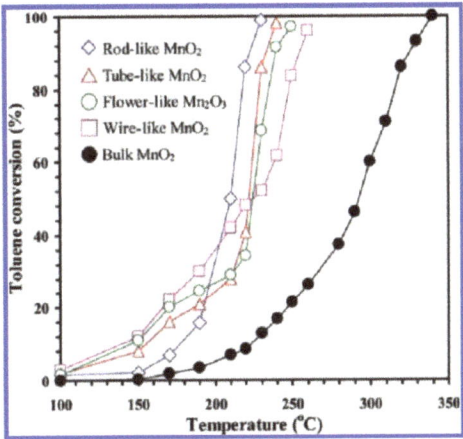

Figure 15. Toluene conversion as a function of reaction temperature over the catalysts under the conditions of toluene concentration = 1000 ppm, toluene/O$_2$ molar ratio = 1/400, and SV = 20,000 mL/(gh). Reprinted with permission from Ref. [101]. Copyright 2012 American Chemical Society.

In reference [102], the synthesis of hollow and solid polyhedral MnO$_x$ was reported. The hollow-structure materials proved to have outstanding performances in low-temperature toluene combustion (100% conversion at temperatures up to 240 °C) and high thermal stability (Figures 16 and 17). Catalytic activity was correlated with cavity nature, the higher content of active oxygen, and the higher average oxidation state of Mn. From the TPR measurements, the solid sample presented two main reduction peaks centered at lower temperatures than for the hollow one, but the H_2 consumption in the second case was far higher (Figure 16). This is an indication that the hollow sample had much more active oxygen and might have been more reducible and active than the solid one (the consumption of hydrogen was directly related to the concentration of active oxygen).

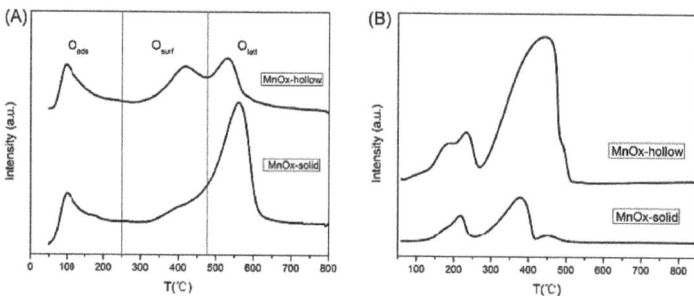

Figure 16. (**A**) O_2-TPD and (**B**) H_2-TPR profiles of MnOx-hollow and MnOx-solid. Reprinted with permission from Ref. [102]. Copyright 2017 Elsevier.

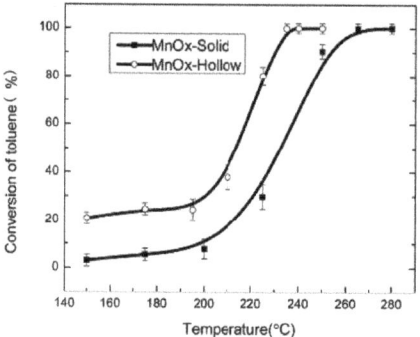

Figure 17. Toluene conversion over MnOx catalysts. Reaction conditions: catalysts weight = 0.3 g, toluene concentration = 1000 ppm, WHSV = 32,000 mL $g^{-1}h^{-1}$. Reprinted with permission from Ref. [102]. Copyright 2017 Elsevier.

A parameter of paramount importance for catalytic activity is the surface area of a catalyst, since the number of active sites increases as the exposed area increases. Thus, many authors have tried and succeeded in developing different methods for the synthesis of manganese oxides with increased surface areas [40,55,103]. In reference [55], MnOx with a high surface area and excellent activity in propane combustion was obtained (T_{90} = 265 °C) using a wet combustion procedure in the presence of organic acids, which produced a significant increase in surface area. The best results were obtained with glyoxilic acid, the surface area increasing considerably from 3 m^2/g (in absence of organic acids) up to 43 m^2/g.

Reference [40] describes the successful preparation of mesoporous α-MnO_2 microspheres with high specific surface areas, with control of the microstructures by adjustment of the synthesis temperature (Figure 18). When the temperature rose, the microspheres went from dense-assembly to loose-assembly, and the nanorod length on the surface of the α-MnO_2 microspheres increased gradually. Among the catalysts prepared at different temperatures, the best performance was exhibited by the catalyst with the larger surface area (Figure 19). A large surface area implies a higher concentration of surface Mn^{3+} and consequently a higher concentration of oxygen vacancies and adsorbed surface oxygen species, which could be incorporated in the lattice or used themselves as active oxygen.

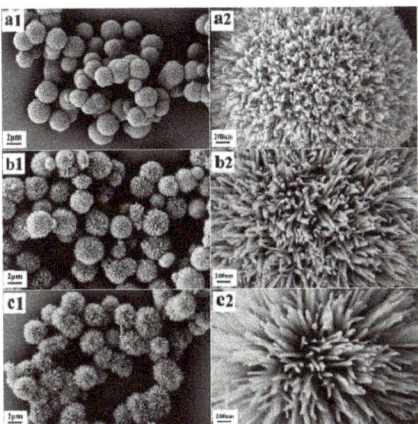

Figure 18. SEM images of MnO_2-40 (**a1,a2**), MnO_2-60 (**b1,b2**), and MnO_2-80 (**c1,c2**). Reprinted with permission from Ref. [40]. Copyright 2016 Elsevier.

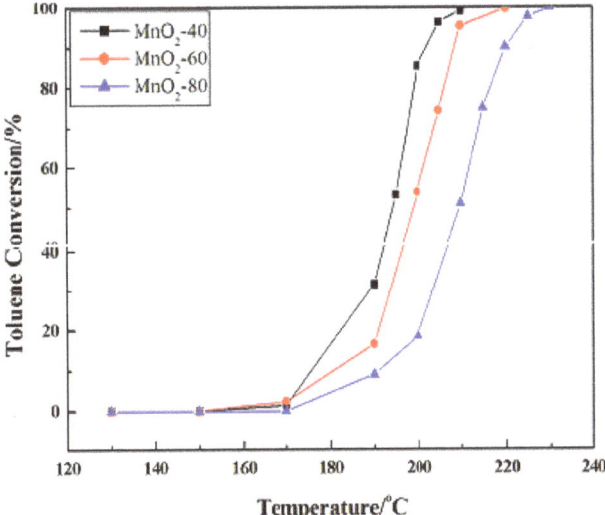

Figure 19. Catalytic combustion of toluene activity over α-MnO_2 obtained at different temperatures. Reprinted with permission from Ref. [40]. Copyright 2016 Elsevier.

In another study [103], a new method was developed to obtain a material with a three-dimensional macroporous and mesoporous morphology and a γ-MnO_2-like structure by selectively removing La cations from three-dimensional ordered macroporous $LaMnO_3$. The as-prepared material had a high specific surface area (245.7 m^2/g) and excellent catalytic performance (T_{90} = 252 °C, at GHSV = 120,000 mL (g h)$^{-1}$), which was related to both the three-dimensional macroporous and mesoporous morphology, which increased the accessibility of toluene to the active sites, and the γ-MnO$_2$-like structure, which improved the O$_{latt}$ mobility of the material.

In Table 2, some results obtained for the oxidation of hydrocarbons over single manganese oxides are summarized, the temperature when 90% conversion (T$_{90}$) was reached being the measurement parameter for catalytic performance [40,88,90,93,95,96,99–102,104–108]. In several cases, the average oxygen state (AOS) of Mn ions instead of the Mn^{4+}/Mn^{3+} ratio was listed. The value of AOS offers direct information about the reduction level of the oxide: the lower the AOS compared to the stoichiometric value, the more metallic ions in a reduced oxidation state exist in the oxide.

Table 2. Oxidation of hydrocarbons over single manganese oxides.

MnO$_x$	Oxidized VOC	Reaction Conditions	BET Surface (m^2/g)	Mn^{4+}/Mn^{3+}/O$_{ads}$/O$_{latt}$	TPR	T$_{90}$ (°C)	Ref.
Mn$_2$O$_3$	Naphtalene	100 ppm VOC in air 50 mL/min GHSV: 60.000 h^{-1}	76	-	247,392	240	[104]
Mn$_2$O$_3$	Propane	VOC:air 0.5:99.5 50 mL/min GHSV: 15.000 h^{-1}	76	-	247,392	335 [a]	[105]
Mn$_2$O$_3$	Toluene	500 ppm VOC, 20% O$_2$ in N$_2$ 100 mL/min	47.1	0.35/0.90	208,273,448	254	[96]
Mn$_2$O$_3$	Ethylene	1000 ppm VOC in air GHSV: 19.100 h^{-1}	43	0.63/0.32	365,489	330 [b]	[95]
Mn$_2$O$_3$	Toluene	1000 ppm VOC 30% O$_2$ in Ar 50 mL/min	7.7	0.90/0.63	379,434	270 [b]	[106]
Mn$_3$O$_4$	Ethylene	1000 ppm VOC in air GHSV: 19.100 h^{-1}	46	0.56/0.37	320,667	260 [b]	[95]
Mn$_3$O$_4$	Propane Propene	2% VOC; 12% O$_2$ in He 350 mL/min	24	- -	- -	300 [b] 270 [b]	[90]
Mn$_3$O$_4$ hollow Mn$_3$O$_4$ solid	Toluene	1000 ppm VOC in air WHSV: 32.000 mL/(g$_{cat}$·h)	90 21	3.66 [c]/4.47 3.10 [c]/2.30	230,444 218,377	240 [b] 280 [b]	[102]
Mn$_2$O$_3$+MnO$_2$	Ethylene	1000 ppm VOC in air GHSV: 19.100 h^{-1}	43	1.28/0.25	350,370,489	350	[95]
α-Mn$_2$O$_3$+β-MnO$_2$	Toluene	4000 mgC/m^3 GHSV: 16.000 h^{-1}	46	3.78 [c]/0.31	320,360,430, 480	280	[107]
MnO$_2$ rod MnO$_2$ wire MnO$_2$ tube Mn$_2$O$_3$ flower Bulk MnO$_2$	Toluene	1000 ppm VOC VOC/O$_2$ 1/400 in N$_2$ WHSV: 20.000 mL/(g$_{cat}$·h)	83.0 83.2 44.8 162.3 10	1.72/1.50 3.22/0.78 2.04/1.15 0.24/0.92 -	265; 285 275 268 260; 332 -	225 245 233 238 322	[101]
α–MnO$_2$	Toluene	4 g/m^3 VOC GHSV: 10.000 h^{-1}	150	0.8/0.84	232	202	[40]
α–MnO$_2$	Toluene	1000 ppm VOC in air 66.6 mL/min WHSV: 60.000 mL/(g$_{cat}$·h)	166	-/0.72	294,348	267	[40]
γ–MnO$_2$ β–MnO$_2$	N-hexane	250 ppm in air 100 mL/min	100 50			200 260	[93]
γ–MnO$_2$	N-hexane	125 ppm VOC, 20% O$_2$ in N$_2$ WHSV: 72.000 h^{-1}	103	3.9 [c]	-	170	[88]

Table 2. Cont.

MnO$_x$	Oxidized VOC	Reaction Conditions	BET Surface (m^2/g)	Mn^{4+}/Mn^{3+}/O$_{ads}$/O$_{latt}$	TPR	T$_{90}$ (°C)	Ref.
δ–MnO$_2$	Toluene	1000 ppm VOC, 21% O$_2$ in N$_2$ 150 mL/min WHSV: 90.000 mL/(g$_{cat}$·h)	231	1.12/1.192	362	200	[108]
α–MnO$_2$ β–MnO$_2$ γ–MnO$_2$ δ–MnO$_2$	Toluene	2000 ppm VOC 20% O$_2$ in N$_2$ 100 mL/min GHSV: 120.000 mL/(g$_{cat}$·h)	245.7 163.4 11.3 75.9	1.02/0.97 1.67/0.34 1.30/0.25 1.78/0.14	349,401,469 364,406,475 349.401.469 305,435	252 258 304 272	[103]
α–MnO$_2$	Toluene	1000 ppm VOC in air 500 mL/min GHSV: 30.000 h^{-1}	137.33	0.58/0.40	238,311,365, 376	203	[99]
α–MnO$_2$ β–MnO$_2$ γ–MnO$_2$ δ–MnO$_2$	Propane	250 ppm VOC in air SV: 72.000 h^{-1}	33.2 8.4 64.3 141	- - - -	300 322 324 278	290 373 280 -	[100]

^{a} T$_{80}$. ^{b} T$_{100}$. ^{c} Average oxidation state (AOS).

Although an important factor to consider in the development of a high-performance catalyst is the stability of the catalyst, in this review we have not explored this vast topic, choosing instead to emphasize the redox properties. The poor resistance to SO$_2$ and H$_2$O has remained a challenge for some MnO$_x$-based catalysts, especially unsupported MnO$_x$ catalysts [109]. Tang et al. [41], by a chemical leaching method, synthesized a Co$_3$O$_4$ material with superior catalytic stability for propane oxidation at a high WHSV: 240.000 mL/(g h). Compared with a 1%Pd/Al$_2$O$_3$ sample (when the conversion decreased by more than 10%), for the Co$_3$O$_4$, any deactivation occurred within 60 h of measurement. This catalyst also had a very good water/sulfur-resistance property. At 280 °C, the conversion of propane remained complete for a long time when either 5 ppm SO$_2$ or 3.0% water was introduced. However, the water vapor effect on the catalytic oxidation of VOCs should be taken into consideration because it may adsorb on the active sites of porous materials and inhibit the adsorption of VOCs, suppressing their further catalytic oxidation [110,111]. The addition of water beyond an optimum level results in a loss of activity due to the sintering of the catalyst. Zhang et al. studied the stability of a Co$_3$O$_4$ nanocatalyst in the oxidation of toluene, with and without moisture [112]. The toluene conversions remained constant, without any deactivation under dry conditions during the stability test, but suffered a slight decrease (being 10% lower) when 1.5 vol.% H$_2$O was introduced into the stream. The addition of a higher concentration (5 vol.%) of moisture produced a sharp decrease in conversion, from 100 to 70%, though this rapidly came back to its original level once the moisture was stopped. This indicates that there was competitive adsorption between water and VOC molecules on the surface active sites, causing the deactivation of the catalyst, but it exhibited good regeneration ability and stability. Similar results regarding the effect of the concentration of water were also obtained by Li et al. [113]. Contrary to water, carbon dioxide is not an inhibitor, but the formation of carbonate species might lead to the deactivation of the catalyst when the reaction is performed at low temperatures [114]. On the other hand, the origin of carbonaceous material (coke) was ascribed to adsorbed carbon monoxide on the catalyst. The deposition of coke on the outer surface of catalysts or inside pores is another problem, because it can poison the active center and cause the deactivation of catalysts. Working in excess-oxygen conditions could prevent this problem.

In this review, we have limited ourselves to the study of CoO$_x$ and MnO$_x$ single oxides, considering this a starting point for such a vast field of research. MnO$_x$ particles supported on activated carbon, carbon nanotubes, Ce−Zr solid solutions, Al$_2$O$_3$, etc., are used as catalysts for heterogeneous catalytic reactions [31]. Al_2O_3 and zeolite supports are considered inactive and enlarge surface areas, dispersing active sites. CeO_2 and Ce–Zr solid solutions are active supports because they often serve as adsorption sites for reactions. Non-noble metal catalysts include mixed-metal-oxide catalysts and perovskite catalysts. Perovskite oxides (with the formula ABO_3) containing transition metals, such as Mn and Co, at B sites have high oxidation abilities toward VOCs [111]. $LaMnO_3$ (manganites) and $LaCoO_3$ (cobaltites) are considered active catalyst materials. A category of mixed oxides with improved properties is $CoMnO_x$. Han et al. prepared a $CoMnO_x$ catalyst with a high amount of surface-chemically adsorbed oxygen [115]; Dong et al. synthesized nanoflower spinel $CoMn_2O_4$ with a large surface area, high mobility of oxygen species, and cationic vacancies [116]. Various metals were used for composites with Mn or Co, because of the synergistic effect between the two metallic components, which can improve the catalytic properties of the final materials.

5. Conclusions

Due to the rapid growth of industrialization and urbanization, volatile organic compounds (VOCs) have become the main pollutants in air. Their abatement is imperative due to their harmful effects on human health and to their precursor role in the formation of photochemical smog. In this context, catalytic oxidation has proved to be a useful technology, as the pollutants can be totally oxidized at moderate operating temperatures under 500 °C.

High-performance noble-metal-based catalysts for complete oxidation of VOCs at low temperatures have been developed. However, these catalyst types present high manufacturing costs and poor resistance to poisoning. Hence, there is a need to replace these materials with abundant ones which are cheaper, less susceptible to supply fluctuations, and more environmentally friendly. Thus, a better alternative, widely explored, is the use of transition metal oxides as catalysts. They have the advantages of much lower costs, higher thermal stabilities, and resistance to poisoning, which are ultimately reflected in a decrease in the total cost of the depollution technology.

The oxidation of VOCs is a complex process that involves the consideration of many different parameters: firstly, there is a large variety of polluting organic compounds which interact differently with metallic oxides, depending on the structural and electronic properties specific to each class; on the other hand, interactions also depend on the structural and physicochemical properties of the catalyst itself. In this review, we targeted the oxidation of hydrocarbons, and the focus was on the characteristic properties of catalysts which influence their performance in total-oxidation reactions.

Based on the redox mechanism, which is accepted as the one that describes how the oxidation reaction of hydrocarbons on transition metal oxides generally proceeds, the catalytic activity of these oxides strongly depends on the following factors: (i) their specific surface areas and morphologies; (ii) their redox properties; and (iii) their abilities to adsorb oxygen in the form of high electrophilic species, O_2^-, O^-, which are also very active species for total oxidation.

Generally, the higher specific surface area, the larger the number of exposed active sites. In addition, a porous structure permits the diffusion of the molecules of reactants inside the pores, which leads to the easier accessibility of the active centers.

Redox properties are dependent on: (i) the presence of metallic ions in higher oxidation states (Co^{3+}, Mn^{3+}, or Mn^{4+}) which make the oxide easier to be reduced; (ii) the presence of oxygen vacancies, which act as the active centers for oxygen activation (necessary for re-oxidation of oxides after interaction with hydrocarbons); (iii) the presence of metallic ions in reduced oxidation states, which result in more oxygen vacancies on the surface and also produce the weakening of the bond length of M-O and contribute to the easier release of surface lattice oxygen; and (iv) the high mobility of lattice oxygen.

Cobalt and manganese oxides are abundant materials which are non-toxic and have been proved to be very active in the total oxidation of hydrocarbons, being good candidates for the replacement of catalysts based on noble metals. In the case of cobalt oxides, Co_3O_4 was proved to be the most active, and the improvement of its catalytic performance has been achieved through controlled synthesis so as to obtain specific morphologies which present large surfaces and porous structures and which predominantly expose facets with high reactivity and defective structures (which contain large numbers of oxygen vacancies).

In the case of manganese oxides, the multitude of oxides with different oxidation states, which in turn can present several crystallographic phases, establishing the factors that determine their catalytic activities is more difficult. However, the increased reducibility, the presence of adsorbed oxygen, and the Mn^{3+}/Mn^{4+} couple generally have a beneficial effect.

Briefly, in this review, recent progresses made in the development of single-oxide catalysts (CoO_x and MnO_x) for the total oxidation of hydrocarbons has been summarized in order to establish the parameters that influence catalytic performances. Starting from this, new and high-performance catalytic materials with improved characteristics can be developed, either by obtaining specific morphologies, by depositing them on supports or by obtaining mixed oxides. Therefore, this research could be useful for tailoring advanced and high-performance catalysts for the total oxidation of VOCs.

Author Contributions: Conceptualization, V.B., P.C. and C.H.; writing—original draft preparation, V.B., A.V., P.C. and C.H.; writing—review and editing, V.B., A.V. and P.C.; supervision, V.B., A.V. and C.H. All authors have read and agreed to the published version of the manuscript.

Funding: This research received no external funding.

Conflicts of Interest: The authors declare no conflict of interest.

References

1. He, C.; Cheng, J.; Zhang, X.; Douthwaite, M.; Pattisson, S.; Hao, Z. Recent advances in the catalytic oxidation of volatile organic compounds: A review based on pollutant sorts and sources. *Chem. Rev.* **2019**, *119*, 4471–4568. [CrossRef] [PubMed]
2. Rusu, A.O.; Dumitriu, E. Destruction of volatile organic compounds by catalytic oxidation. *Environ. Eng. Manag. J.* **2003**, *2*, 273–302. [CrossRef]
3. Guo, Y.; Wen, M.; Li, G.; An, T. Recent advances in VOC elimination by catalytic oxidation technology onto various nanoparticles catalysts: A critical review. *Appl. Catal. B Environ.* **2021**, *281*, 119447. [CrossRef]
4. Lee, J.E.; Ok, Y.S.; Tsang, D.C.; Song, J.; Jung, S.C.; Park, Y.K. Recent advances in volatile organic compounds abatement by catalysis and catalytic hybrid processes: A critical review. *Sci. Total Environ.* **2020**, *719*, 137405. [CrossRef] [PubMed]
5. Liang, X.; Chen, X.; Zhang, J.; Shi, T.; Sun, X.; Fan, L.; Wang, L.; Ye, D. Reactivity-based industrial volatile organic compounds emission inventory and its implications for ozone control strategies in China. *Atmos. Environ.* **2017**, *162*, 115–126. [CrossRef]
6. Atkinson, R.; Arey, J. Gas-phase tropospheric chemistry of biogenic volatile organic compounds: A review. *Atmos. Environ.* **2003**, *37*, 197–219. [CrossRef]
7. Yuan, B.; Hu, W.W.; Shao, M.; Wang, M.; Chen, W.T.; Lu, S.H.; Zeng, L.M.; Hu, M. VOC emissions, evolutions and contributions to SOA formation at a receptor site in eastern China. *Atmos. Chem. Phys.* **2013**, *13*, 8815–8832. [CrossRef]
8. Fischer, E.V.; Jacob, D.J.; Yantosca, R.M.; Sulprizio, M.P.; Millet, D.B.; Mao, J.; Pandey Deolal, S. Atmospheric peroxyacetyl nitrate (PAN): A global budget and source attribution. *Atmos. Chem. Phys.* **2014**, *14*, 2679–2698. [CrossRef]
9. Zhong, L.; Haghighat, F. Photocatalytic air cleaners and materials technologies—Abilities and limitations. *Build. Environ.* **2015**, *91*, 191–203. [CrossRef]
10. Zhang, J.; Nosaka, Y. Mechanism of the OH radical generation in photocatalysis with TiO_2 of different crystalline types. *J. Phys. Chem. C* **2014**, *118*, 10824–10832. [CrossRef]
11. Okal, J.; Zawadzki, M. Catalytic combustion of butane on $Ru/\gamma-Al_2O_3$ catalysts. *Appl. Catal. B Environ.* **2009**, *89*, 22–32. [CrossRef]
12. Moretti, E.C. Reduce VOC and HAP emissions. *Chem. Eng. Prog.* **2002**, *98*, 30–40. Available online: http://pascal-francis.inist.fr/vibad/index.php?action=getRecordDetail&idt=13701676 (accessed on 25 August 2021).
13. Kouotou, P.M.; Pan, G.; Tian, Z. CVD-Made Spinels: Synthesis, Characterization and Applications for Clean Energy. In *Magnetic Spinels—Synthesis, Properties and Applications*; IntechOpen: London, UK, 2017. [CrossRef]
14. Kamal, M.S.; Razzak, S.A.; Hossain, M.M. Catalytic oxidation of volatile organic compounds (VOCs)—A review. *Atmos. Environ.* **2016**, *140*, 117–134. [CrossRef]
15. Peng, R.S.; Li, S.J.; Sun, X.B.; Ren, Q.M.; Chen, L.M.; Fu, M.L.; Wu, J.L.; Ye, D.Q. Size Effect of Pt Nanoparticles on the Catalytic Oxidation of Toluene over Pt/CeO_2 Catalysts. *Appl. Catal. B Environ.* **2018**, *220*, 462–470. [CrossRef]
16. Huang, S.Y.; Zhang, C.B.; He, H. Complete Oxidation of o-Xylene over Pd/Al_2O_3 Catalyst at Low Temperature. *Catal. Today* **2008**, *139*, 15–23. [CrossRef]
17. Liotta, L.F. Catalytic Oxidation of Volatile Organic Compounds on Supported Noble Metals. *Appl. Catal. B Environ.* **2010**, *100*, 403–412. [CrossRef]

18. Bratan, V.; Chesler, P.; Hornoiu, C.; Scurtu, M.; Postole, G.; Pietrzyk, P.; Gervasini, A.; Auroux, A.; Ionescu, N.I. In situ electrical conductivity study of Pt-impregnated VO_x/γ-Al_2O_3 catalysts in propene deep oxidation. *J. Mater. Sci.* **2020**, *55*, 10466–10481. [CrossRef]
19. Bratan, V.; Munteanu, C.; Chesler, P.; Negoescu, D.; Ionescu, N.I. Electrical characterization and the catalytic properties of SnO_2/TiO_2 catalysts and their Pd-supported equivalents. *Rev. Roum. Chim.* **2014**, *59*, 335–341. Available online: http://revroum.lew.ro/wp-content/uploads/2014/5/Art%2007.pdf (accessed on 24 March 2015).
20. Gelles, T.; Krishnamurthy, A.; Adebayo, A.; Rownaghi, A.; Rezaei, F. Abatement of gaseous volatile organic compounds: A material perspective. *Catal. Today* **2020**, *350*, 3–18. [CrossRef]
21. Vedrine, J.C. Heterogeneous catalysis on metal oxides. *Catalysts* **2017**, *7*, 341. [CrossRef]
22. Vasile, A.; Caldararu, M.; Hornoiu, C.; Bratan, V.; Ionescu, N.I.; Yuzhakova, T.; Rédey, Á. Elimination of gas pollutants using SnO_2-CeO_2 catalysts. *Environ. Eng. Manag. J.* **2012**, *11*, 481–486. Available online: http://www.eemj.icpm.tuiasi.ro/pdfs/vol11/no2/31_700_Vasile_11.pdf (accessed on 29 May 2012). [CrossRef]
23. Bratan, V.; Chesler, P.; Vasile, A.; Todan, L.; Zaharescu, M.; Caldararu, M. Surface properties and catalytic oxidation on V_2O_5-CeO_2 catalysts. *Rev. Roum. Chim.* **2011**, *56*, 1055–1065. Available online: http://revroum.lew.ro/wp-content/uploads/2011/RRCh_10-11_2011/Art%2017.pdf (accessed on 24 March 2015).
24. Vedrine, J.C. Metal oxides in heterogeneous oxidation catalysis: State of the art and challenges for a more sustainable world. *ChemSusChem* **2019**, *12*, 577–588. [CrossRef] [PubMed]
25. Urda, A.; Herraiz, A.; Redey, A.; Marcu, I.C. Co and Ni ferrospinels as catalysts for propane total oxidation. *Catal. Commun.* **2009**, *10*, 1651–1655. [CrossRef]
26. Huang, H.; Xu, Y.; Feng, Q.; Leung, D.Y.C. Low temperature catalytic oxidation of volatile organic compounds: A review. *Catal. Sci. Technol.* **2015**, *5*, 2649–2669. [CrossRef]
27. Vasile, A.; Hornoiu, C.; Bratan, V.; Negoescu, D.; Caldararu, M.; Ionescu, N.I.; Yuzhakova, T.; Redey, A. In situ electrical conductivity of propene interaction with SnO_2-CeO_2 mixed oxides. *Rev. Roum. Chim.* **2013**, *58*, 365–369. Available online: http://revroum.lew.ro/wp-content/uploads/2013/4/Art%2005.pdf (accessed on 24 February 2022).
28. Blazowski, W.S.; Walsh, D.E. Catalytic Combustion: An Important Consideration for Future Applications. *Combust. Sci. Technol.* **1975**, *10*, 233–244. [CrossRef]
29. Dixon, G.M.; Nicholls, D.; Steiner, H. Activity pattern in the disproportionation and dehydrogenation of cyclohexene to cyclohexane and benzene over the oxides of the first series of transitiona elements. *Proc. Third Intern. Congr. Catal.* **1965**, *2*, 81.
30. Moro-Oka, Y.; Morikawa, Y.; Ozaki, A. Regularity in the catalytic properties of metal oxides in hydrocarbon oxidation. *J. Catal.* **1967**, *7*, 23–32. [CrossRef]
31. Xu, H.; Yan, N.; Qu, Z.; Liu, W.; Mei, J.; Huang, W.; Zhao, S. Gaseous heterogeneous catalytic reactions over Mn-based oxides for environmental applications: A critical review. *Environ. Sci. Technol.* **2017**, *51*, 8879–8892. [CrossRef]
32. Baldi, M.; Finocchio, E.; Milella, F.; Busca, G. Catalytic combustion of C3 hydrocarbons and oxygenates over Mn_3O_4. *Appl. Catal. B Environ.* **1998**, *16*, 43–51. [CrossRef]
33. Todorova, S.; Blin, J.L.; Naydenov, A.; Lebeau, B.; Kolev, H.; Gaudin, P.; Dotzeva, A.; Velinova, R.; Filkova, D.; Ivanova, I.; et al. Co_3O_4-MnO_x oxides supported on SBA-15 for CO and VOCs oxidation. *Catal. Today* **2020**, *357*, 602–612. [CrossRef]
34. Li, K.; Chen, C.; Zhang, H.; Hu, X.; Sun, T.; Jia, J. Effects of phase structure of MnO_2 and morphology of δ-MnO_2 on toluene catalytic oxidation. *Appl. Surf. Sci.* **2019**, *496*, 143662. [CrossRef]
35. Deng, J.; Zhang, L.; Dai, H.; Xia, Y.; Jiang, H.; Zhang, H.; He, H. Ultrasound-assisted nanocasting fabrication of ordered mesoporous MnO_2 and Co_3O_4 with high surface areas and polycrystalline walls. *J. Phys. Chem. C* **2010**, *114*, 2694–2700. [CrossRef]
36. Salek, G.; Alphonse, P.; Dufour, P.; Guillemet-Fritsch, S.; Tenailleau, C. Low-temperature carbon monoxide and propane total oxidation by nanocrystalline cobalt oxides. *Appl. Catal. B Environ.* **2014**, *147*, 1–7. [CrossRef]
37. Lu, T.; Su, F.; Zhao, Q.; Li, J.; Zhang, C.; Zhang, R.; Liu, P. Catalytic oxidation of volatile organic compounds over manganese-based oxide catalysts: Performance, deactivation and future opportunities. *Sep. Purif. Technol.* **2022**, *296*, 121436. [CrossRef]
38. Kang, S.B.; Hazlett, M.; Balakotaiah, V.; Kalamaras, C.; Epling, W. Effect of Pt: Pd ratio on CO and hydrocarbon oxidation. *Appl. Catal. B Environ.* **2018**, *223*, 67–75. [CrossRef]
39. Lahousse, C.; Bernier, A.; Grange, P.; Delmon, B.; Papaefthimiou, P.; Ioannides, T.; Verykios, X. Evaluation of γ-MnO_2 as a VOC removal catalyst: Comparison with a noble metal catalyst. *J. Catal.* **1998**, *178*, 214–225. [CrossRef]
40. Lin, T.; Yu, L.; Sun, M.; Cheng, G.; Lan, B.; Fu, Z. Mesoporous α-MnO_2 microspheres with high specific surface area: Controlled synthesis and catalytic activities. *Chem. Eng. J.* **2016**, *286*, 114–121. [CrossRef]
41. Tang, W.; Xiao, W.; Wang, S.; Ren, Z.; Ding, J.; Gao, P.X. Boosting catalytic propane oxidation over PGM-free Co_3O_4 nanocrystal aggregates through chemical leaching: A comparative study with Pt and Pd based catalysts. *Appl. Catal. B Environ.* **2018**, *226*, 585–595. [CrossRef]
42. Ma, L.; Geng, Y.; Chen, X.; Yan, N.; Li, J.; Schwank, J.W. Reaction mechanism of propane oxidation over Co_3O_4 nanorods as rivals of platinum catalysts. *Chem. Eng. J.* **2020**, *402*, 125911. [CrossRef]
43. Langmuir, I. Part II.—"Heterogeneous reactions". Chemical reactions on surfaces. *Trans. Faraday Soc.* **1922**, *17*, 607–620. [CrossRef]
44. Ertl, G. Reactions at well-defined surfaces. *Surf. Sci.* **1994**, *299/300*, 742–754. [CrossRef]
45. Zhang, Z.; Jiang, Z.; Shangguan, W. Low-temperature catalysis for VOCs removal in technology and application: A state-of-the-art review. *Catal. Today* **2016**, *264*, 270–278. [CrossRef]

46. Liang, R.; Hu, A.; Hatat-Fraile, M.; Zhou, N. Fundamentals on Adsorption, Membrane Filtration, and Advanced Oxidation Processes for Water Treatment. In Nanotechnology for Water Treatment and Purification, Lecture Notes in Nanoscale Science and Technology. Hu, A., Apblett, A., Eds.; Springer: Cham, Switzerland, 2014; Volume 22. [CrossRef]
47. Pease, R.N.; Taylor, H.S. The catalytic formation of water vapor from hydrogen and oxygen in the presence of copper and copper oxide. *J. Am. Chem. Soc.* **1922**, *44*, 1637–1647. [CrossRef]
48. Senseman, C.E.; Nelson, O.A. Catalytic Oxidation of Anthracene to Anthraquinone. *Ind. Eng. Chem.* **1923**, *15*, 521–524. [CrossRef]
49. Weiss, J.M.; Downs, C.R.; Burns, R.M. Oxide Equilibria in Catalysis. *Ind. Eng. Chem.* **1923**, *15*, 965–967. [CrossRef]
50. Mars, P.; Krevelen, D.W.V. Spec. Suppl. *Chem. Eng. Sci.* **1954**, *3*, 41–59. [CrossRef]
51. Wang, Q.; Yeung, K.L.; Bañares, M.A. Ceria and its related materials for VOC catalytic combustion: A review. *Catal. Today* **2020**, *356*, 141–154. [CrossRef]
52. Vasile, A.; Bratan, V.; Hornoiu, C.; Caldararu, M.; Ionescu, N.I.; Yuzhakova, T.; Rédey, Á. Electrical and catalytic properties of cerium–tin mixed oxides in CO depollution reaction. *Appl. Catal. B Environ.* **2013**, *140*, 25–31. [CrossRef]
53. Bratan, V.; Ionescu, N.I. Oscillations in the system carbon monoxide-TiO_2. *Rev. Roum. Chim.* **2016**, *61*, 665–669. Available online: https://revroum.lew.ro/wp-content/uploads/2016/08/Art%2008.pdf (accessed on 9 November 2021).
54. Bielanski, A.; Najbar, M. Adsorption species of oxygen on the surfaces of transition metal oxides. *J. Catal.* **1972**, *25*, 398–406. [CrossRef]
55. Puértolas, B.; Smith, A.; Vázquez, I.; Dejoz, A.; Moragues, M.; Garcia, T.; Solsona, B. The different catalytic behaviour in the propane total oxidation of cobalt and manganese oxides prepared by a wet combustion procedure. *Chem. Eng. J.* **2013**, *229*, 547–558. [CrossRef]
56. Golodets, G.I. On Principles of Catalyst Choice for Selective Oxidation. In *Studies in Surface Science and Catalysis*; Elsevier: Amsterdam, The Netherlands, 1990; Volume 55, pp. 693–700. [CrossRef]
57. Boreskov, G.K. Catalytic Activation of Dioxygen. In *Catalysis. CATALYSIS—Science and Technology*; Anderson, J.R., Boudart, M., Eds.; Springer: Berlin/Heidelberg, Germany, 1982; Volume 3. [CrossRef]
58. Wang, X.; Tian, W.; Zhai, T.; Zhi, C.; Bando, Y.; Golberg, D. Cobalt (II, III) oxide hollow structures: Fabrication, properties and applications. *J. Mater. Chem.* **2012**, *22*, 23310–23326. [CrossRef]
59. Liu, Q.; Wang, L.C.; Chen, M.; Cao, Y.; He, H.Y.; Fan, K.N. Dry citrate-precursor synthesized nanocrystalline cobalt oxide as highly active catalyst for total oxidation of propane. *J. Catal.* **2009**, *263*, 104–113. [CrossRef]
60. Zhang, W.; Diez-Ramirez, J.; Anguita, P.; Descorme, C.; Valverde, J.L.; Giroir-Fendler, A. Nanocrystalline Co_3O_4 catalysts for toluene and propane oxidation: Effect of the precipitation agent. *Appl. Catal. B Environ.* **2020**, *273*, 118894. [CrossRef]
61. Solsona, B.; Davies, T.E.; Garcia, T.; Vázquez, I.; Dejoz, A.; Taylor, S.H. Total oxidation of propane using nanocrystalline cobalt oxide and supported cobalt oxide catalysts. *Appl. Catal. B Environ.* **2008**, *84*, 176–184. [CrossRef]
62. Garcia, T.; Agouram, S.; Sánchez-Royo, J.F.; Murillo, R.; Mastral, A.M.; Aranda, A.; Vázquez, I.; Dejoz, A.; Solsona, B. Deep oxidation of volatile organic compounds using ordered cobalt oxides prepared by a nanocasting route. *Appl. Catal. A Gen.* **2010**, *386*, 16–27. [CrossRef]
63. Xia, Y.; Dai, H.; Jiang, H.; Zhang, L. Three-dimensional ordered mesoporous cobalt oxides: Highly active catalysts for the oxidation of toluene and methanol. *Catal. Commun.* **2010**, *11*, 1171–1175. [CrossRef]
64. Xie, S.; Liu, Y.; Deng, J.; Yang, J.; Zhao, X.; Han, Z.; Zhang, K.; Dai, H. Insights into the active sites of ordered mesoporous cobalt oxide catalysts for the total oxidation of o-xylene. *J. Catal.* **2017**, *352*, 282–292. [CrossRef]
65. Liu, Z.; Cheng, L.; Zeng, J.; Hu, X.; Zhangxue, S.; Yuan, S.; Bo, Q.; Zhang, B.; Jiang, Y. Boosting catalytic oxidation of propane over mixed-phase CoO-Co_3O_4 nanoparticles: Effect of CoO. *Chem. Phys.* **2021**, *540*, 110984. [CrossRef]
66. Li, Y.; Chen, T.; Zhao, S.; Wu, P.; Chong, Y.; Li, A.; Zhao, X.; Chen, G.; Jin, X.; Qiu, Y.; et al. Engineering cobalt oxide with coexisting cobalt defects and oxygen vacancies for enhanced catalytic oxidation of toluene. *ACS Catal.* **2022**, *12*, 4906–4917. [CrossRef]
67. Wang, C.A.; Li, S.; An, L. Hierarchically porous Co_3O_4 hollow spheres with tunable pore structure and enhanced catalytic activity. *Chem. Commun.* **2013**, *49*, 7427–7429. [CrossRef]
68. Xiao, X.; Liu, X.; Zhao, H.; Chen, D.; Liu, F.; Xiang, J.; Hu, Z.; Li, Y. Facile shape control of Co_3O_4 and the effect of the crystal plane on electrochemical performance. *Adv. Mater.* **2012**, *24*, 5762–5766. [CrossRef]
69. Yao, J.; Shi, H.; Sun, D.; Lu, H.; Hou, B.; Jia, L.; Xiao, Y.; Li, D. Facet-dependent activity of Co_3O_4 catalyst for C_3H_8 combustion. *ChemCatChem* **2019**, *11*, 5570–5579. [CrossRef]
70. Jian, Y.; Tian, M.; He, C.; Xiong, J.; Jiang, Z.; Jin, H.; Zheng, L.; Albilali, R.; Shi, J.W. Efficient propane low-temperature destruction by Co_3O_4 crystal facets engineering: Unveiling the decisive role of lattice and oxygen defects and surface acid-base pairs. *Appl. Catal. B Environ.* **2021**, *283*, 119657. [CrossRef]
71. Xie, X.; Shen, W. Morphology control of cobalt oxide nanocrystals for promoting their catalytic performance. *Nanoscale* **2009**, *1*, 50–60. [CrossRef]
72. Niu, H.; Wu, Z.; Hu, Z.T.; Chen, J. Imidazolate-mediated synthesis of hierarchical flower-like Co_3O_4 for the oxidation of toluene. *Mol. Catal.* **2021**, *503*, 111434. [CrossRef]
73. Xie, X.; Li, Y.; Liu, Z.Q.; Haruta, M.; Shen, W. Low-temperature oxidation of CO catalysed by Co_3O_4 nanorods. *Nature* **2009**, *458*, 746–749. [CrossRef]
74. Bielanski, A.; Haber, J. *Oxygen in Catalysis*; CRC Press: Boca Raton, FL, USA, 1991. [CrossRef]

75. Finocchio, E.; Busca, G.; Lorenzelli, V.; Escribano, V.S. FTIR studies on the selective oxidation and combustion of light hydrocarbons at metal oxide surfaces. Part 2.—Propane and propene oxidation on Co_3O_4. *J. Chem. Soc. Faraday Trans.* **1996**, *92*, 1587–1593. [CrossRef]
76. Zhong, J.; Yikui, Z.; Zhang, M.; Feng, W.; Xiao, D.; Wu, J.; Chen, P.; Fu, M.; Ye, D. Toluene oxidation process and proper mechanism over Co_3O_4 nanotubes: Investigation through in-situ DRIFTS combined with PTR-TOF-MS and quasi in-situ XPS. *Chem. Eng. J.* **2020**, *397*, 125375. [CrossRef]
77. Liu, W.; Liu, R.; Zhang, H.; Jin, Q.; Song, Z.; Zhang, X. Fabrication of Co_3O_4 nanospheres and their catalytic performances for toluene oxidation: The distinct effects of morphology and oxygen species. *Appl. Catal. A Gen.* **2020**, *597*, 117539. [CrossRef]
78. Ma, C.Y.; Mu, Z.; Li, J.J.; Jin, Y.G.; Cheng, J.; Lu, G.Q.; Hao, Z.P.; Qiao, S.Z. Mesoporous Co_3O_4 and Au/Co_3O_4 catalysts for low-temperature oxidation of trace ethylene. *J. Am. Chem. Soc.* **2010**, *132*, 2608–2613. [CrossRef] [PubMed]
79. Ren, Q.; Feng, Z.; Mo, S.; Huang, C.; Li, S.; Zhang, W.; Chen, L.; Fu, M.; Wu, J.; Ye, D. 1D-Co_3O_4, 2D-Co_3O_4, 3D-Co_3O_4 for catalytic oxidation of toluene. *Catal. Today* **2019**, *332*, 160–167. [CrossRef]
80. Li, C.; Liu, X.; Wang, H.; He, Y.; Song, L.; Deng, Y.; Cai, S.; Li, S. Metal-organic framework derived hexagonal layered cobalt oxides with {1 1 2} facets and rich oxygen vacancies: High efficiency catalysts for total oxidation of propane. *Adv. Powder Technol.* **2022**, *33*, 103373. [CrossRef]
81. Zhang, W.; Hu, L.; Wu, F.; Li, J. Decreasing Co_3O_4 particle sizes by ammonia-etching and catalytic oxidation of propane. *Catal. Lett.* **2017**, *147*, 407–415. [CrossRef]
82. Zhang, W.; Wu, F.; Li, J.; You, Z. Dispersion–precipitation synthesis of highly active nanosized Co_3O_4 for catalytic oxidation of carbon monoxide and propane. *Appl. Surf. Sci.* **2017**, *411*, 136–143. [CrossRef]
83. Zhang, W.; Lassen, K.; Descorme, C.; Valverde, J.L.; Giroir-Fendler, A. Effect of the precipitation pH on the characteristics and performance of Co_3O_4 catalysts in the total oxidation of toluene and propane. *Appl. Catal. B Environ.* **2021**, *282*, 119566. [CrossRef]
84. Lei, J.; Wang, S.; Li, J. Mesoporous Co_3O_4 derived from Co-MOFs with different morphologies and ligands for toluene catalytic oxidation. *Chem. Eng. Sci.* **2020**, *220*, 115654. [CrossRef]
85. Ghosh, S.K. Diversity in the family of manganese oxides at the nanoscale: From fundamentals to applications. *ACS Omega* **2020**, *5*, 25493–25504. [CrossRef]
86. Menezes, P.W.; Indra, A.; Littlewood, P.; Schwarze, M.; Göbel, C.; Schomäcker, R.; Driess, M. Nanostructured manganese oxides as highly active water oxidation catalysts: A boost from manganese precursor chemistry. *ChemSusChem* **2014**, *7*, 2202–2211. [CrossRef] [PubMed]
87. Dong, Y.; Li, K.; Jiang, P.; Wang, G.; Miao, H.; Zhang, J.; Zhang, C. Simple hydrothermal preparation of α-, β-, and γ-MnO_2 and phase sensitivity in catalytic ozonation. *RSC Adv.* **2014**, *4*, 39167–39173. [CrossRef]
88. Cellier, C.; Ruaux, V.; Lahousse, C.; Grange, P.; Gaigneaux, E.M. Extent of the participation of lattice oxygen from γ-MnO_2 in VOCs total oxidation: Influence of the VOCs nature. *Catal. Today* **2006**, *117*, 350–355. [CrossRef]
89. Busca, G.; Finocchio, E.; Lorenzelli, V.; Ramis, G.; Baldi, M. IR studies on the activation of C–H hydrocarbon bonds on oxidation catalysts. *Catal. Today* **1999**, *49*, 453–465. [CrossRef]
90. Finocchio, E.; Busca, G. Characterization and hydrocarbon oxidation activity of coprecipitated mixed oxides Mn_3O_4/Al_2O_3. *Catal. Today* **2001**, *70*, 213–225. [CrossRef]
91. Baldi, M.; Escribano, V.S.; Amores, J.M.G.; Milella, F.; Busca, G. Characterization of manganese and iron oxides as combustion catalysts for propane and propene. *Appl. Catal. B Environ.* **1998**, *17*, L175–L182. [CrossRef]
92. Baldi, M.; Milella, F.; Gallardo-Amores, J.M. A study of Mn-Ti oxide powders and their behaviour in propane oxidation catalysis. *J. Mater. Chem.* **1998**, *8*, 2525–2531. [CrossRef]
93. Lahousse, C.; Bernier, A.; Gaigneaux, E.; Ruiz, P.; Grange, P.; Delmon, B. Activity of manganese dioxides towards VOC total oxidation in relation with their crystallographic characteristics. In *Studies in Surface Science and Catalysis*; Elsevier: Amsterdam, The Netherlands, 1997; Volume 110, pp. 777–785. [CrossRef]
94. Kim, S.C.; Shim, W.G. Catalytic combustion of VOCs over a series of manganese oxide catalysts. *Appl. Catal. B Environ.* **2010**, *98*, 180–185. [CrossRef]
95. Piumetti, M.; Fino, D.; Russo, N. Mesoporous manganese oxides prepared by solution combustion synthesis as catalysts for the total oxidation of VOCs. *Appl. Catal. B Environ.* **2015**, *163*, 277–287. [CrossRef]
96. Zhang, X.; Zhao, H.; Song, Z.; Liu, W.; Zhao, J.; Zhao, M.; Xing, Y. Insight into the effect of oxygen species and Mn chemical valence over MnO_x on the catalytic oxidation of toluene. *Appl. Surf. Sci.* **2019**, *493*, 9–17. [CrossRef]
97. Lamaita, L.; Peluso, M.A.; Sambeth, J.E.; Thomas, H.J. Synthesis and characterization of manganese oxides employed in VOCs abatement. *Appl. Catal. B Environ.* **2005**, *61*, 114–119. [CrossRef]
98. Huang, N.; Qu, Z.; Dong, C.; Qin, Y.; Duan, X. Superior performance of α@β-MnO_2 for the toluene oxidation: Active interface and oxygen vacancy. *Appl. Catal. A Gen.* **2018**, *560*, 195–205. [CrossRef]
99. Chen, L.; Liu, Y.; Fang, X.; Cheng, Y. Simple strategy for the construction of oxygen vacancies on α-MnO_2 catalyst to improve toluene catalytic oxidation. *J. Hazard. Mater.* **2021**, *409*, 125020. [CrossRef] [PubMed]
100. Xie, Y.; Yu, Y.; Gong, X.; Guo, Y.; Guo, Y.; Wang, Y.; Lu, G. Effect of the crystal plane figure on the catalytic performance of MnO_2 for the total oxidation of propane. *CrystEngComm* **2015**, *17*, 3005–3014. [CrossRef]
101. Wang, F.; Dai, H.; Deng, J.; Bai, G.; Ji, K.; Liu, Y. Manganese oxides with rod-, wire-, tube-, and flower-like morphologies: Highly effective catalysts for the removal of toluene. *Environ. Sci. Technol.* **2012**, *46*, 4034–4041. [CrossRef] [PubMed]

102. Liao, Y.; Zhang, X.; Peng, R.; Zhao, M.; Ye, D. Catalytic properties of manganese oxide polyhedra with hollow and solid morphologies in toluene removal. *Appl. Surf. Sci.* **2017**, *405*, 20–28. [CrossRef]
103. Si, W.; Wang, Y.; Peng, Y.; Li, X.; Li, K.; Li, J. A high-efficiency γ-MnO_2-like catalyst in toluene combustion. *Chem. Commun.* **2015**, *51*, 14977–14980. [CrossRef]
104. Garcia, T.; Solsona, B.; Taylor, S.H. Naphthalene total oxidation over metal oxide catalysts. *Appl. Catal. B Environ.* **2006**, *66*, 92–99. [CrossRef]
105. Solsona, B.E.; Garcia, T.; Jones, C.; Taylor, S.H.; Carley, A.F.; Hutchings, G.J. Supported gold catalysts for the total oxidation of alkanes and carbon monoxide. *Appl. Catal. A Gen.* **2006**, *312*, 67–76. [CrossRef]
106. Zhang, X.; Lv, X.; Bi, F.; Lu, G.; Wang, Y. Highly efficient Mn_2O_3 catalysts derived from Mn-MOFs for toluene oxidation: The influence of MOFs precursors. *Mol. Catal.* **2020**, *482*, 110701. [CrossRef]
107. Santos, V.P.; Pereira, M.F.R.; Órfão, J.J.M.; Figueiredo, J.L. The role of lattice oxygen on the activity of manganese oxides towards the oxidation of volatile organic compounds. *Appl. Catal. B Environ.* **2010**, *99*, 353–363. [CrossRef]
108. Lyu, Y.; Li, C.; Du, X.; Zhu, Y.; Zhang, Y.; Li, S. Catalytic oxidation of toluene over MnO_2 catalysts with different Mn (II) precursors and the study of reaction pathway. *Fuel* **2020**, *262*, 116610. [CrossRef]
109. Gao, L.; Li, C.; Li, S.; Zhang, W.; Du, X.; Huang, L.; Zeng, G. Superior performance and resistance to SO_2 and H_2O over CoO_x-modified MnO_x/biomass activated carbons for simultaneous Hg^0 and NO removal. *Chem. Eng. J.* **2019**, *371*, 781–795. [CrossRef]
110. Yang, C.; Miao, G.; Pi, Y.; Xia, Q.; Wu, J.; Li, Z.; Xiao, J. Abatement of various types of VOCs by adsorption/catalytic oxidation: A review. *Chem. Eng. J.* **2019**, *370*, 1128–1153. [CrossRef]
111. Neha; Prasad, R.; Singh, S.V. Catalytic abatement of CO, HCs and soot emissions over spinel-based catalysts from diesel engines: An overview. *J. Environ. Chem. Eng.* **2020**, *8*, 103627. [CrossRef]
112. Zhang, Q.; Mo, S.; Chen, B.; Zhang, W.; Huang, C.; Ye, D. Hierarchical Co_3O nanostructures in-situ grown on 3D nickel foam towards toluene oxidation. *Mol. Catal.* **2018**, *454*, 12–20. [CrossRef]
113. Li, G.; Zhang, C.; Wang, Z.; Huang, H.; Peng, H.; Li, X. Fabrication of mesoporous Co_3O_4 oxides by acid treatment and their catalytic performances for toluene oxidation. *Appl. Catal. A Gen.* **2018**, *550*, 67–76. [CrossRef]
114. Bion, N.; Can, F.; Courtois, X.; Duprez, D. Transition Metal Oxides for Combustion and Depollution Processes. In *Metal Oxides in Heterogeneous Catalysis*; Elsevier: Amsterdam, The Netherlands, 2018; pp. 287–353. [CrossRef]
115. Zhao, L.; Zhang, Z.; Li, Y.; Leng, X.; Zhang, T.; Yuan, F.; Zhu, Y. Synthesis of Ce_aMnO_x hollow microsphere with hierarchical structure and its excellent catalytic performance for toluene combustion. *Appl. Catal. B Environ.* **2019**, *245*, 502–512. [CrossRef]
116. Dong, C.; Qu, Z.; Qin, Y.; Fu, Q.; Sun, H.; Duan, X. Revealing the highly catalytic performance of spinel $CoMn_2O_4$ for toluene oxidation: Involvement and replenishment of oxygen species using in situ designed-TP techniques. *ACS Catal.* **2019**, *9*, 6698–6710. [CrossRef]

Article

Iron-Modified Titanate Nanorods for Oxidation of Aqueous Ammonia Using Combined Treatment with Ozone and Solar Light Irradiation

Silviu Preda [1], Polona Umek [2], Maria Zaharescu [1], Crina Anastasescu [1,*], Simona Viorica Petrescu [1], Cătălina Gîfu [3], Diana-Ioana Eftemie [1], Razvan State [1], Florica Papa [1,*] and Ioan Balint [1,*]

1 "Ilie Murgulescu" Institute of Physical-Chemistry of the Romanian Academy, 202 Splaiul Independentei, 060021 Bucharest, Romania; predas@icf.ro (S.P.); mzaharescu@icf.ro (M.Z.); simon_pet@yahoo.com (S.V.P.); deftemie@yahoo.com (D.-I.E.); rstate@icf.ro (R.S.)
2 Jožef Stefan Institute, Jamova Cesta 39, 1000 Ljubljana, Slovenia; polona.umek@ijs.si
3 National Research and Development Institute for Chemistry and Petrochemistry, 060021 Bucharest, Romania; gifu_ioanacatalina@yahoo.com
* Correspondence: canastasescu@icf.ro (C.A.); frusu@icf.ro (F.P.); ibalint@icf.ro (I.B.); Tel.: +4-021-318-85-95 (I.B.)

Abstract: Sodium titanate nanorods were synthesized by a hydrothermal method and subsequently modified with an iron precursor. For comparison, Fe_2O_3 nanocubes were also obtained through a similar hydrothermal treatment. Pristine, Fe-modified nanorods and Fe_2O_3 nanocubes were suspended in diluted ammonia solutions (20 ppm) and exposed to ozone and simulated light irradiation. Ammonia abatement, together with the resulting nitrogen-containing products (NO_3^-), was monitored by ion chromatography measurements. The generation of reactive oxygen species (·OH and O_2^-) in the investigated materials and their photoelectrochemical behaviour were also investigated. Morphological and structural characterizations (SEM, XRD, XRF, UV–Vis, H_2-TPR, NH_3-TPD, PL, PZC) of the studied catalysts were correlated with their activity for ammonia degradation with ozone- and photo-assisted oxidation. An increase in ammonia conversion and a decreasing amount of NO_3^- were achieved by combining the above-mentioned processes.

Keywords: titanate nanorods; Fe-modified titanate; Fe_2O_3 nanocubes; ammonia catalytic ozonation assisted by solar light

1. Introduction

The decontamination of waste waters, including the removal of nitrogen-containing pollutants (NH_3), is mandatory in order to sustain the increased global demand of drinking water since natural resources are limited. Therefore, biological water treatments, in addition to advanced catalytic oxidation processes (AOPs) are pathways that are nowadays largely explored by research studies and already validated depollution technologies [1]. High-performance water treatment processes with a view to removing ammonia play a beneficial role in the environment because they contribute to reducing water acidification and eutrophication, contributing to the sustainable use of water resources. In the last decade, many ammonia removal processes have been studied: biological treatment [2], bio filtration treatment [3], air/steam stripping [4] break-point chlorination [5], chemical precipitation [6], ion exchange [7], photo and catalytic ozone oxidation [8,9].

Catalytic ozonation is an advanced oxidation process and has become greatly significant in recent years. During the ozonation process, catalysts favour the decomposition of O_3, generating reactive oxygen species [10]. Many studies [8] indicated the efficient removal of NH_4^+ by oxidation with ozone using oxide-based catalytic systems of transition metals MO_x (M = Co, Ni, Mn, Sn, Cu, Mg and Al). MgO has a high catalytic activity but low selectivity to N_2 gas, while Co_3O_4 has good selectivity to N_2 gas but a lower activity.

A promising, effective catalyst for NH_4^+ removal, which is selective to nitrogen gas, was reported by Chen et al. [11] using a MgO/Co_3O_4 (molar ratio 2:8) composite catalyst. Noble metals supported on metal oxides are also widely investigated and used for aqueous ammonia oxidation into safe gaseous compounds [12–15].

Clearly, many efforts have been made to sustain the implementation of depollution technologies that generate non harmful degradation end-products, but recently, many of these efforts also focused on the usage of green and regenerable energies, such as solar light. For instance, an increased number of photocatalysts were developed and tested in order to achieve the selective degradation of aqueous NH_3 into N_2 under UV and solar light irradiation [16–18]. Most of these are materials based on TiO_2, displaying a large scale of morphologies and modifiers.

Since titanates with a 1D morphology were extensively and successfully investigated, both for their electronic and optical properties [19–23] it is reasonable to assume that ammonia degradation on these materials can bring significant advancements for environmental preservation. In order to favour the ammonia photodegradation under solar light irradiation, pristine titanate nanorods were modified with iron for the improvement of light absorption and separation of photogenerated charges. These were also compared with bare Fe_2O_3. Additionally, iron-based compounds are largely used for water depollution processes because they are non-toxic and cheap.

The aim of this work was to optimize the aqueous ammonia oxidation process in order to obtain a high ammonia conversion but also increase selectivity to gaseous nitrogen-containing products. This was successfully achieved by combining the ozone oxidation of aqueous ammonia with its photodegradative abatement under solar irradiation.

2. Results

2.1. Catalysts Synthesis

Sodium titanate nanorods were prepared starting from commercial TiO_2, anatase). Then, 8 g of TiO_2 was homogenized for 30 min at 3000 rpm in 80 mL 10 M NaOH aqueous solution. The mixture was ultrasonicated for 40 min, and then a definite amount (18 mL) was transferred to the PTFE-lined pressure vessels (Parr Instruments, Moline, IL, USA), with a filling degree of 80%. The pressure vessel was kept at 175 °C for 72 h. The as-obtained mixture was dispersed in distilled water, ultrasonicated for 5 min, filtered to remove excess solution, then dried overnight at 100 °C.

The as-prepared nanorods (denoted TiR) were washed several times using ultrapure water slightly acidified with HCl (pH 6) and further modified with Fe, according to the following procedure: 0.1 g of titanate nanorods were suspended in 3×10^{-2} M $FeCl_3 \cdot 6H_2O$ solution and gently shaken for 24 h. The filtered powder was subjected to hydrothermal treatment in 1 M NaOH solution at 160 °C for 3 h. The sample is denoted as FeTiR.

Fe_2O_3 nanocube synthesis follows the above-mentioned hydrothermal procedure, starting from 0.25 g $FeCl_3 \cdot 6H_2O$ and filtering, washing and drying at 80 °C. The sample is denoted as Fe_2O_3.

2.2. Catalysts Characterisation

2.2.1. Scanning Electron Microscopy (SEM)

In order to identify the morphological properties of the materials of interest, SEM images were recorded in addition to EDAX analysis.

Figure 1a,b reveals a morphological similarity between the as-prepared sodium titanate nanorods and Fe-modified nanorods, their lengths ranging from tens of nanometers to micrometers. A slight loss of transparency, together with an incipient surface roughness, could be perceived in the FeTiR sample, relative to the as-prepared TiR sample. Additionally, EDAX spectra confirmed the presence of iron in the modified nanorods (1.84 wt%) and a smaller sodium amount (7.14 wt%) than for pure nanorods (12.27 wt%).

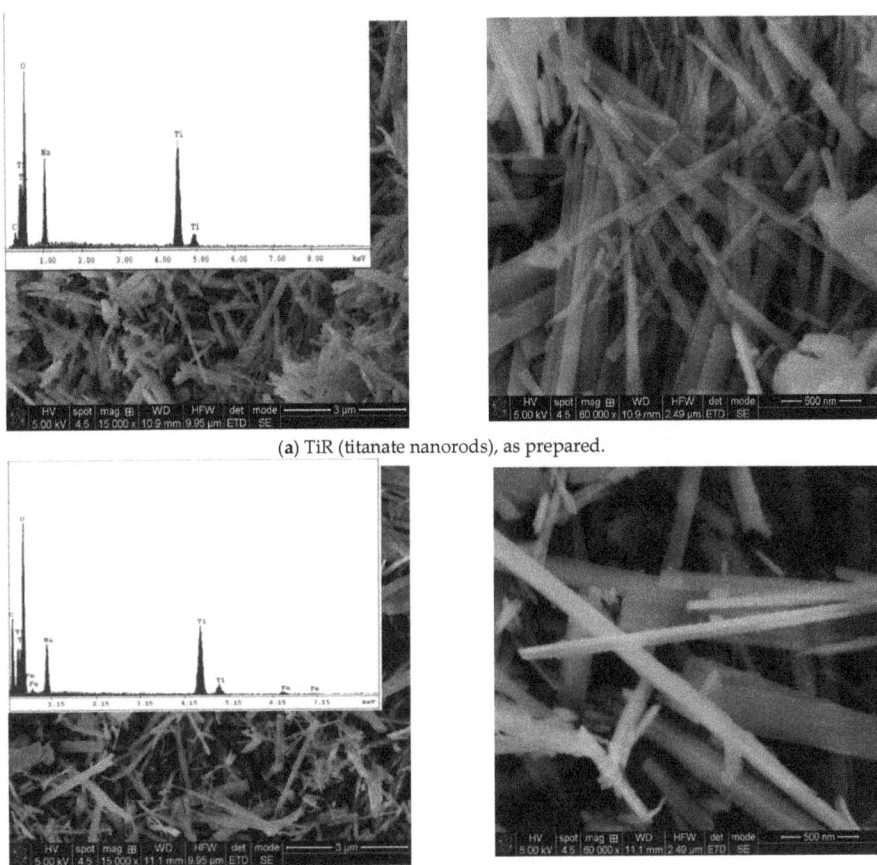

(a) TiR (titanate nanorods), as prepared.

(b) FeTiR (Fe-modified titanate nanorods).

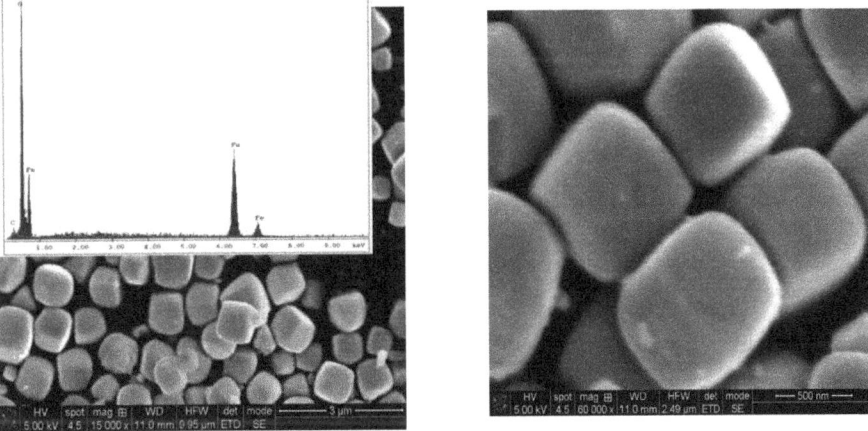

(c) Fe_2O_3, as prepared.

Figure 1. SEM images and EDAX spectra: (**a**) TiR as prepared, (**b**) FeTiR, (**c**) Fe_2O_3 nanocubes.

Figure 1c shows well-defined Fe_2O_3 nanocubes with a narrow size distribution around 700–800 nm and smooth surfaces. Ma et al. [24] reported the obtaining of Fe_2O_3 microcubes and small nanoparticles depending on the hydrothermal treatment parameters.

2.2.2. X-ray Diffraction (XRD) and X-ray Fluorescence (XRF)

The identification of crystalline phases and elemental characterization was performed in order to explain the catalytic performance of the target materials in the investigated media.

The XRD pattern of the sample TiR, presented in Figure 2a (upper-side), presents typical reflections, which indicates the formation of sodium titanate with nanorods morphology, as also noticed by other groups [25,26]. The phase composition of the sodium titanate nanorods was identified as $NaTi_3O_6(OH) \cdot 2H_2O$, according to PDF file no. 00-210-4964. The crystal structure of $NaTi_3O_6(OH) \cdot 2H_2O$ was described as a layered structure, similar to $Na_2Ti_3O_7$, belonging to monoclinic space group C2/m. The water molecules of crystallization rendered the structure more open, which is essential for cation-exchange behaviour [25]. The XRF measurement detected the Na/Ti weight ratio of 15/85. The deviation against theoretical ratio was a consequence of the thorough washing procedure after hydrothermal synthesis was completed and the vulnerability of this structure to cation-exchange behaviour.

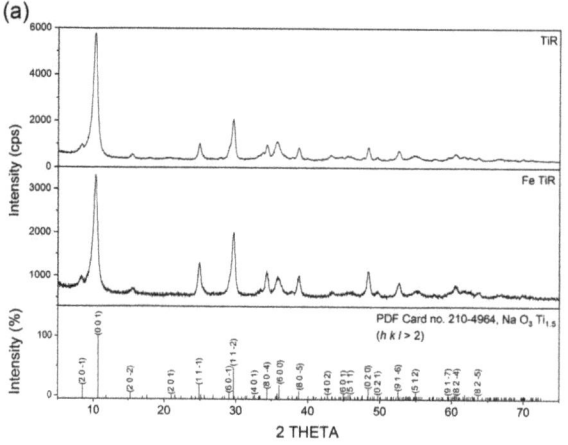

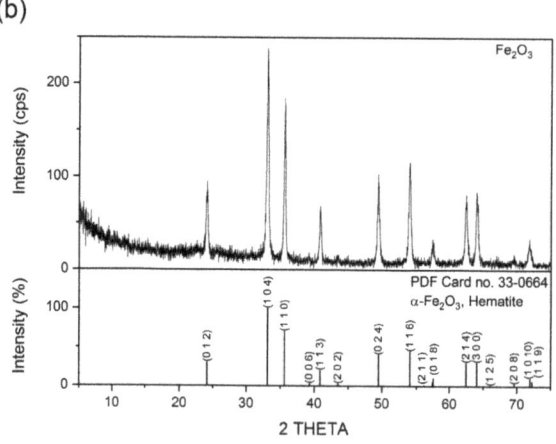

Figure 2. XRD pattern of (**a**) titanate nanorods: pristine TiR and FeTiR (**b**) Fe_2O_3.

Figure 2b (bottom side) presents the sample FeTiR, which contains sodium titanate nanorods submitted for the cation-exchange procedure. Sodium was partially exchanged by iron, as the XRF measurements detected. The weight ratio Fe/Na/Ti was 6/10/84. The percentage of Fe relative to the overall composition, measured by XRF, was 2.9611 mass %. These results are in agreement with the EDAX elemental analysis. Besides the sodium titanate nanorods, no iron-based compounds were detected by XRD. Accordingly, the reasonable assumptions are: the amorphous phase of iron compounds located on the nanorod surface, a partial replacement of sodium by iron cations during the ion-exchange procedure (supported by the XRF results). Furthermore, a shifting of the (001) reflection to smaller 2θ, (for TiR relative to FeTiR sample, respectively, to a larger interlayer spacing (d-value) supports the ion-exchange approach (including sodium–proton exchange).

The XRD pattern of the sample Fe_2O_3 is presented in Figure 2b. Single-phase α-Fe_2O_3 (hematite) was identified in the sample, according to PDF file no. 00-033-0664. Even single-phase hematite crystallizes, a ~5% amorphous phase can still be detected.

The structure parameters of the three samples are listed in Table 1.

Table 1. The structure parameters.

Sample	(001) Crystal Plane		Unit Cell Parameters					
	2θ (°)	d-Value (Å)	a (Å)	b (Å)	c (Å)	α (°)	β (°)	γ (°)
TiR	10.195(4)	8.670(4)	21.461(10)	3.757(7)	12.113(7)	90	135.59(2)	90
FeTiR	10.291(5)	8.589(4)	21.497(9)	3.7305(16)	12.051(6)	90	135.592(18)	90
Fe_2O_3	-	-	5.038(2)	5.038(2)	13.780(6)	90	90	120

2.2.3. UV–Vis Spectroscopy

The optical properties of the materials were revealed by the recorded diffuse reflectance (DR) spectra, the light absorption characteristics of the catalysts being correlated with their photoactivity.

Figure 3a illustrates a maximum absorption peak located at 260 nm, smaller for TIR and much higher for Fe-modified TiR sample. By enlarging the representation for TiR sample in Figure 3b, a small peak centred at 350 nm can be observed. This appears to be shifted to 390 nm and increased for the Fe-modified nanorods, which unlike the unmodified nanorods, also have a strong absorption in the visible range.

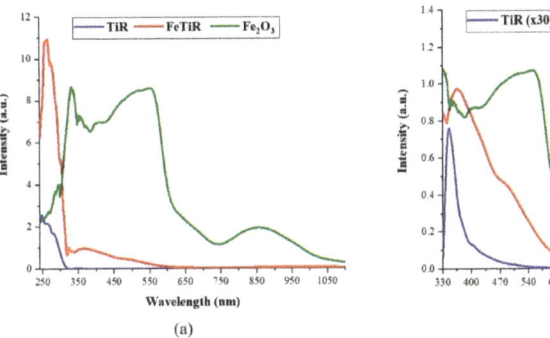

Figure 3. Comparative UV–Vis spectra are as follows: (a) Absorbance of TiR, FeTiR and Fe_2O_3 registered in 250–1100 nm range. (b) Absorbance of TiR (multiplied 30 times), FeTiR and Fe_2O_3 (decreased 8 times) recorded in 330–1100 nm range.

On the other hand, Fe_2O_3 emphasizes a broad light absorption, spanning on the whole spectral range. In the 750–1100 nm domain a broad absorption band can be observed, a similar but discrete shape is perceived in the inset of Figure 3b for the FeTiR sample.

In conclusion, Figure 3a,b clearly shows the improvement of light absorption in the visible range for Fe TiR nanorods relative to pristine TiRs.

2.2.4. Photoluminescence Measurements

Generally, photoluminescence measurements carried out using semiconductors are meant to act as photocatalysts that assess the photogenerated electron–hole pairs recombination and high PL signal, which indicates a reduced photocatalytic activity [27]. By modifying the photocatalyst and its encountering media, different photoluminescence signals can be obtained.

For the TiR and FeTiR samples, Figure 4 shows the same PL emission maxima (λ_{em} = 357 and 425 nm), which slightly decrease for the Fe-modified nanorods in diluted ammonia solution. This indicates a beneficial photo-mediated interaction between the catalyst surface and the adsorbed reactant (NH_4^+), which consumes part of the photo-generated charges, lowering their recombination. Consequently, the ammonia photodecomposition is expected to take place over the FeTiR sample. No significant PL signals were obtained for Fe_2O_3.

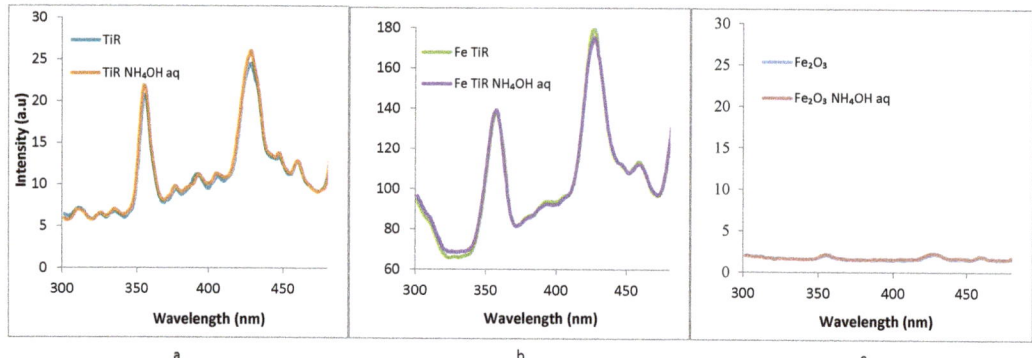

Figure 4. Room-temperature photoluminescence spectra of the powders suspended in water and diluted ammonia solutions, collected for λ_{exc} = 260 nm: (**a**) TiR, (**b**) FeTiR, (**c**) Fe_2O_3.

2.2.5. Reactive Oxygen Species Generation: Hydroxyl Radical (·OH) and Superoxide Anion (O_2^-)

Hydroxyl Radical (·OH)

According to our previously reported data [28], the generation of ·OH radicals was evaluated, taking into account the presence of fluorescent coumarin derivative (namely umbelliferone), caused by its interaction with photo-generated ·OH radicals over the investigated materials.

The characteristic broad peak is located around 470 nm and is depicted in Figure 5a,b. If ammonia photo-oxidation is mediated by ·OH radicals, this is likely to be caused by both TiR and FeTiR samples since they generate this oxygen species.

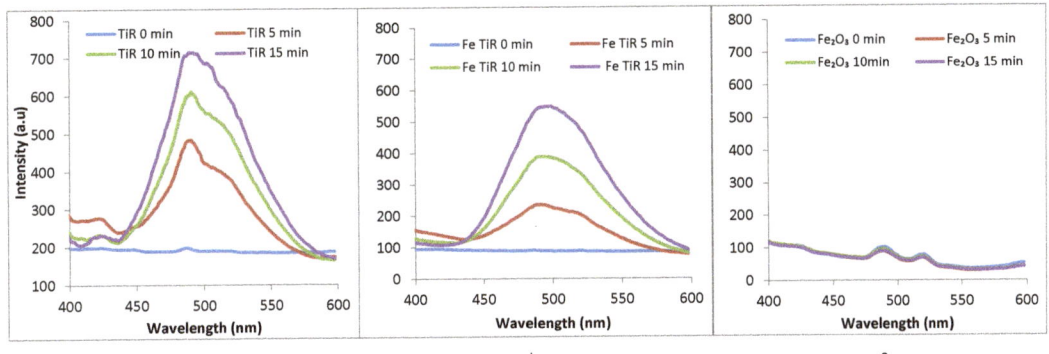

Figure 5. ·OH radicals trapping by coumarin under solar irradiation over TiR (**a**), FeTiR (**b**) and Fe$_2$O$_3$ (**c**).

Superoxide Anion (O$_2^-$)

In order to evaluate superoxide anion (O$_2^-$) generation, the XTT Formazan complex formation (due to the reaction of XTT with the photogenerated O$_2^-$) is evaluated based on spectrophotometric monitoring.

According to Figure 6a–c, there is no characteristic peak with maxima located at 475 nm for any of the investigated samples, only traces produced by Fe$_2$O$_3$ after 15 min of irradiation.

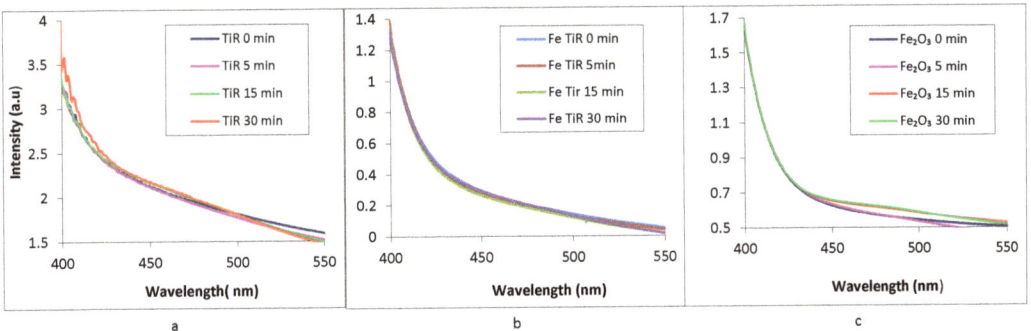

Figure 6. Investigation of O$_2^-$ generation over TiR (**a**), FeTiR (**b**), Fe$_2$O$_3$ cubes (**c**) under simulated solar light irradiation.

Consequently, it is reasonable to assume that the photogenerated electrons are not captured by O$_2$ but possibly involved in H$_2$ production.

2.2.6. Electrokinetic Potential Measurements

Electrokinetic potential measurements can reveal the surface properties of the interest powders suspended in water and diluted ammonia aqueous solution.

Figure 7 reveals the negative electrokinetic charges of all investigated samples. Nonetheless, a small difference in electrokinetic potential is observed for aqueous suspensions of titanate nanorods −49.54 mV and −45.24 for TiR and FeTiR, respectively. This could indicate more positive charges provided by the Fe presence in the titanate. Additionally, Fe$_2$O$_3$ cubes appear to be less negatively charged (electrokinetic potential being −29.2 mV). These above-mentioned values are clearly shifted toward the positive scale in the presence of ammonia. The difference is strongly related to the NH$_4^+$ adsorption on the investigated surfaces. These data confirm the electrostatically driven adsorption of NH_4^+ onto the catalyst surface for all samples.

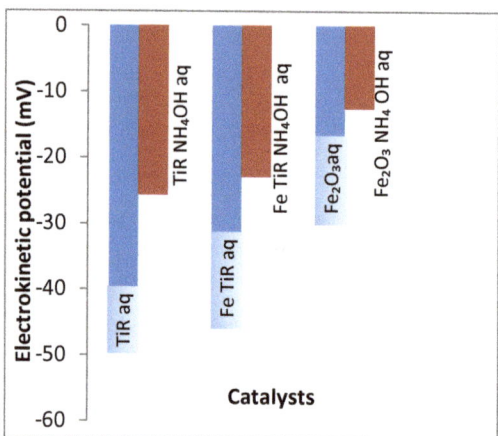

Figure 7. The electrokinetic potential of TiRs, Fe-modified TiRs and Fe_2O_3 cubes suspended in deionised water and diluted ammonia solution.

2.2.7. Photoelectrochemical Measurements

Current–potential dependence was registered for all samples under simulated solar light irradiation, before and after ammonia was added to the electrolyte.

The investigated powder suspensions were deposited by drop casting on transparent conductive oxide (TCO)-coated glass substrates. Figure 8 presents the plotted current density ($\mu A/cm^2$) versus potential (0–1 V) for the above-mentioned working electrodes under solar irradiation, before and after NH_4OH was added to the electrolyte solution. Lower values of the registered photocurrent density and open circuit voltage (V_{OC}) can be observed for the FeTiR relative to the TiR sample in both investigated media. Unlike the nanorods, the Fe_2O_3 cubes (green curve in Figure 8a) generate a much higher photocurrent that is significantly lowered by the ammonia addition.

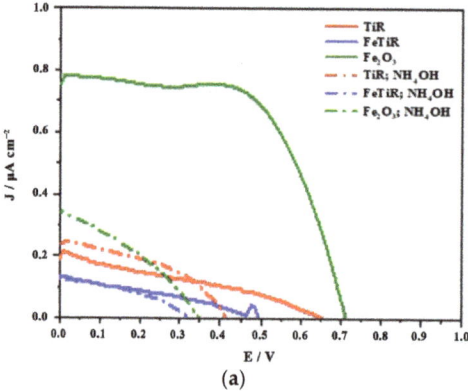

(a)

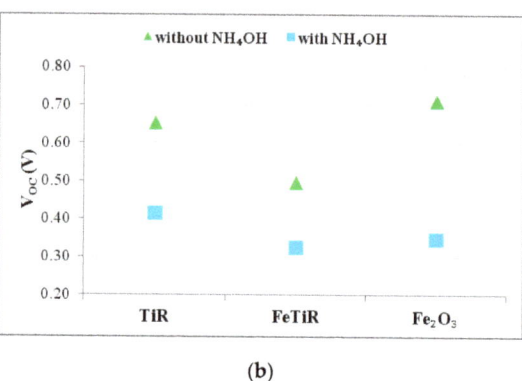

(b)

Figure 8. (a) J/V curves recorded for the investigated materials in pure electrolyte and diluted ammonia containing electrolyte under simulated solar irradiation; (b) V_{oc} representation for the investigated samples.

PCE (power conversion efficiency), η, was determined using the following equation:

$$PCE = \eta = \frac{J_{SC} V_{OC} FF}{P_{in}}$$

where: J_{sc}—short-circuit current density, V_{oc}—open-circuit voltage, FF—fill factor and P_{in}—power of incident light on the working electrode.

Figure 9 emphasizes the higher power conversion efficiency of the Fe_2O_3 sample in water that strongly decreases after the ammonia addition. On the contrary, the behaviour of the nanorods is scarcely affected by ammonia addition.

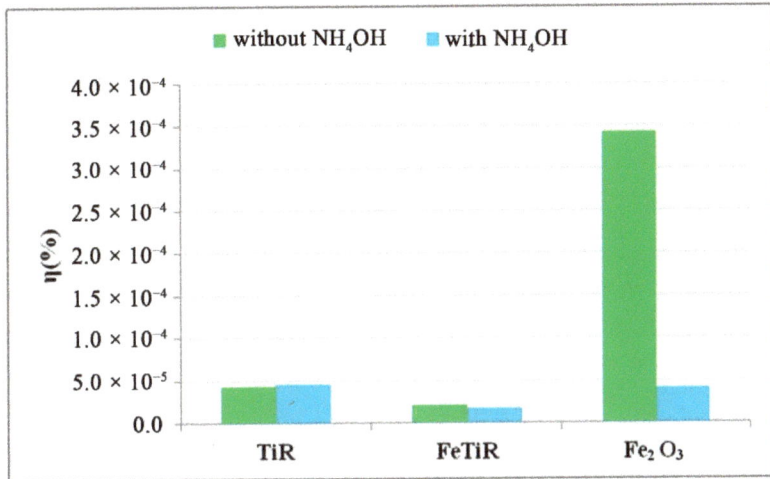

Figure 9. Comparative power conversion efficiency of the investigated samples measured in pure and ammonia containing electrolyte.

2.2.8. Temperature-Programmed Reduction Measurements (H_2-TPR)

Hydrogen temperature-programmed reduction (H_2-TPR) measurements were carried out to determine the redox behaviour of the involved chemical species of catalyst. As can be seen from Figure 10, the TiR sample is not significantly reduced in this temperature range. For the FeTiR and Fe_2O_3 samples, Figure 10 presents three similar but slightly shifted (to the higher temperature) peaks above 500 °C. These correspond to the successive reduction stages of Fe_2O_3 to Fe^0. The shift of the maxima to the higher temperatures in the case of Fe-modified nanorods could be due to the different encountering of the Fe species in titanate and Fe_2O_3 network, respectively. Unlike the Fe-modified nanorods, Fe_2O_3 cubes generate a distinct (TPR) peak around 450 °C, which the previous literature assigns to the presence of iron hydroxide [29]. The next overlapped peaks from 500 to 600 °C can be ascribed to the process of $Fe_2O_3 \rightarrow Fe_3O_4$. The largest hydrogen consumption at a high temperature 600–800 °C can be assigned to the reduction of Fe_3O_4 to metallic iron via wüstite (FeO) $Fe_3O_4 \rightarrow Fe^0$ [30].

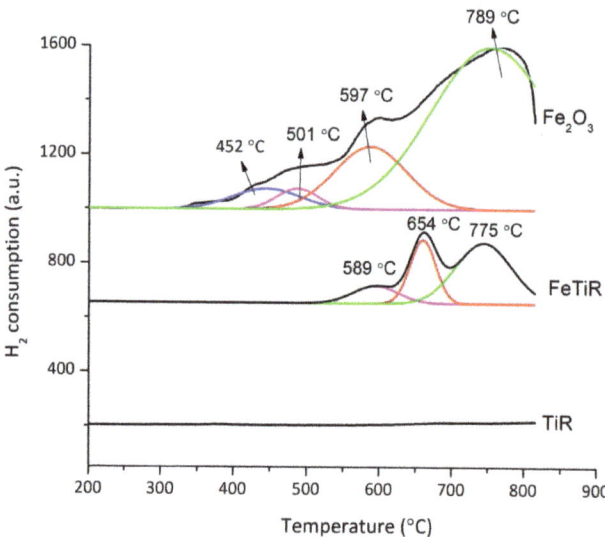

Figure 10. Temperature-programmed reduction behaviour of the investigated samples.

The amount of H_2 (μmol) consumed by each catalyst is listed in Table 2. It is clear that TiR sample is not significantly reduced in this temperature range.

Table 2. The H_2-TPR results for TiR, FeTiR and Fe_2O_3 catalysts.

Catalyst	H_2 Consumption TPR (μmol/g)				Total H_2 Consumption (μmol/g)
	I FeO(OH)	II $Fe_2O_3 \rightarrow Fe_3O_4$	III $Fe^{2+,3+} \rightarrow Fe^0$	IV	
TiR	0	0	0	0	
FeTiR	0	128	244	232	604
Fe_2O_3	1625	3092	4098	9352	18,167

In conclusion, in the FeTiR and Fe_2O_3 samples, the presence of Fe^{3+} as a single species can be identified using the TPR analysis. In the case of the Fe_2O_3 sample, besides hematite, the additional presence of an amorphous hydroxide phase is suggested. In fact, the XRD analysis indicates the presence of an amorphous phase (5%) for the Fe_2O_3 sample. The theoretical amount of H_2 consumption needed for the complete reduction of $Fe^{3+} \rightarrow Fe^0$ in pure Fe_2O_3 hematite is 18,750 μmol·g_{cat}^{-1}. The measured value was 18,167 μmol·g_{cat}^{-1}, which is close to the theoretical value. Concerning the synthesis procedure, the amorphous phase presence can be explained as follows. Despite the fact that the hydrothermal treatment applied for the obtaining of Fe_2O_3 cubes and FeTiR is similar, there is a significant difference regarding the overall synthesis procedure. The iron source of Fe_2O_3 nanocubes sample, obtained by hydrothermal synthesis, was $FeCl_3·6H_2O$ in the aqueous solution. The sodium titanate nanorods were first subjected to an ion-exchange procedure in the presence of $FeCl_3·6H_2O$ and were then hydrothermally treated.

2.2.9. Temperature-Programmed Desorption Measurements (NH_3-TPD)

The total acidity and distribution of the catalyst acid sites were calculated from the peak integration.

NH_3-TPD were performed in order to study the surface acidity of the FeTiR catalyst by comparing with TiR and Fe_2O_3. According to Figure 11 and Table 3, NH_3-TPD profiles of FeTiR, TiR and Fe_2O_3 catalysts were mainly composed of three desorption zones at the following temperatures: 50–300 °C (weak acid groups), 300–600 °C (moderate acid sites) and 600–800 °C (strong acid sites), respectively.

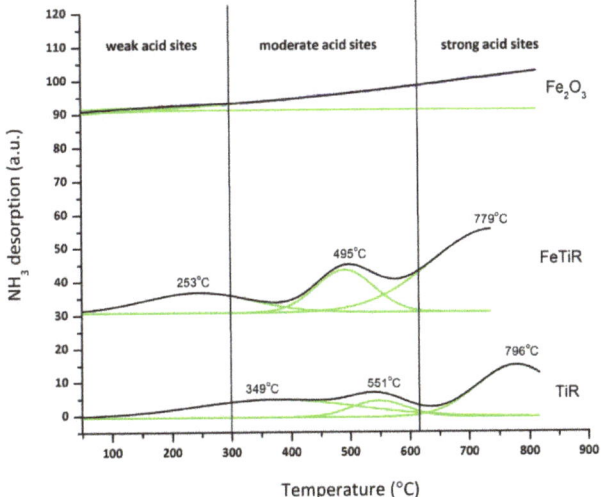

Figure 11. NH_3-TPD profile.

Table 3. The total acidity and distribution of the acid sites of the TiR, Fe_2O_3 and FeTiR catalysts.

Catalysts	Weak Acid Sites µmol·g^{-1} T < 300 °C	Moderate Acid Sites µmol·g^{-1} 300 °C < T < 600 °C	Strong Acid Sites µmol·g^{-1} T > 600 °C	Total Acid Sites µmol·g^{-1}
TiR	0.185	0.682	9.764	10.631
FeTiR	0.714	2.204	11.989	14.907
Fe_2O_3	0.329	0.964	1.306	2.599

The thermal stability of the nanorods reached up to 350 °C, and the comparative evaluation of all investigated sample was perfectly valid for this temperature, which was similar to the structure of catalysts in our catalytic media. High-temperature investigations are relevant for comparisons with the literature data.

The total acidity of TiR sample increases after iron addition (FeTiR), while Fe_2O_3 does not absorb ammonia. Additionally, a shift of low-temperature desorption (from 349 to 253 °C) is observed after the Fe addition to TiR, indicating change in the surface acidity of the FeTiR catalyst. Wang et al. [31] showed that the NH_3 species adsorbed at a low temperature T < 500 °C can be related to the presence of Brønsted acid sites, while adsorption at a high temperature (>600 °C) was correlated with the adsorption of NH_3 species on Lewis acid sites.

3. Catalytic Assays

Prior to ozone exposure and light irradiation, the steady state of the catalytical system is achieved (the suspended catalyst powder in aqueous ammonia solution is stirred for 30 min). It is expected that the reactant adsorption on the catalyst surface (certified by the electrokinetic potential measurements) and the working pH of the diluted ammonia solution are around 10. For these parameters, equilibrium occurs between free ammonia (NH_3) and ammonium ions (NH_4^+).

3.1. Ammonia Oxidation with Ozone

The following steps are presumably involved in aqueous ammonia catalytic oxidation with ozone:

– An equilibrium reaction (Equation (1)):

$$NH_4^+ + OH^- \rightleftarrows NH_3 + H_2O \qquad (1)$$

– Direct oxidation with ozone, especially for low pH (pH < Pka):

$$NH_4^+ + 3O_3 \rightarrow NO_2^- + H_2O + 3O_2 + 2H^+ \qquad (2)$$

– The reaction of ·OH radicals resulting from the decomposition of O_3 (Equation (3)) (pH > Pka):

$$6 \cdot OH + NH_3 \rightarrow NO_2^- + H^+ + 4H_2O \qquad (3)$$

$$NO_2^- + O_3 \rightarrow NO_3^- + O_2 \qquad (4)$$

$$NO_2^- + 2 \cdot OH \rightarrow NO_3^- + H_2O \qquad (5)$$

In both cases (Equations (4) and (5)), the oxidation of NO_2^- to NO_3^- occurs rapidly due to the strong oxidizing agents.

Figures 12 and 13 present the recorded catalytic results for the following working parameters: NH_4^+ initial concentration: 20.00 ppm, starting pH: 10.2, catalyst amount: 0.15 g/L, reaction temperature: 25 °C, and reaction time: 3 h.

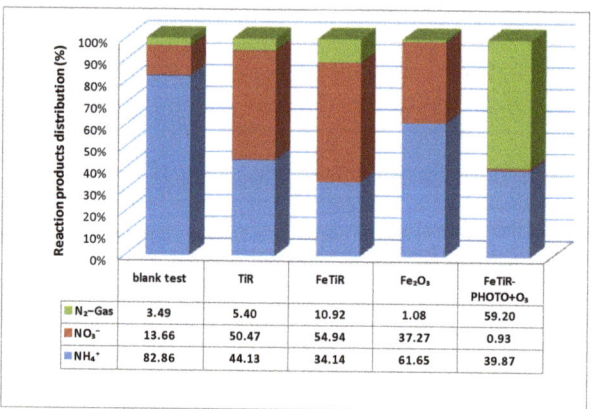

Figure 12. Reaction products distribution for aqueous ammonia oxidation with ozone and a photo-assisted process (carried out for FeTiR catalysts).

The amount of converted NH_4^+ in the absence of the catalyst (denoted as blank test in Figure 13) is low; a higher NH_4^+ conversion for the catalytic oxidation of ammonia with ozone is recorded over the FeTiR catalyst. The main degradation product in the solution is nitrate, and the nitrite ions are not present. A rapid oxidation of NO_2^- to NO_3^- in the presence of ozone occurs.

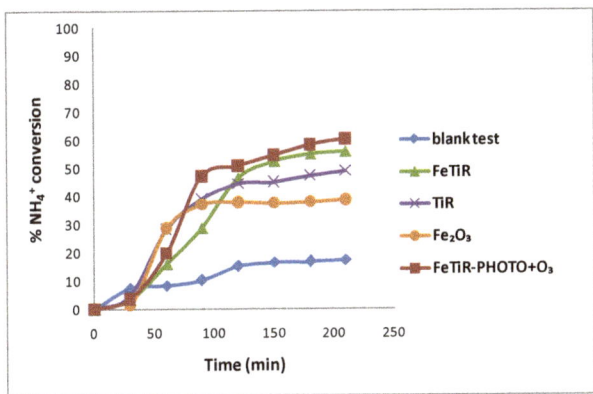

Figure 13. NH$_4^+$ conversion for catalytic oxidation with ozone and combined photo-assisted degradation pathway (for FeTiR catalyst).

Taking these data into account, the FeTiR sample was tested in a combined degradation pathway: ozone and solar light irradiation. An increase in conversion was recorded (Figure 13), together with a lower amount of NO$_3^-$. Based on these results, the increased selectivity of ammonia degradation to gaseous-nitrogen-containing end products is expected.

3.2. Photocatalytic Assays

The above-mentioned working parameters were preserved for the photocatalytic tests of the TiR, FeTiR, Fe$_2$O$_3$ samples. Similarly, the comparative evaluation of the combined photocatalytic ammonia oxidation with ozone was performed on the FeTiR sample and is presented in Figures 14 and 15.

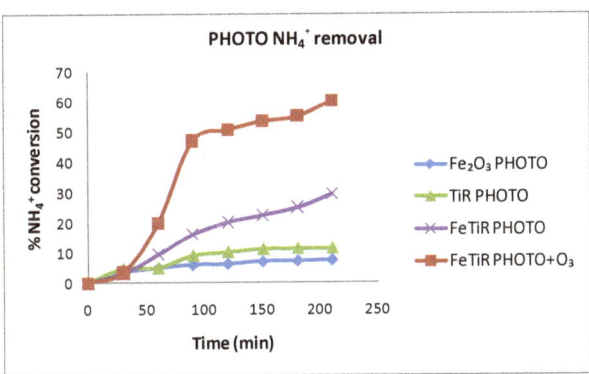

Figure 14. Comparative NH$_4^+$ photodegradation for the investigated samples.

After solar light irradiation of the system, the following steps are taken:

$$\text{Catalyst} + h\nu \rightarrow e^-_{CB} + h^+_{VB}$$

$$e^-_{CB} + O_2 \rightarrow O_2^- \tag{6}$$

$$h^+_{VB} + H_2O \rightarrow \cdot OH \tag{7}$$

$$h^+_{VB} + OH^-_{surf} \rightarrow \cdot OH \tag{8}$$

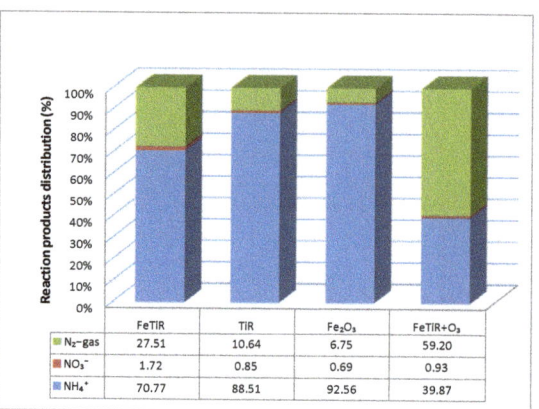

Figure 15. Reaction products distribution for aqueous ammonia photo-oxidation and ozone photo-assisted process (carried out for FeTiR catalyst).

According to the reported data [32], in the presence of oxygen, the following reactions are possible:

$$4NH_3 + 3O_2 \rightarrow 2N_2 + 6H_2O \tag{9}$$

$$2NH_3 + 2O_2 \rightarrow N_2O + 3H_2O \tag{10}$$

$$2NH_3 + 3O_2 \rightarrow 2NO_2^- + 2H^+ + 2H_2O \tag{11}$$

$$2NO^-_2 + O_2 \rightarrow 2NO_3^- \tag{12}$$

Figure 14 shows the highest NH_4^+ conversion over the course of 180 min under solar irradiation and bubbled O_2 for the Fe-modified nanorods. The selectivity to gaseous nitrogen compounds is high (Figure 15) since the measured NO_3^- amount is very low and nitrite is also missing. By adding ozone to the photocatalytic system, an increased catalytic activity of this sample can be obtained. Figure 15 also emphasizes the improvement of NH_4^+ conversion induced by ozone, adding to the photo-driven degradative process and reducing NO_3^- formation (1.72–0.93%).

Figure 16 shows the nitrate yield versus time of the ozonation process and photo-oxidation process. Due to the fact that the gaseous reaction products cannot be separately quantified, the yield of NO_3^- formation was measured. The lowest value was registered for the ozone photo-assisted degradation of ammonia for the FeTiR sample.

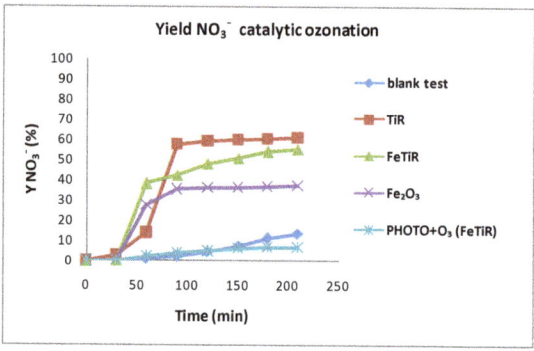

Figure 16. Yield of NO_3^- vs. time for ammonia ozonation in addition to photo-oxidation.

Taking into account the previously mentioned structural and functional characterization of the investigated catalysts, the following can be concluded for the photodegradative oxidation of the ammonia:

- The investigated ROS (both hydroxyl radicals and superoxide anions) are not fully involved in the degradative pathway.
- The photogenerated charges on the catalyst surface play a major role in ammonia oxidation.
- The highest photocatalytic activity of the Fe-modified titanate nanorods can be related to the increased light absorption relative to the bare samples.

4. Materials and Methods

4.1. Synthesis of Materials

The sodium titanate nanorods (TiR), Fe-modified titanate nanorods (FeTiR) and Fe_2O_3 were prepared, starting from commercial TiO_2, anatase (Aldrich), $FeCl_3 \cdot 6H_2O$ (Fluka), NaOH 97% (Alpha Aesar), and HCl 37.5% (Alfa Aesar).

4.2. Structural and Functional Characterization of the Obtained Materials

4.2.1. SEM—Scanning Electron Microscopy

In order to identify the morphological properties of the materials of interest, scanning electron microscopy (SEM) was used with a high-resolution microscope, FEI Quanta3 DFEG model, Brno, Czech Republic, working at 5 kV voltage, in high-vacuum mode with Everhart–Thornley secondary electron (SE) detector, coupled with energy-dispersive X-ray (EDAX) spectrometer.

4.2.2. X-ray Diffraction (XRD) and X-ray Fluorescence (XRF)

Elemental analysis of the samples was carried out using a Rigaku ZSX Primus II spectrometer, Rigaku Corp., Tokyo, Japan with wavelength dispersion in vacuum atmosphere. The spectrometer was equipped with 4.0 kW X-ray Rh tube. The XRF results were analysed using EZ-scan combined with Rigaku SQX fundamental parameters software (standard less), which is capable of automatically correcting all matrix effects, including line overlaps.

Powder X-ray diffraction patterns were recorded using Rigaku's Ultima IV multipurpose diffraction system, Rigaku Corp., Tokyo, Japan, a Cu target tube (λ = 1.54060 Å) and a graphite (002) monochromator, with working conditions of 30 mA and 40 kV. The data were collected at room temperature between 3 and 75° in 2θ, with a 0.02° step size and a scanning rate of 1°/min. Phase identification was performed using Rigaku's PDXL software, connected to the ICDD PDF-2 database. The lattice constants were refined using diffraction line position.

4.2.3. UV–Vis Spectrophotometry

In order to investigate the optical properties of the investigated samples, the diffuse reflectance (DR) spectra were recorded with a UV–Vis spectrophotometer, Analytik Jena Specord 200 Plus, Jena, Germany. The diffuse reflectance UV–Vis data were converted into absorbance using the Kubelka–Munk function.

4.2.4. Photoluminescence (PL) Measurements

Photoluminescence (PL) data were obtained using a Carry Eclipse fluorescence spectrometer, Agilent Technologies, Kuala Lumpur, Malaysia, for the following working parameters: 120 nm min^{-1} scan rate, 0.5 nm spectral resolution, and 10 nm slits in excitation and emission. Powder samples (0.001 g) were suspended in ultrapure water. The measurements were performed at room temperature for λ_{exc} = 260 nm.

4.2.5. Identification of Reactive Oxygen Species (ROS) Generation under Solar Irradiation

For ·OH radicals trapping, 1 mg of each catalyst was suspended in 10 mM coumarin (Merck) solution and exposed to simulated solar light irradiation for fluorescent umbelliferone formation. This was conducted with Carry Eclipse fluorescence spectrometer for λ_{exc} = 330 nm.

For O_2^- formation, 3 mg of each sample was suspended in 3 mM solution of XTT sodium salt (2, 3-bis(2-methoxi-4-nitro-5sulfophenyl)-2H-tetrazolium-5-carboxanilide) (Sigma-Aldrich) and exposed to simulated solar light, following XTT reduction by the photogenerated O_2^-. These results in the formation of XTT formazan complex were identified with a UV–Vis spectrophotometer, Analytik Jena Specord 200 Plus, a broad peak located at 470 nm being indicative.

4.2.6. Electrokinetic Potential Measurements

Electrokinetic potential measurements were carried out on a Malvern Nano ZS Zetasizer, Model ZEN 3600, Malvern, UK, at room temperature. The electrokinetic potential values were calculated using Helmholtz–Smoluchowski equation. The tests were conducted in triplicate.

4.2.7. Photoelectrochemical (PEC) Tests

Photoelectrochemical measurements were carried out in an electrochemical cell equipped with a quartz window using a two-electrode configuration by means of a Zahner IM6 potentiostat, Zahner-Elektrik GmbH, Kronach—Gundelsdorf, Germany. The reference and counter electrode leads were connected to a platinum wire, and working electrode lead was attached to prepared samples (geometric area of ~3 cm^2). An aqueous suspension of interest powder (0.01 g in 2 mL ultrapure water) was sonicated and coated on TCO conductive glass (Solaronix, Aubonne, Switzerland) by successive depositions and subsequent drying. The as-obtained coatings were heat treated at 300 °C in air for 1 h. The experiments were conducted in 0.5 M Na_2SO_4 electrolyte, pure or containing NH_4OH (12 µL/120 mL) under simulated solar light irradiation (AM 1.5, Peccell-L01, Yokohama, Japan). Photocurrent–voltage curves were recorded by linear sweep voltammetry at a scan rate of 10 mV/s.

4.2.8. H_2-TPR-Hydrogen Temperature-Programmed Reduction

Hydrogen temperature-programmed reduction (H_2-TPR) was carried out to determine the reducibility of catalytic species using the CHEMBET 3000 Quantachrome apparatus, Boynton Beach, Fl, USA. The sample (20 mg) was heated from room temperature up to 1073 K with 10 K/min heating rate, using a stream of 5% H_2/Ar. The hydrogen consumption was estimated from the area of the recorded peaks.

The calibration of the TCD signal was performed by injecting a known quantity of hydrogen (typically 50 µL) in the carrier gas (Ar). The experimentally obtained peak surface (mV·s) was thus converted into micromoles of hydrogen.

4.2.9. NH_3-TPD Temperature-Programmed Desorption

Quantity and distribution of acid sites were performed by temperature-programmed desorption of ammonia (NH_3-TPD) technique with the same apparatus mentioned above. Before the measurement, the catalyst was saturated using 200 µL injected pulse NH_3 gas at 50 °C. The NH_3-TPD experiment was carried out in a flow system with thermal conductivity detectors (TCD) by heating with 10 K/min (from room temperature up to 1073 K). This was used as a carrier gas (70 mL·min^{-1}). The NH_3 desorption was estimated from the area of the recorded peaks. The calibration of the TCD signal was performed by injecting a known quantity of NH_3. The surface of the as-obtained peak (mV·s) was converted into micromoles of ammonia.

4.3. Catalytic Ozonation and Photocatalytic Tests

Ozonation of ammonia in water was performed by using a semi-batch reactor connected to a gas flow line. Aside from the reactor, the catalytic setup contained a pH meter, stirring system, and a recirculator to maintain a constant temperature (25 °C) and an ozone-generating source. Typically, 0.015 g of catalyst was added into 100 mL of the ammonia solution containing 20 ppm NH_4^+. The suspension was vigorously stirred in a stream of O_3/O_2 (10 cm^3·min^{-1}). O_3 was generated from O_2 using an ozone generator (OzoneFIX, Mureș, Romania).

Photocatalytic tests of aqueous ammonia oxidation were carried out in a photoreactor (120 mL) provided with a quartz window and thermostated at 18 °C. Catalyst (0.015 g) was suspended in ammonia solution containing 20 ppm NH_4^+ obtained by adding of ammonia solution (NH_4OH 25%) in ultrapure water. Light irradiation of the reaction medium (AM 1.5) was realised by exposing it to AM 1.5 light provided by a solar simulator (Peccell-L01) with a xenon short-arc lamp (150 W). A gaseous mixture of argon (10 cm^3·min^{-1}) and oxygen (1 cm^3·min^{-1}) was bubbled into the aqueous suspension. The liquid aliquots were collected every 30 min and analysed with ion chromatographs (Dionex ICS 900, Sunnyvale, CA, USA) for the monitoring of NH_4^+ and the resulting anions (NO_3^-).

Ammonia oxidation with ozone, assisted by solar light irradiation, was performed using the above-mentioned set-up, also equipped with an ozone generator:

$$X_{NH_4^+}\% = \frac{[NH_4^+]_t}{[NH_4^+]_i} * 100 \tag{13}$$

$$S_{N_2-gas}\% = 100 - S_{NO_3^-}\% \tag{14}$$

$$S_{NO_3^-}\% = \frac{[NO_3^-]_t}{[NH_4^+]_c} * 100 \tag{15}$$

$$Y_{NO_3^-}\% = \frac{[NO_3^-]_t}{[NH_4^+]_i} * 100 \tag{16}$$

where: $X_{NH_4^+}$ is the NH_4^+ conversion, $Y_{NO_3^-}$ Yis the yield of NO_3^-, $[NH_4^+]_t$ is the ammonium ion concentration consumed at time t, $[NH_4^+]_i$ is initial ammonium ion concentration, and $[NO_3^-]_t$ is the nitrate concentration formed at time t.

5. Conclusions

Hydrothermally synthesized titanate nanorods were successfully modified by the addition of Fe precursor.

α-Fe_2O_3 with a well-defined morphology (nanocubes) was obtained by similar hydrothermal treatment and tested for comparative evaluation.

The comparison of the structural and functional characterization data of both titanate-based samples showed that the modification of titanate nanorods with iron did not affect the nanorod morphology and significantly improved optical and photocatalytical properties.

Catalytic investigations for aqueous ammonia oxidation showed the best reactivity for the Fe-modified sample (FeTiR) under solar light irradiation and ozonation. The NH_4^+ conversion and selectivity to gaseous end products was further improved by combining ozonation with the photo-assisted catalytic oxidation of aqueous ammonia.

An environmentally friendly, non-expensive and innovative depollution technology can be developed based on these materials.

Author Contributions: Conceptualization, S.P., M.Z., C.A. and F.P.; methodology, C.A., I.B. and F.P.; formal analysis, investigation, S.P., P.U., C.A., F.P., S.V.P., C.G. and R.S.; data curation, D.-I.E. I.B. and C.A.; writing—original draft preparation, C.A., F.P., S.V.P. and R.S.; writing—review and editing, C.A., R.S. and F.P.; supervision, I.B.; funding acquisition, C.A. and F.P. All authors have read and agreed to the published version of the manuscript.

Funding: This research was funded by Unitatea Executiva pentru Finantarea Invatamantului Superior, a Cercetarii, Dezvoltarii si Inovarii (UEFISCDI), grant number: PN-III-P2-2.1-PTE-2019-0222, 26PTE/2020 DENOX.

Conflicts of Interest: The authors declare no conflict of interest.

References

1. Oller, I.; Malato, S.; Sánchez-Pérez, J.A. Combination of Advanced Oxidation Processes and biological treatments for wastewater decontamination—A review. *Sci. Total Environ.* **2011**, *409*, 4141–4166. [CrossRef] [PubMed]
2. Shiskowski, D.M.; Mavinic, D.S. Biological treatment of a high ammonia leachate: Influence of external carbon during initial startup. *Water Res.* **1998**, *32*, 2533–2541. [CrossRef]
3. Malone, R.F.; Pfeiffer, T.J. Rating fixed film nitrifying biofilters used in recirculating aquaculture systems. *Aquacult. Eng.* **2006**, *34*, 389–402. [CrossRef]
4. Yuan, M.-H.; Chen, Y.-H.; Tsai, J.-Y.; Chang, C.-Y. Removal of ammonia from wastewater by air stripping process in laboratory and pilot scales using a rotating packed bed at ambient temperature. *J. Taiwan Inst. Chem. Eng.* **2016**, *60*, 488–495. [CrossRef]
5. Devi, P.; Dalai, A.K. Implications of breakpoint chlorination on chloramines decay and disinfection by-products formation in brine solution. *Desalination* **2021**, *504*, 114961. [CrossRef]
6. Bhuiyan, M.I.H.; Mavinic, D.; Beckie, R. Nucleation and growth kinetics of struvite in a fluidized bed reactor. *J. Cryst. Growth* **2008**, *310*, 1187–1194. [CrossRef]
7. Pansini, M. Natural zeolites as cation exchangers for environmental protection. *Miner. Depos.* **1996**, *31*, 563–575. [CrossRef]
8. Ichikawa, S.I.; Mahardiani, L.; Kamiya, Y. Catalytic oxidation of ammonium ion in water with ozone over metal oxide catalysts. *Catal. Today* **2014**, *232*, 192–197. [CrossRef]
9. Krisbiantoro, P.A.; Togawa, T.; Kato, K.; Zhang, J.; Otomo, R.; Kamiya, Y. Ceria-supported palladium as a highly active and selective catalyst for oxidative decomposition of ammonium ion in water with ozone. *Catal. Commun.* **2021**, *149*, 106204. [CrossRef]
10. Chen, Y.; Wu, Y.; Liu, C.; Guo, L.; Nie, J.; Chen, Y.; Qiu, T. Low-temperature conversion of ammonia to nitrogen in water 2 with ozone over composite metal oxide catalyst. *J. Environ. Sci.* **2018**, *66*, 265–273. [CrossRef]
11. Taguchi, J.; Okuhara, T. Selective oxidative decomposition of ammonia in neutral water to nitrogen over titania-supported platinum or palladium catalyst. *Appl. Catal. A Gen.* **2000**, *194*, 89–97. [CrossRef]
12. Ukropec, R.; Kuster, B.F.M.; Schouten, J.C.; van Santen, R.A. Low temperature oxidation of ammonia to nitrogen in liquid phase. *Appl. Catal. B Environ.* **1999**, *23*, 45–47. [CrossRef]
13. Wang, Y.; Xu, W.; Li, C.; Yang, Y.; Geng, Z.; Zhu, T. Effects of IrO_2 nanoparticle sizes on Ir/Al_2O_3 catalysts for the selective oxidation of ammonia. *Chem. Eng. J.* **2022**, *437*, 135398. [CrossRef]
14. Dobrescu, G.; Papa, F.; State, R.; Balint, I. Characterization of bimetallic nanoparticles by fractal analysis. *Powder Technol.* **2018**, *338*, 905–914. [CrossRef]
15. State, R.; Papa, F.; Dobrescu, G.; Munteanu, C.; Atkinson, I.; Balint, I.; Volceanov, A. Green synthesis and characterization of gold nanoparticles obtained by a direct reduction method and their fractal dimension. *Environ. Eng. Manag. J.* **2015**, *14*, 587–593. [CrossRef]
16. Shavisi, Y.; Sharifnia, S.; Mohamadi, Z. Solar-Light-Harvesting Degradation of Aqueous Ammonia by CuO/ZnO Immobilized on Pottery Plate: Linear Kinetic Modeling for Adsorption and Photocatalysis Process. *J. Environ. Chem. Eng.* **2016**, *4*, 2736–2744. [CrossRef]
17. Bahadori, E.; Conte, F.; Tripodi, A.; Ramis, G.; Rossetti, I. Photocatalytic Selective Oxidation of Ammonia in a Semi-Batch Reactor: Unravelling the Effect of Reaction Conditions and Metal Co-Catalysts. *Catalysts* **2021**, *11*, 209. [CrossRef]
18. Wang, I.; Edwards, J.G.; Davies, J.A. Photooxidation of aqueous ammonia with titania-based heterogeneous catalysts. *Sol. Energy* **1994**, *52*, 459–466. [CrossRef]
19. Bavykin, D.V.; Friedrich, J.M.; Walsh, F.C. Protonated Titanates and TiO_2 Nanostructured Materials: Synthesis, Properties and Applications. *Adv. Mater.* **2006**, *18*, 2807–2824. [CrossRef]
20. Amy, L.; Favre, S.; Gau, D.L.; Faccio, R. The effect of morphology on the optical and electrical properties of sodium titanate nanostructures. *Appl. Surf. Sci.* **2021**, *555*, 149610. [CrossRef]
21. Sayahi, H.; Aghappor, K.; Mohsenzadeh, F.; Morad, M.M.; Darabi, H.R. TiO_2 nanorods integrated with titania nanoparticles: Large specific surface area 1D nanostructures for improved efficiency of dye-sensitized solar cells (DSSCs). *Sol. Energy* **2021**, *215*, 311–320. [CrossRef]
22. Kerkez, O.; Boz, I. Photo(electro)catalytic Activity of Cu^{2+}-Modified TiO_2 Nanorod, Array Thin Films under Visible Light Irradiation. *J. Phys. Chem. Solids* **2014**, *75*, 611–618. [CrossRef]
23. Preda, S.; Anastasescu, C.; Balint, I.; Umek, P.; Sluban, M.; Negrila, C.; Angelescu, D.G.; Bratan, V.; Rusu, A.; Zaharescu, M. Charge separation and ROS generation on tubular sodium titanates exposed to simulated solar light. *Appl. Surf. Sci.* **2019**, *470*, 1053–1063. [CrossRef]
24. Ma, J.; Lian, J.; Duan, X.; Liu, X.; Zheng, W. α-Fe_2O_3: Hydrothermal Synthesis, Magnetic and Electrochemical Properties. *J. Phys. Chem. C* **2010**, *114*, 10671–10676. [CrossRef]
25. Andrusenko, I.; Mugnaioli, E.; Gorelika, T.E.; Koll, D.; Panthöfer, M.; Tremelb, W.; Kolb, U. Structure analysis of titanate nanorods by automated electron diffraction tomography. *Acta Crystallogr. Sect. B Struct. Sci.* **2011**, *67*, 218–225. [CrossRef]
26. Shirpour, M.; Cabana, J.; Doef, M. New materials based on a layered sodium titanate for dual electrochemical Na and Li intercalation systems. *Energy Environ. Sci.* **2013**, *6*, 2538–2547. [CrossRef]

27. Liqiang, J.; Yichun, Q.; Baiqi, W.; Shudan, L.; Baojiang, J.; Libin, Y.; Wei, F.; Honggang, F.; Jiazhong, S. Review of photoluminescence performance of nano-sized semiconductor materials and its relationships with photocatalytic activity. *Sol. Energy Mater. Sol. Cells* **2006**, *90*, 1773–1787. [CrossRef]
28. Anastasescu, C.; Negrila, C.; Angelescu, D.G.; Atkinson, I.; Anastasescu, M.; Spataru, N.; Zaharescu, M.; Balint, I. Distinct and interrelated facets bound to photocatalysis and ROS generation on insulators and semiconductors: Cases of SiO_2, TiO_2 and their composite SiO_2-TiO_2. *Catal. Sci. Technol.* **2018**, *8*, 5657–5668. [CrossRef]
29. Liu, B.; Geng, S.; Zheng, J.; Jia, X.; Jiang, F.; Liu, X. Unravelling the new roles of Na and Mn promoter in CO_2 hydrogenation over Fe_3O_4-Based catalysts for enhanced selectivity to light α-olefins. *ChemCatChem* **2018**, *10*, 4718–4732. [CrossRef]
30. Stoicescu, C.S.; Culita, D.; Stanica, N.; Papa, F.; State, R.N.; Munteanu, G. Temperature programmed reduction of a core-shell synthetic magnetite: Dependence on the heating rate of the reduction mechanism. *Termochim. Acta.* **2022**, *709*, 179146. [CrossRef]
31. Wang, X.; Zhao, Z.; Xu, Y.; Li, Q. Promoting effect of Ti addition on three-dimensionally ordered macroporous Mn-Ce catalysts for NH_3-SCR reaction: Enhanced N_2 selectivity and remarkable water resistance. *Appl. Surf. Sci.* **2021**, *569*, 151047. [CrossRef]
32. Shibuya, S.; Aoki, S.; Sekine, Y.; Mikami, I. Influence of oxygen addition on photocatalytic oxidation of aqueous ammonia over platinum-loaded TiO_2. *Appl. Catal. B Environ.* **2013**, *138–139*, 294–298. [CrossRef]

Article

Highly Efficient and Sustainable ZnO/CuO/g-C$_3$N$_4$ Photocatalyst for Wastewater Treatment under Visible Light through Heterojunction Development

Md. Abu Hanif [1,*], Jeasmin Akter [2], Young Soon Kim [1], Hong Gun Kim [1,3], Jae Ryang Hahn [2] and Lee Ku Kwac [1,3,*]

1. Institute of Carbon Technology, Jeonju University, Jeonju 55069, Korea; kyscjb@jj.ac.kr (Y.S.K.); hkim@jj.ac.kr (H.G.K.)
2. Department of Chemistry, Research Institute of Physics and Chemistry, Jeonbuk National University, Jeonju 54896, Korea; tina44445@gmail.com (J.A.); jrhahn@jbnu.ac.kr (J.R.H.)
3. Graduate School of Carbon Convergence Engineering, Jeonju University, Jeonju 55069, Korea
* Correspondence: hanif4572@gmail.com (M.A.H.); kwac29@jj.ac.kr (L.K.K.); Tel.: +82-63-220-3157 (M.A.H.); +82-63-220-3063 (L.K.K.)

Abstract: Dye-containing pollutants are currently a threat to the environment, and it is highly challenging to eliminate these dyes photocatalytically under visible light. Herein, we designed and prepared a ZnO/CuO/g-C$_3$N$_4$ (ZCG) heterostructure nanocomposite by a co-crystallization procedure and applied it to eliminate pollutants from wastewater via a photocatalytic scheme. The structural and morphological features of the composite confirmed the formation of a ZCG nanocomposite. The photocatalytic capability of the ZCG photocatalyst was investigated via the decomposition of methylene blue dye. The outstanding activity level of 97.46% was reached within 50 min. In addition, the proficiency of the ZCG composite was 753%, 392%, 156%, and 130% higher than photolysis, g-C$_3$N$_4$, CuO, and ZnO, respectively. Furthermore, the photodeterioration activity on Congo red was also evaluated and found to be excellent. The enhanced catalytic achievement is attributed to the construction of heterojunctions among the constituent compounds. These properties boost the charge transfer and decrease the recombination rate. Moreover, the reusability of the ZCG product was explored and a negligible photoactivity decline was detected after six successful runs. The outcomes suggest the as-prepared nanocomposite can be applied to remove pollutants, which opens a new door to practical implementation.

Keywords: ZnO/CuO/g-C$_3$N$_4$; co-crystallization; heterojunction; photocatalytic degradation; organic pollutant

Citation: Hanif, M.A.; Akter, J.; Kim, Y.S.; Kim, H.G.; Hahn, J.R.; Kwac, L.K. Highly Efficient and Sustainable ZnO/CuO/g-C$_3$N$_4$ Photocatalyst for Wastewater Treatment under Visible Light through Heterojunction Development. *Catalysts* **2022**, *12*, 151. https://doi.org/10.3390/catal12020151

Academic Editors: Florica Papa, Anca Vasile and Gianina Dobrescu

Received: 29 December 2021
Accepted: 21 January 2022
Published: 25 January 2022

Publisher's Note: MDPI stays neutral with regard to jurisdictional claims in published maps and institutional affiliations.

Copyright: © 2022 by the authors. Licensee MDPI, Basel, Switzerland. This article is an open access article distributed under the terms and conditions of the Creative Commons Attribution (CC BY) license (https://creativecommons.org/licenses/by/4.0/).

1. Introduction

In recent years, environmental pollution has become a severe problem for human health and aquatic life due to industrialization and the use of a massive amount of chemicals. The pollutants that arise from industries, including food coloring, dyeing, printing, and textiles, which contain various types of untreated dyes, are released directly into water bodies. Among several dyes used, methylene blue (MB) is a cationic dye, and it is applied on a vast scale as a colorant material in industries. People who come into contact with MB face acute illness, especially abdominal pain, nausea, headache, corneal injury, dizziness, and anemia, as well as it being detrimental for aquatic and other organisms [1]. In addition, Congo red (CR) is also a noxious anionic azo dye. CR is widely used in textile industries, as a pH indicator, and in the diagnosis of amyloidosis and benzidine metabolization, etc. This toxic dye is responsible for dangerous diseases such as gastrointestinal and eye disorders and skin irritations, and causes allergic and respiratory problems [2]. Hence, it is necessary to eliminate the harmful dyes from wastewater before mixing them into the aquatic environment. However, the decolorization technique is very challenging due to the complex structure of the dyes. This is why numerous research teams are trying to find a sustainable method to resolve this serious issue. The traditional approaches, including adsorption, sand filtration, coagulation/flocculation, precipitation, biodegradation, etc., are familiar methods used to eradicate the contaminants from wastewater [3,4]. Nevertheless, the sustainability of most of these conventional methods is in doubt because they require a huge amount of catalyst. Additionally, such procedures may generate huge amounts of secondary pollutants or convey one form to another form. Conversely, advanced oxidation processes such as ozonation, photocatalysis, Fenton oxidation might be the next generation step for wastewater treatment systems. In particular, the photocatalytic method is an appropriate process in respect of sustainability. Moreover, it is more appropriate than others due to its better performance, lack of secondary pollutant formation, low cost, easy operation, and eco-friendliness. The harmful pollutants are converted into various non-toxic products such as carbon dioxide and water with the assistance of reactive oxygen species in the photocatalysis system [1]. However, selection of a suitable photocatalyst is a vital concern for enhancing deterioration proficiency.

Currently, semiconductor-based photocatalysts have been explored extensively in various fields using methods such as water splitting, supercapacitors, removal of pollutants from wastewater by photocatalysis, H_2 production, CO_2 reduction, and so on. The increasing attention being paid to the use of semiconductor materials is due to their electronic structure. Among the widely used semiconductor photocatalysts, ZnO and CuO are most significantly applied due to being low cost and non-toxic, with notable thermal and electrical conductivity and outstanding photocatalytic efficiency. Nonetheless, these two materials, when used alone, have some inconvenient limitations in visible light photocatalytic applications. In general, the photodecomposition proficiency of ZnO and CuO alone is minimal due to insufficient visible light harvesting and the high recombination rate of photoinduced electron-hole (e^--h^+) pairs, low surface area, and low electron charge separation [5–8]. In contrast, ZnO and CuO enhance their catalytic activity by working as a supporting material through heterojunction construction [7,9]. These heterojunctions and hierarchical nanostructures diminish the recombination rate and enhance the photocatalytic activity of the composite materials compared with pure oxides [10–13]. Although the ZnO/CuO composite showed greater photodegradation proficiency than the separated ZnO or CuO compounds, we found that the catalytic activity of ZnO/CuO material did not gain a satisfactory level for practical implementation via visible light irradiation [9,14–16]. Therefore, the ZnO/CuO composite needs to be further coupled with an electrically conducting material to increase its photodeterioration ability under visible light irradiation. Thus, it remains a challenge to discover the appropriate photocatalytic compound which will improve the VL photocatalytic activity.

Based on the above hypothesis, graphitic carbon nitride (g-C_3N_4) might be a suitable compound for coupling with the ZnO/CuO composite because of its low cost and large-scale manufacturing, unique electronic and optical properties, good chemical and thermal stability, facile preparation, and non-toxicity [17,18]. In addition, the photocatalytic activity of g-C_3N_4-related composites has been extensively studied [19–23]. However, the photocatalytic efficiency towards pollutant degradation of pristine g-C_3N_4 is low due to its quick e^--h^+ pair recombination ability [18,24]. Therefore, to counteract this drawback, various methods have been applied in attempts to increase the visible light photocatalysis, particularly coupling with other semiconductors or metals, non-metal doping, control of morphology, etc.

In this work, we focused on the preparation of a ZnO/CuO/g-C_3N_4 (ZCG) nanocomposite through an efficient co-crystallization method and executed the photocatalytic process for the treatment of dye-containing wastewater followed by visible light irradiation. Our synthesis process may open a new way to prepare other g-C_3N_4-based semiconductor composites in the near future. The physicochemical properties of the as-synthesized ZCG nanocomposite were systematically investigated by using various spectroscopic and microscopic techniques. The ZCG nanocomposite showed excellent photocatalytic MB deterioration activity compared with g-C_3N_4, ZnO, CuO, and with photolysis under visible light illumination. Additionally, the photocatalytic ability of the ZCG heterostructure for the photodegradation of CR dye was also evaluated and showed remarkable efficiency. The enhanced activity that was achieved may be due to the heterojunction construction among g-C_3N_4, ZnO, and CuO compounds, which suppressed the recombination rate of the photoinduced e^--h^+ pairs and enhanced the flow of e^- transfer. In addition, in order to understand the recyclability and stability of the as-synthesized ZCG composite, we explored the recycling ability up to six times along with XRD and FTIR analysis before and after photodegradation. There were no significant changes in photocatalytic activity and structural properties during recycling, which strongly implies the possibility for the practical implementation of our work. The probable photocatalytic degradation mechanism was also discussed.

2. Results and Discussion

2.1. Structural Characterization

The crystal structure of the as-prepared g-C_3N_4, CuO-NPs, ZnO-NPs, and ZCG composite was inspected by X-ray diffraction (XRD) and is presented in Figure 1. The XRD pattern of the g-C_3N_4 confirms two peaks (Figure 1a): one is at 12.77° corresponding to the (100) plane of the inter-planar stacking of a tri-s-triazine unit and the other one is at 27.74° and belongs to the (002) plane due to the interplanar stacking of the conjugated aromatic system (JCPDS 87-1526). The diffraction peaks of CuO-NPs are positioned at 32.43°, 35.46°, 38.63°, 48.68°, 53.36°, 58.28°, 61.45°, 66.14°, and 67.96° (Figure 1b) and correspond to the crystal planes (110), ($\bar{1}$11), (111), ($\bar{2}$02), (020), (202), ($\bar{1}$13), ($\bar{3}$11), and (113), respectively, with monoclinic morphology (JCPDS 96-901-5569). The ZnO-NPs diffraction peaks were detected at 31.51°, 34.18°, 35.96°, 47.34°, 56.39°, 62.66°, 66.22°, 67.81°, and 67.81° for the (100), (002), (101), (102), (110), (113), (200), (112), and (201) facets, respectively (Figure 1c). All of the diffraction peaks of the ZnO-NPs were matched with the ZnO hexagonal wurtzite structure (JCPDS 36-1451). All the characteristic diffraction peaks of g-C_3N_4, CuO-NPs, and ZnO-NPs appeared in the ZCG nanocomposite, confirming its successful formation (Figure 1d). No additional peaks were found in the composite, indicating the as-prepared ZCG nanocomposite is highly pure.

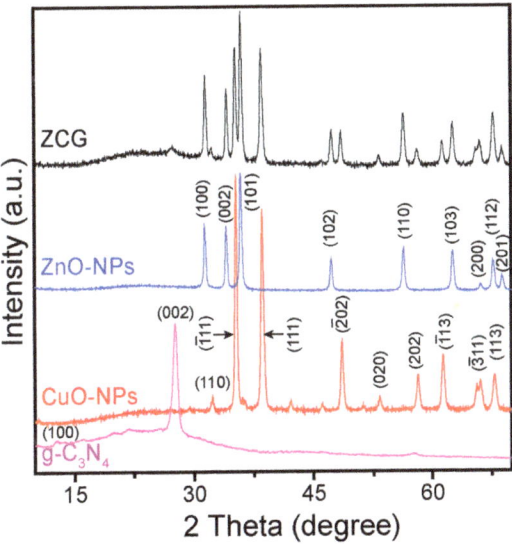

Figure 1. XRD spectra of the (**a**) g-C_3N_4, (**b**) CuO-NPs, (**c**) ZnO-NPs, and (**d**) ZCG nanocomposite.

The crystallite size (D) of the CuO-NPs, ZnO-NPs, and ZCG nanocomposite was calculated applying Scherrer's formula ($D = \frac{k\lambda}{\beta \cos\theta}$, where k denotes non-dimensional shape factor (0.94), β is the full width at half maximum, λ is the X-ray wavelength, and θ is the Bragg angle). In this case, all the above-mentioned planes were used to calculate the average D of the materials. The average D values of the CuO-NPs, ZnO-NPs, and ZCG nanocomposite were 22.46, 21.26, and 20.37 nm, respectively. Further, the D values of the CuO-NPs, ZnO-NPs, and ZCG nanocomposite were verified by using the modified Scherrer equation and the corresponding curves are shown in Figure S1. The calculated D values of the CuO-NPs, ZnO-NPs, and ZCG nanocomposite were 27.01, 26.45, and 25.32 nm, respectively (described in Supplementary File in detail). There is little difference in the results and the order of crystallite size (CuO-NPs > ZnO-NPs > ZCG) is the same.

The binding energy, valence state, and purity of the ZCG nanocomposite were analyzed by X-ray photoelectron spectroscopy (XPS) and are shown in Figure 2. Figure 2a indicates the survey spectrum of the composite and affirms the presence of Zn, Cu, O, N, and C elements in the composite. The high-resolution Zn 2p plot was observed at binding energies of 1021.37 and 1044.37 nm for Zn-$2p_{3/2}$ and Zn-$2p_{1/2}$ chemical environments, respectively (Figure 2b). These results imply the existence of a Zn^{2+} oxidation state in the ZCG nanocomposite. In addition, the difference between the mentioned peaks is 23.0 eV, confirming the oxidation state of the Zn in the composite is +2 [25]. Figure 2c displays the XPS features of the Cu 2p. The peaks at 952.64 and 932.68 eV were allotted for Cu-$2p_{1/2}$ and Cu-$2p_{3/2}$ of CuO. In addition, the satellite peaks were found to be 940.63, 943.32, and 961.71 eV, confirming the presence of a partially filled d orbital ($3d^9$) of Cu^{2+} in the nanocomposite [1]. The two deconvoluted peaks of O 1s were positioned at 529.31, and 530.79 eV, respectively (Figure 2d). These characteristic peaks are observed due to the lattice oxygen (O^{2-}) and adsorbed molecular oxygen on the surface of the ZCG composite [26,27]. The core-level spectra of N 1s are shown in Figure 2e. The three deconvoluted bands at binding energies of 400.75, 399.70, and 398.31 eV correspond to the amino functions (C–N–H), tertiary nitrogen N–$(C)_3$, and sp^2-bonded triazine rings (C–N=C), respectively [28,29]. The deconvoluted high-resolution XPS spectrum of C-1s (Figure 2f) was exhibited at 284.39, 285.46, and 287.86 eV. The peak at 287.86 eV is attributed to sp^2-coordinated carbon-containing aromatic rings (N–C=N) [30]. The peak located at 285.46 eV is due to C–O species on the surface of g-C_3N_4 [28,31]. The peak centered at a binding energy of 284.6 eV is ascribed to sp^2-hybridized C–C bonds. Therefore, XPS investigation suggested that the ZCG nanocomposite was composed of g-C_3N_4, ZnO, and CuO, and confirmed their successful combination in the composite.

The nature of the chemical bonding and functional groups of g-C_3N_4, ZnO-NPs, CuO-NPs, and the ZCG nanocomposite were studied through Fourier transform infrared spectroscopy (FTIR) and are demonstrated in Figure 3. Figure 3a shows the characteristic peaks of g-C_3N_4. The peak that appears at 805 cm^{-1} is due to the breathing mode of the s-triazine ring [28]. A series of bands at approximately 1218, 1315, 1411, and 1551 cm^{-1} are related to the stretching vibrations of aromatic C–N heterocycles, and the peak at 2118 cm^{-1} is observed for an asymmetric stretching mode of cyano groups (C≡N) [32]. The absorption band around 449 cm^{-1} corresponds to the Zn–O stretching vibration (Figure 3b). The peak centered between 400 and 555 cm^{-1} is attributed to the ZnO hexagonal structure [25]. Figure 3c represents the FTIR curve of CuO-NPs and the most significant peaks are found at 585 and 531 cm^{-1}, corresponding to the stretching vibrational mode of Cu–O bonds associated with the plane ($\bar{1}01$), and (101), respectively [33]. In addition, the wide band between 3000–3500 cm^{-1} is ascribed for all the compounds as a result of the stretching mode of O–H from the atmosphere. All the demonstrative bands are evident in the spectrum of the ZCG composite (Figure 3d), further confirming the coexistence of ZnO, CuO, and g-C_3N_4.

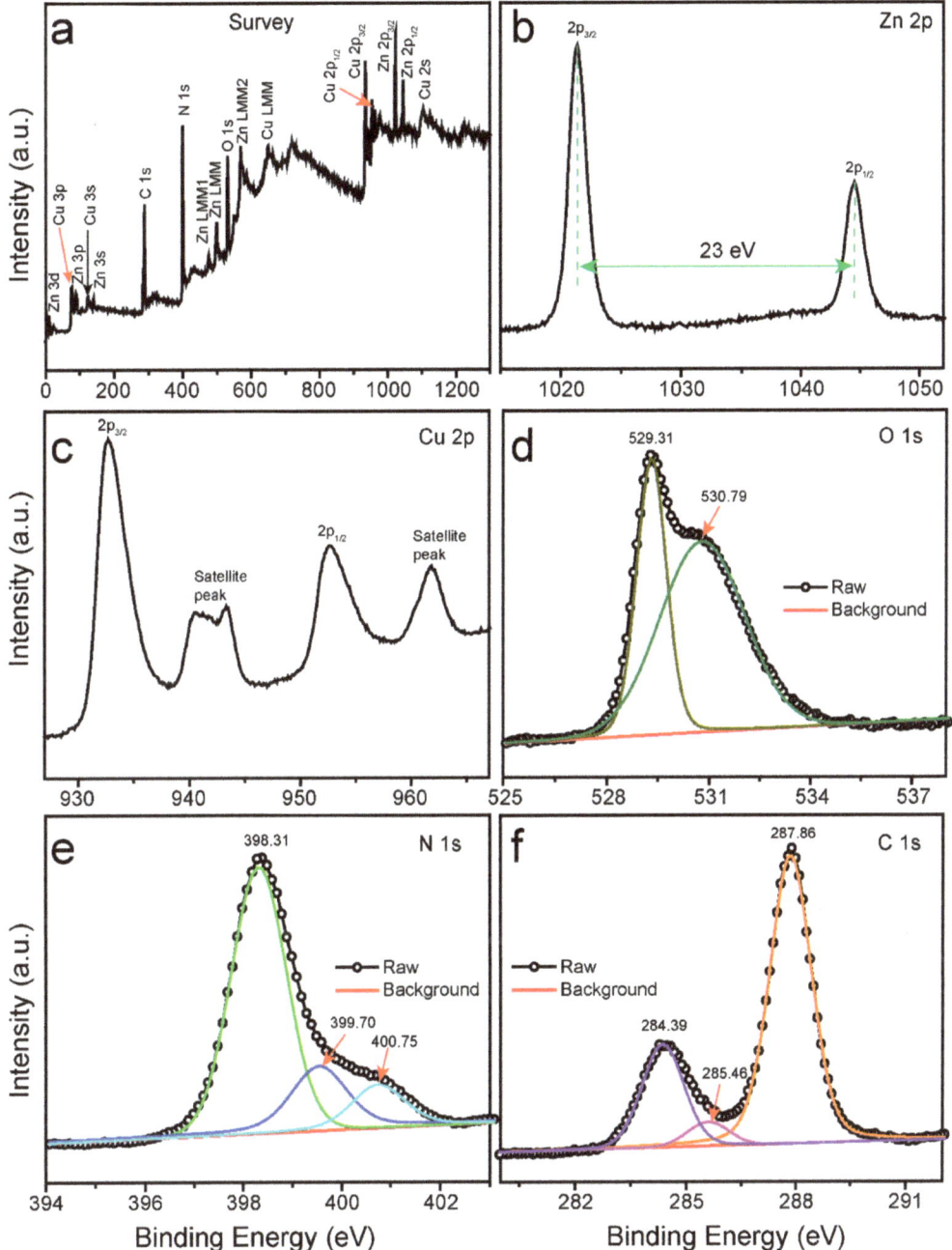

Figure 2. XPS spectra of ZCG nanocomposite. (**a**) Survey; (**b**–**f**) High-resolution spectra of Zn 2p, Cu 2p, O 1s, N 1s, and C 1s, respectively.

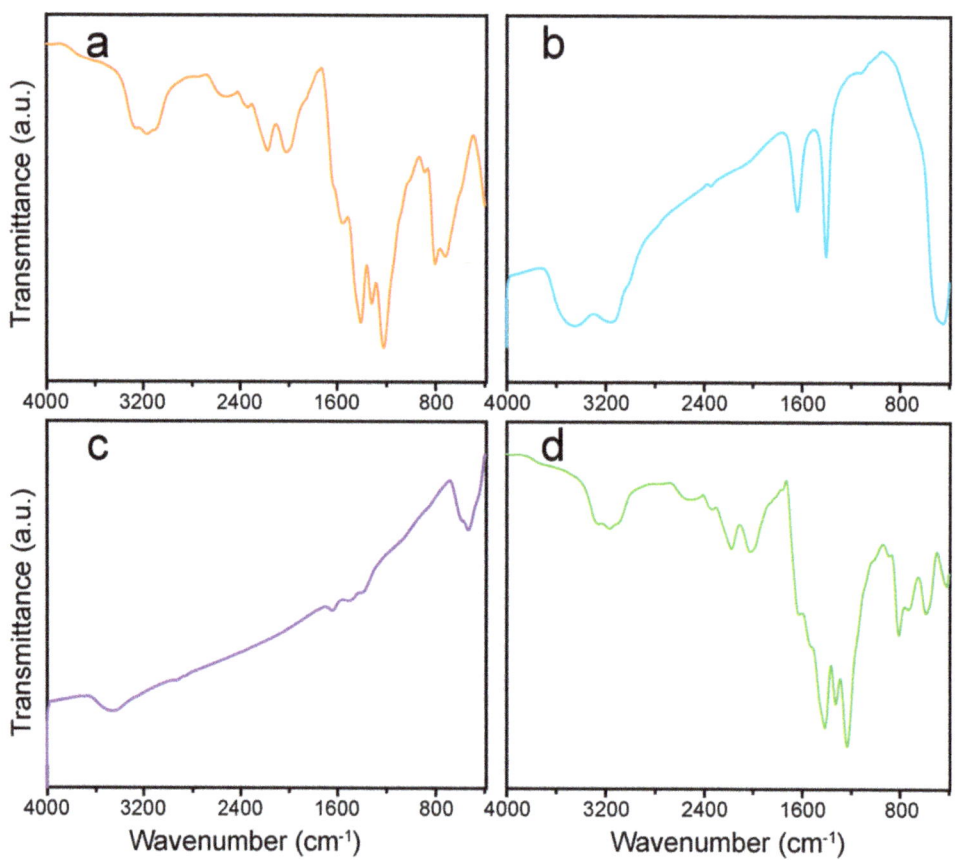

Figure 3. FTIR spectra of the (**a**) g-C$_3$N$_4$, (**b**) ZnO-NPs, (**c**) CuO-NPs, and (**d**) ZCG nanocomposite.

2.2. Optical Characterization

The optical characteristics of the material are a crucial parameter in the application as a photocatalyst. UV–vis spectroscopy was applied to determine the optical features of the g-C$_3$N$_4$, ZnO-NPs, CuO-NPs, and ZCG nanocomposite. The absorption edge of the g-C$_3$N$_4$, ZnO-NPs, CuO-NPs, and ZCG nanocomposite indicates that all the samples can be used in the visible region of light. In addition, their corresponding bandgap energy (E_g) was calculated by using Tauc plots (Figure 4). The E_g of the materials was measured via extrapolation of the $(\alpha h\nu)^2$ vs. E, where $h\nu$ refers to photon energy ($1239/\lambda$, in eV) and α represents the absorption coefficient. The intercept of the E axis where $(\alpha h\nu)^2 = 0$ gives the E_g value. The calculated E_g values of the g-C$_3$N$_4$, ZnO-NPs, CuO-NPs, and ZCG nanocomposite were 2.33, 2.25, 1.46, and 1.69 eV, respectively. Interestingly, after the construction of the nanocomposite, the E_g value was noticeably lower than that of the individual compounds except CuO-NPs. The reduced E_g of the composite facilitates the e^--h^+ pair generation and minimizes their recombination rate. In general, the lower the E_g value, the higher the photocatalytic activity of the material. Although the bandgap of the pure CuO is smaller than those of the nanocomposites, the pure CuO showed low photocatalytic efficiency. This is due to the photocatalytic efficiency also being sensitive to several factors, such as crystalline size, heterojunction formation, and charge recombination effect. The ZCG composite showing higher photocatalytic activity than the single CuO could be due to the modification of surface morphology such as heterojunction development, and the small crystallite size of the nanocomposite.

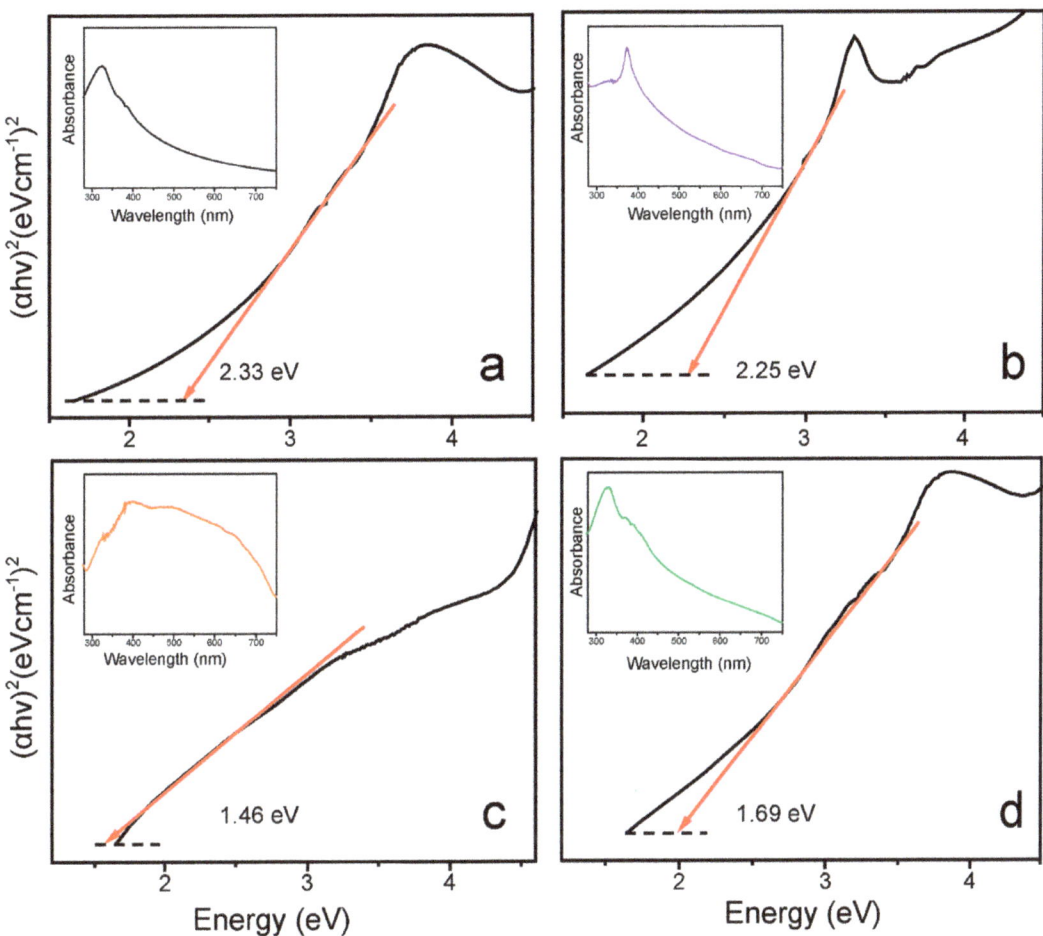

Figure 4. Tauc plots of the (**a**) g-C$_3$N$_4$, (**b**) ZnO-NPs, (**c**) CuO-NPs, and (**d**) ZCG nanocomposite. Each inset displays the representative UV–vis absorption data of the compound.

2.3. Morphological Investigation

The morphological features of the as-prepared g-C$_3$N$_4$, CuO-NPs, ZnO-NPs, and ZCG nanocomposite were analyzed by field-emission scanning electron microscopy (FE-SEM) and are depicted in Figure 5. Figure 5a shows the structure of as-synthesized g-C$_3$N$_4$ is aggregated, thin, and a flat sheet. The as-constructed ZnO-NPs are nanorod-shaped structures and the size and shape of the nanorods are almost uniform throughout the compounds (Figure 5b). Figure 5c displays the spherical-shaped CuO-NPs. The size of the CuO-NPs lies between 64 and 180 nm and the average size is approximately 98 nm (forty-five particles are counted). The ZCG nanocomposite is composed of three types of morphology: sheet-like g-C$_3$N$_4$, nanorod-like ZnO, and sphere-like CuO (Figure 5d). These results indicate the successful formation of a ZCG heterostructure nanocomposite with a C$_3$N$_4$, ZnO, and CuO chemical environment. Such types of heterostructure enhance the electron flow and reduce the recombination rate and, therefore, improve the photocatalytic efficiency of the product [34].

Figure 5. FE-SEM images of the as-synthesized (**a**) g-C$_3$N$_4$, (**b**) ZnO-NPs, (**c**) CuO-NPs, and (**d**) ZCG nanocomposite.

For a better understanding of the heterostructure, we further used high-resolution transmission electron microscopy (HR-TEM) to explain the presence of g-C$_3$N$_4$, ZnO, and CuO nanoparticles in the ZCG composite. Figure 6a,b show the low and high magnification micrographs of the compound. As can be clearly seen from both the figures, the ZCG nanocomposite is constructed by g-C$_3$N$_4$, ZnO, and CuO compounds and their individual morphology is intact after forming the composite. The outcomes are also balanced with the findings of FE-SEM.

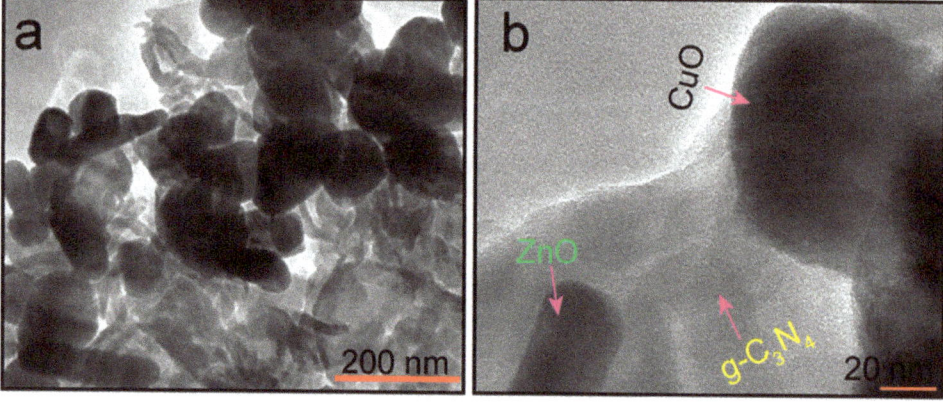

Figure 6. (**a**) Low and (**b**) high resolution micrographs of the ZCG nanocomposite.

The chemical confirmation and purity of the as-prepared ZCG nanocomposite were analyzed through energy-dispersive X-ray spectroscopy (EDS) and are shown in Figure 7a,b. Figure 7a from the FE-SEM image was used to determine the EDS analysis. The result shows that the five constituent elements Zn, Cu, O, C, and N exist in the composite and their atomic and weight percentages (inset) are indicated in Figure 7b. The prepared nanocomposite is free from impurities due to the absence of additional peaks. These findings are in line with the XRD outcomes. The distribution of the elements in the composite was further inspected using elemental mapping images of the ZCG composite (Figure 7c,d). The outcomes confirm that the Zn, Cu, O, N, and C elements were homogeneously distributed in the nanocomposite.

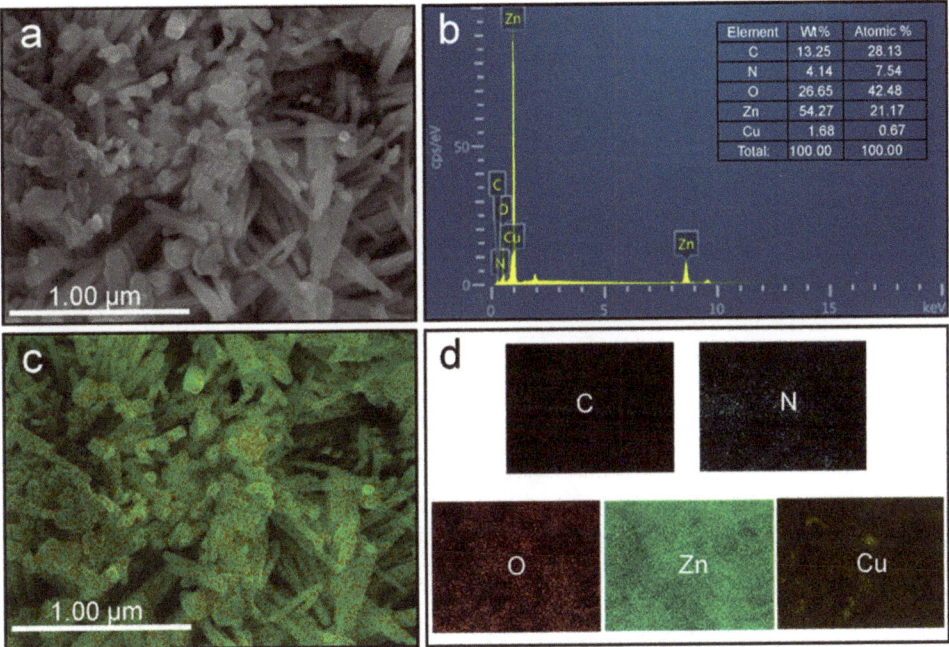

Figure 7. (**a**,**b**) EDS and (**c**,**d**) elemental mapping analysis of the ZCG nanocomposite.

2.4. Photocatalytic Assessment of the Compounds

The photodegradation activity of g-C_3N_4, ZnO-NPs, CuO-NPs, ZCG, and photolysis (PL) was evaluated for the deterioration of MB as a representative organic pollutant under visible light illumination for 50 min. The adsorption ability was also inspected under dark conditions and is displayed in Figure 8a. The results show the adsorption proficiency of the compounds is negligible compared with photodegradation and increase in sequence PL < g-C_3N_4 < CuO-NPs < ZnO-NPs < ZCG. The photocatalytic activity of the aforementioned materials was calculated through the following Equation (1) [35]:

$$\eta(\%) = \left(\frac{C_0 - C_t}{C_0}\right) \times 100 \tag{1}$$

where η refers to the degradation ability, and C_t and C_0 express the absorbance at t and 0 min, respectively. Figure 8a shows the photodegradation efficiency of all the investigated samples along with PL. The as-constructed ZCG composite shows the highest efficiency (97.46%) within 50 min of visible light stimulation. The catalytic activity was enhanced in the order PL (12.95%) < g-C_3N_4 (24.88%) < CuO-NPs (62.09%) < ZnO-NPs (75.37%) <

ZCG (97.46%). The photocatalytic activity of the ZCG nanocomposite was 753%, 392%, 156%, 130% higher than that of PL, g-C$_3$N$_4$, CuO-NPs, and ZnO-NPs, respectively. In addition, as can be realized from Figure 8a, the activity of the ZCG composite gradually increased after the introduction of g-C$_3$N$_4$, CuO-NPs, and ZnO-NPs materials. Further, the reduced catalytic ability of g-C$_3$N$_4$, CuO-NPs, and ZnO-NPs is observed because of the quick recombining impact of photoinduced e^--h^+ pairs in these substances. Conversely, the enhanced efficiency of the ZCG composite followed due to the heterostructure development in the composite which suppressed the recombination effect.

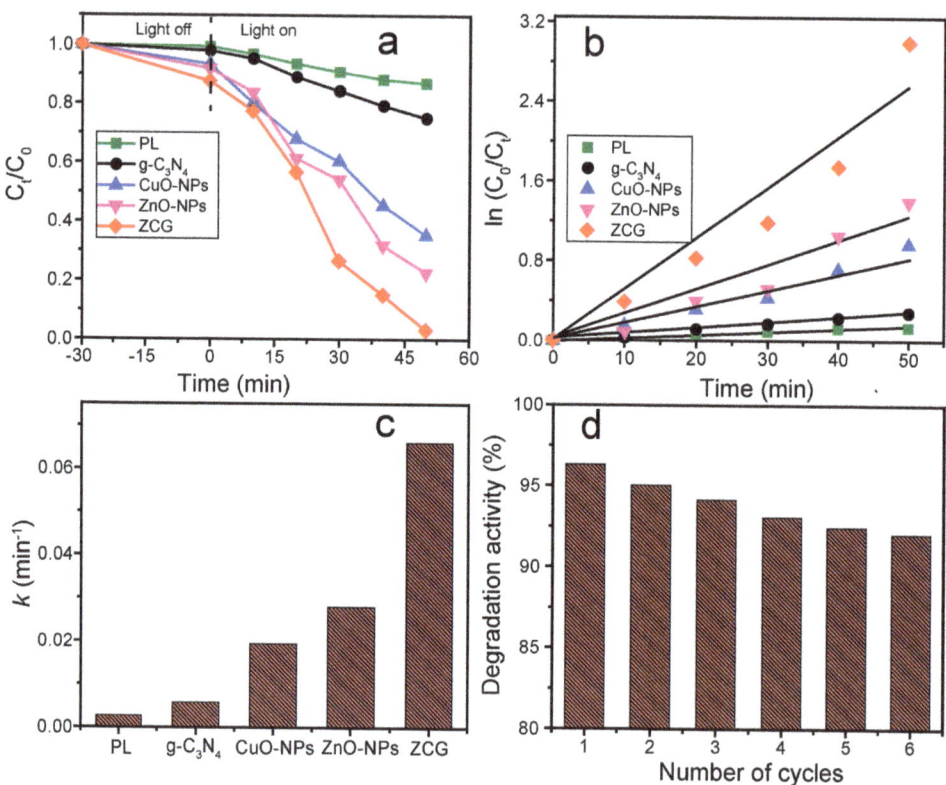

Figure 8. (**a**) Degradation proficiency as a function of time, (**b**) Corresponding kinetics curves, (**c**) Estimated rate constant, (**d**) Recyclability of the ZCG nanocomposite.

Further, a kinetics study was undertaken to better understand the product's catalytic activity, as shown in Figure 8b. The linear plots of ln (C_0/C_t) vs. illumination time for MB photodegradation show pseudo-first-order reaction kinetics. The deterioration rate constant (k) was calculated employing Equation (2) [35]:

$$\ln\left(\frac{C_0}{C_t}\right) = kt \qquad (2)$$

where t represents the irradiation time of the reaction. Figure 8c shows the k values for PL, g-C$_3$N$_4$, CuO-NPs, ZnO-NPs, and ZCG were 0.0028, 0.0057, 0.0193, 0.0280, and 0.0660 min^{-1}, respectively. The ZCG composite demonstrated the highest k value compared to other inspected materials. This value is 23.6 times greater than without the catalyst (PL). Even the k value of the ZCG nanocomposite is 11.6, 3.4, and 2.4 times larger than the g-C$_3$N$_4$, CuO-NPs, and ZnO-NPs compounds, respectively. The k value of the as-prepared

ZCG nanocomposite is enhanced compared with those of the single materials g-C_3N_4, CuO-NPs, and ZnO-NPs used for MB photodegradation via visible light irradiation. Hence, the composite materials display a vital property for boosting the photocatalytic activity and rate constant compared with single materials applied in the reaction.

To reduce wastewater treatment costs and prevent the generation of secondary pollutants, photocatalysts with excellent reusability and high stability are highly recommended for research. The recyclability and constancy of the ZCG catalyst were investigated in as many as six successive runs of MB photodegradation upon 50 min visible light irradiation. As can be observed from Figure 8d, the photocatalytic ability of the ZCG nanocomposite exhibited a negligible decreasing efficiency after six cycles, which might be due to the damage caused by the catalyst throughout the recycling procedure. In addition, the structural changes of the ZCG composite were evaluated by XRD and FTIR instruments before and after (six cycles) catalytic reaction (Figure 9). The outcomes confirm that there are no substantial structural changes on the nanocomposite after the reaction. Therefore, our as-synthesized ZCG composite revealed remarkable reusability and stability which can be applied to the decomposition of organic pollutants under visible light.

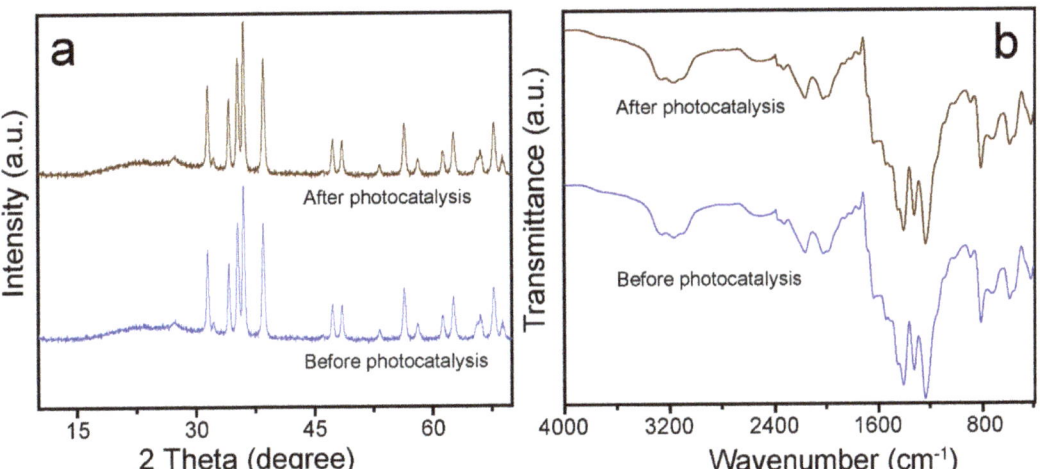

Figure 9. (a) XRD, and (b) FTIR patterns of the ZCG nanocomposite (before, and after photocatalytic analysis).

The photocatalytic efficiency of the as-synthesized ZCG nanocomposite was also studied to evaluate the degradation capability of other organic pollutants. Herein, CR was used to analyze its photodeterioration ability under identical experimental conditions. The catalytic potency of the ZCG composite was found to be 91.73% for the degradation of CR dye within 50 min (Figure 10a). Further, the kinetics analysis was evaluated and is shown in Figure 10b. The result indicates that pseudo-first-order kinetics was followed for the deterioration of CR. The estimated k value of the composite was 0.0371 min^{-1}. Therefore, the photodegradation efficiency of the as-prepared ZCG composite is noticeably higher compared with the previously described composite.

Table 1 displays the photocatalytic activity of the previously reported compounds and as-synthesized ZCG nanocomposite. The assessment indicates that our as-prepared ZCG composite is highly suitable for application in wastewater treatment compared with the other materials.

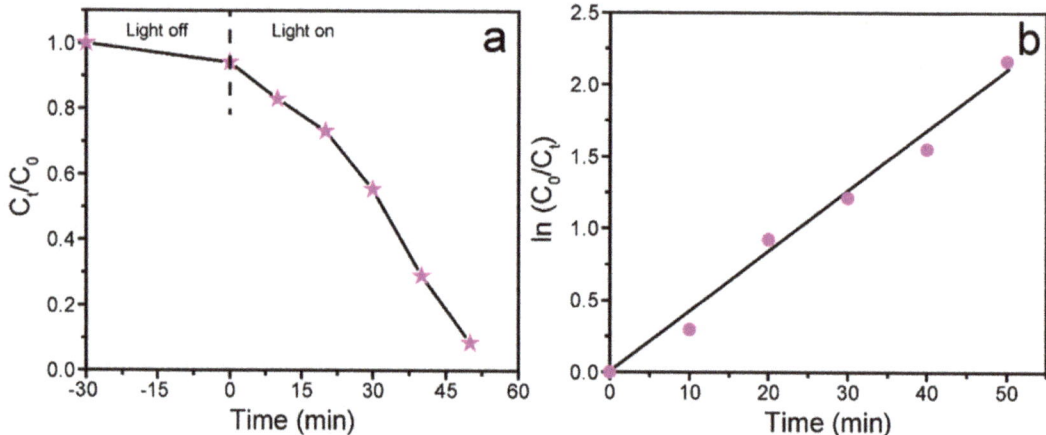

Figure 10. (a) Photocatalytic proficiency of the ZCG nanocomposite for the deterioration of Congo red, and (b) corresponding kinetics plot.

Table 1. Comparison of the photocatalytic decomposition of as-synthesized ZCG composite and ZnO/CuO or g-C_3N_4-based composite.

Catalyst	Pollutant (Dye)	Dye Dose	Catalyst Dose (g)	Degradation (%)	Time (min)	Ref.
ZnO/CuO	MB	10 ppm	0.01	73	120	[9]
ZnO/CuO	MB	3×10^{-5} ML^{-1}	0.5	97.2	120	[14]
ZnO/CuO	MB	6 mg L^{-1}	0.4 g L^{-1}	82	120	[15]
ZnO/CuO	MB	20 mg L^{-1}	0.5 g L^{-1}	93	120	[16]
ZnO/ZnS/g-C_3N_4	MB	100 mL, 10 mg L^{-1}	0.05	75	60	[36]
Fe_2O_3–xS_x/S-doped g-C_3N_4	MB	100 mL, 8 mg L^{-1}	0.07	82	150	[37]
Ba/Alg[1]/CMC[2]/TiO_2	CR	4.3×10^{-5} M	1.2 L^{-1}	91.5	240	[38]
ZnO/CuO/g-C_3N_4 (ZCG)	MB	100 mL, 20 mg L^{-1}	0.02	97.46	50	Our work
	CR	100 mL, 20 mg L^{-1}	0.02	91.73		

[1] Alg = alginate, [2] CMC = carboxymethyl.

2.5. Determination of the Point of Zero Charge (pH$_{pzc}$) and pH Effect

The electrical charge and pH of the catalyst and organic pollutants are vital parameters for understanding photocatalytic efficiency. The pH$_{pzc}$ signifies the charge of the catalyst surface. The pH drift technique was applied to estimate the pH$_{pzc}$ of the as-prepared ZCG composite. The initial pH (pH$_i$) of the solution was adjusted by adding either HCl or NaOH. The final pH (pH$_f$) was then measured. The pH$_{pzc}$ was determined by plotting pH$_i$ vs ΔpH (pH$_f$ − pH$_i$) and pH$_{pzc}$ refers to the point at which the total numbers of positive and negative charges are identical. The pH$_{pzc}$ of the ZCG nanocomposite was 7.36 (Figure 11a). These outcomes indicate that the ZCG catalyst's surface is negative at pH values higher than pH$_{pzc}$ and positive at pH values lower than pH$_{pzc}$. The pH effect of the ZCG composite for the MB deterioration was investigated at different pH values (pH 3, 6, 9, and 12). In addition, the catalyst dose and concentration of the solution in each experiment were identical. The photocatalytic proficiency of the composite was dramatically improved with the increase in the solution's pH (Figure 11b). The highest activity was achieved at a high pH (12, basic medium) within 30 min due to the strong interaction between the cationic MB dye and the anionic (negatively charged) surface of the catalyst. By contrast, the lowest efficiency was observed at pH 3 because at low pH (acidic medium), the catalyst surface behaves with a cationic nature and therefore repulsion takes place between MB and the catalyst surface.

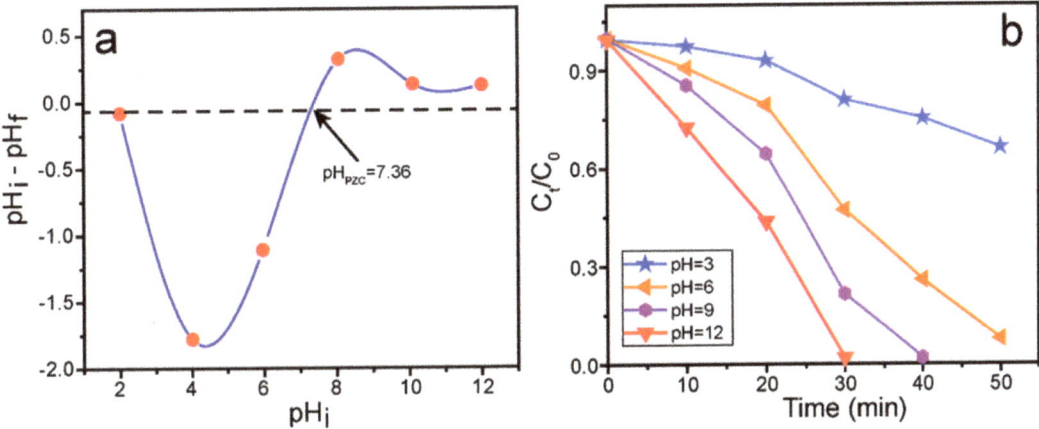

Figure 11. (**a**) Point of zero charge, and (**b**) pH effect on the deterioration of MB under VL using ZCG nanocomposite.

2.6. Proposed Photocatalytic Degradation Mechanism

When the ZCG nanocomposite is exposed to VL irradiation, photoinduced holes (h^+) and electrons (e^-) are produced on its surface by absorption of photons (Equation (3)). Valence band (VB) holds h^+ and conduction band (CB) carries e^-.

$$ZCG + h\nu \rightarrow ZCG\,(e^-_{CB} + h^+_{VB}) \qquad (3)$$

The h^+ in VB plays a key role in generating hydroxyl radicals from H_2O molecules or hydroxyl ions (Equation (4)).

$$h^+_{VB} + H_2O/OH^- \rightarrow HO^\bullet + H^+ \qquad (4)$$

On the other hand, e^- in CB helps generate superoxide radicals from oxygen molecules (Equation (5)).

$$e^-_{CB} + O_2 \rightarrow O_2^{-\bullet} \qquad (5)$$

The produced superoxide radicals also help to generate hydroxyl radicals through a series of reactions (Equations (6)–(10)):

$$H^+ + O_2^{-\bullet} \rightarrow HO_2^\bullet \qquad (6)$$

$$HO_2^\bullet + HO_2^\bullet \rightarrow O_2 + H_2O_2 \qquad (7)$$

$$O_2^{-\bullet} + H_2O_2 \rightarrow HO^\bullet + O_2 + HO^- \qquad (8)$$

$$h\nu + H_2O_2 \rightarrow 2HO^\bullet \qquad (9)$$

$$H_2O_2 + e^- \rightarrow HO^\bullet + HO^- \qquad (10)$$

The resultant reactive oxygen species (ROS; $O_2^{-\bullet}$, $HO^\bullet$) is the key component in the photocatalytic deterioration of pollutants. The dye reacts with ROS and undergoes an oxidation reaction to form degradation products. The h^+ in VB can directly oxidize the contaminant molecules and accelerate the photocatalytic activity of ZCG. Finally, the deterioration of organic pollutants by ZCG occurs as in Equation (11).

$$(HO^{\bullet}, O_2^{-\bullet}, h^+) + Pollutant \rightarrow Degradation\ products \tag{11}$$

3. Materials and Methods

3.1. Materials

Copper(II) sulfate pentahydrate (99%) was received from Showa Chemical Co. Ltd., Japan (Akasaka, Minato-ku, Tokyo). Zinc acetate dihydrate (99.0%), urea (99.0–100.5%), sodium hydroxide (NaOH, 97%), ethyl alcohol (99.5%), and CR were purchased from Sigma-Aldrich (St. Louis, MO, USA). MB was received from Alfa Aesar (Heysham, Lancashire, UK).

3.2. Synthesis of CuO Nanoparticles

The CuO nanoparticles (CuO-NPs) were synthesized via the chemical precipitation method followed by high-pressure annealing. First, 2 g of $CuSO_4.5H_2O$ was dissolved in 10 mL of distilled water. In the clear mixture, aqueous NaOH solution was mixed dropwise up to pH 11. The formation of precipitate was observed. The resultant mixture was stirred using a magnetic stirrer for 3 h at 600 rpm. The color of the precipitate was converted to black from bluish-green. The precipitate was taken out by filtration after washing with distilled water. The sample was desiccated at 70 °C in an oven overnight. The obtained sample was poured into a quartz crucible and stoppered with a lid. The crucible was then positioned in a vacuum chamber which was closed by an airtight gasket made of copper (O_2-free). The sealed chamber was then kept in the furnace and heated at a rate of 7.5 °C min^{-1} at 400 °C for 3h. The calcined compound was ground to obtain the fine powder nanoparticles and reserved for the next use.

3.3. Synthesis of ZnO Nanoparticles and g-C_3N_4 Nanosheet

Both the ZnO nanoparticles (ZnO-NPs) and g-C_3N_4 nanosheet were prepared through a solvent-free modified thermal method. For ZnO-NPs, 5 g of $Zn(CH_3COO)_2.2H_2O$ was poured into a quartz crucible and covered with a lid. The crucible was then positioned in a stainless-steel chamber (SUS314; diameter = 35 mm; length = 60 mm), which was subsequently sealed with an oxygen-free copper gasket (OFHC). The chamber was put into a furnace (KSL-1100X-S-UL-LD, Richmond, CA, USA) and heated at 450 °C for 7 h with a heat-up rate of 5 °C min^{-1}. After thermolysis, the furnace was kept undisturbed to cool slowly to ambient temperature to induce the uniform and smooth formation of ZnO-NPs. Next, the same described procedure for ZnO-NPs was used to prepare a g-C_3N_4 nanosheet; the differences were (i) 2 g of urea was used to prepare g-C_3N_4 nanosheet and (ii) the urea was heated at 520 °C for 380 min with a heating rate of 4 °C min^{-1}.

3.4. Synthesis of ZnO/CuO/g-C_3N_4 Nanocomposite

The ZnO/CuO/g-C_3N_4 nanocomposite was prepared by an efficient co-crystallization process. Initially, as-synthesized CuO, ZnO, and g-C_3N_4 at a 2:2:1 weight ratio were transferred into a beaker containing 40 mL distilled water as a solvent. Then, the resultant mixture was homogenized using a bath sonication for 30 min. After that, the aqueous solution was stirred vigorously for 2 h at 600 rpm using a magnetic stirrer at ambient temperature (26 °C) to produce a nanocomposite in which the constituent components were well distributed. The resultant mixture was then left untouched for 1 day for co-crystallization and kept undisturbed for an additional 2 days to allow the precipitate to settle. The upper 50% of solution (20 mL) was removed, and the remaining mixture was subjected to evaporation at 60 °C in the oven to drive off the solvent molecules completely. The collected, dried powder of the prepared composite was then ground smoothly and stored in a sealed vial for further use. The nanocomposite is denoted as ZCG.

3.5. Instrumentation

An X-ray diffractometer (X'Pert PRO, PANalytical, Lelyweg, Almelo, The Netherlands) equipped with a Cu Kα (λ = 1.5406 Å) radiation source was applied to inspect the crystal structure of the compounds. The chemical composition, oxidation state, and binding energies of the sample were analyzed by X-ray photoelectron spectroscopy (XPS, NEXSA, Thermo Fisher Scientific, East Grinstead, West Sussex, UK). The functional groups of the materials were examined by Fourier transform infrared (FTIR) spectroscopy (Thermo Fisher Scientific, Nicolet iS5). The morphological properties of the composite were evaluated by field-emission scanning electron microscopy (FE-SEM, SU8230, Hitachi, Minato-ku, Tokyo, Japan) and high-resolution transmission electron microscopy (HR-TEM, JEM-2200FS, JEOL, Akishima, Tokyo, Japan). To assess the optical features and bandgaps of the materials, a UV–vis spectrophotometer (Perkin Elmer Lambda 25, Ayer Rajah, Singapore) was used.

3.6. Measurements of Photocatalytic Performance

The photocatalytic proficiency of all the compounds was assessed by measuring the degradation of MB as a target contaminant. The assessments were driven using a xenon lamp (300 W) as a source of visible light. Additionally, a UV cut-off filter ($\geq$400 nm) was engaged to avoid UV irradiation. In a typical experiment, 0.020 g of sample was added into 100 mL of dye solution (20 mg L^{-1}). The resultant suspension was then ultrasonicated for 60 min to confirm the adsorption of dye molecules onto the catalyst surface. Further, this mixture was kept in dark conditions for 30 min to acquire an adsorption–desorption equilibrium. The photodegradation reaction was then started by contact with the assistance of a light source. Before and after the solar irradiation, the change in the dye concentration during the reaction was observed by UV–vis spectrophotometer. For measuring the absorbance of the samples, a constant volume of solution was measured at regular intervals. In addition, the photodegradation proficiency for CR was tested under identical conditions. The photocatalytic investigation of all the samples was executed under uniform experimental conditions. The recyclability of the ZCG nanocomposite was evaluated using a similar procedure in each set. This was achieved by collecting the precipitated sample (after centrifugation) and washing it several times with distilled water. The sample was then dried at 100 °C for approximately 5 h in a preheated furnace for successive evaluation of its recycling performance. Following this, the dry sample was used for the next cycle of the photocatalytic reaction.

4. Conclusions

ZnO/CuO/g-C$_3$N$_4$ nanocomposite was successfully synthesized through a cost-effective, facile co-crystallization method. The prepared nanocomposite was used to treat wastewater containing organic pollutants through visible light illumination and the product's durability and recycling capability was evaluated. The structural and morphological observation confirmed the nanocomposite was perfectly constructed. The top-most photocatalytic activity was observed for the ZCG nanocomposite, which induced 97.46% photodegradation of MB under visible light within 50 min. The ZCG nanocomposite showed 753%, 392%, 156%, and 130% higher photocatalytic efficiency compared with photolysis, g-C$_3$N$_4$, CuO-NPs, and ZnO-NPs, respectively. The photodeterioration of MB followed pseudo-first-order reaction kinetics and the k value of the ZCG composite was 23.6, 11.6, 3.4, and 2.4 times greater than those of photolysis, g-C$_3$N$_4$, CuO-NPs, and ZnO-NPs, respectively. The pH$_{PZC}$ of our as-synthesized ZCG affirmed the anionic nature of the surface at a higher pH and correspondingly showed a cationic nature at lower pH. In addition, the photocatalytic activity and reaction kinetics of the ZCG composite for the deterioration of CR were investigated and the results were found to be remarkable. The outstanding photocatalytic performance was observed due to heterojunction formation among the g-C$_3$N$_4$, CuO-NPs, and ZnO-NPs compounds, which minimized the photogenerated e^--h^+ pair recombination, and increased the electron flow rate. After six consecutive runs, the ZCG product showed excellent stability and recycling performance, confirmed by employing XRD and FTIR analysis of the reused ZCG sample. Therefore, the overall results indicate that the ZCG nanocomposite could be practically used for visible-light-driven wastewater treatment; this sheds new light on the field of catalysts.

Supplementary Materials: The following supporting information can be downloaded at: https://www.mdpi.com/article/10.3390/catal12020151/s1, Figure S1: Linear plots of modified Scherrer equation (a) CuO-NPs, (b) ZnO-NPs, and (c) ZCG nanocomposite.

Author Contributions: Conceptualization, M.A.H. and J.A.; methodology, M.A.H.; software, M.A.H.; formal analysis, M.A.H., J.A. and Y.S.K.; investigation and data curation, M.A.H.; writing—original draft preparation, M.A.H.; writing—review and editing, M.A.H., J.A., Y.S.K., H.G.K., J.R.H. and L.K.K.; visualization, H.G.K., J.R.H. and L.K.K.; supervision, L.K.K.; project administration, L.K.K.; funding acquisition, L.K.K. All authors have read and agreed to the published version of the manuscript.

Funding: This research was supported by the Basic Science Research Program through the National Research Foundation of Korea (NRF) funded by the Ministry of Education (2016R1A6A1A03012069, 2018R1D1A1B07050752). This research was also supported by the Korean Government (NRF-2021R1I1A3045310).

Data Availability Statement: The data are available by corresponding author.

Conflicts of Interest: The authors declare no conflict of interest.

References

1. Akter, J.; Sapkota, K.P.; Hanif, M.A.; Islam, M.A.; Abbas, H.G.; Hahn, J.R. Kinetically controlled selective synthesis of Cu_2O and CuO nanoparticles toward enhanced degradation of methylene blue using ultraviolet and sun light. *Mater. Sci. Semicond. Process.* **2021**, *123*, 105570. [CrossRef]
2. Yang, Y.; Ali, N.; Khan, A.; Khan, S.; Khan, S.; Khan, H.; Xiaoqi, S.; Ahmad, W.; Uddin, S.; Ali, N.; et al. Chitosan-capped ternary metal selenide nanocatalysts for efficient degradation of Congo red dye in sunlight irradiation. *Int. J. Biol. Macromol.* **2021**, *167*, 169–181. [CrossRef] [PubMed]
3. Crini, G.; Lichtfouse, E. Advantages and disadvantages of techniques used for wastewater treatment. *Environ. Chem. Lett.* **2019**, *17*, 145–155. [CrossRef]
4. Akter, J.; Hanif, M.A.; Islam, M.A.; Sapkota, K.P.; Hahn, J.R. Selective growth of $Ti^{3+}/TiO_2/CNT$ and $Ti^{3+}/TiO_2/C$ nanocomposite for enhanced visible-light utilization to degrade organic pollutants by lowering TiO_2-bandgap. *Sci. Rep.* **2021**, *11*, 9490. [CrossRef] [PubMed]
5. He, X.; Yang, D.P.; Zhang, X.; Liu, M.; Kang, Z.; Lin, C.; Jia, N.; Luque, R. Waste eggshell membrane-templated CuO-ZnO nanocomposites with enhanced adsorption, catalysis and antibacterial properties for water purification. *Chem. Eng. J.* **2019**, *369*, 621–633. [CrossRef]
6. Bharathi, P.; Harish, S.; Archana, J.; Navaneethan, M.; Ponnusamy, S.; Muthamizhchelvan, C.; Shimomura, M.; Hayakawa, Y. Enhanced charge transfer and separation of hierarchical CuO/ZnO composites: The synergistic effect of photocatalysis for the mineralization of organic pollutant in water. *Appl. Surf. Sci.* **2019**, *484*, 884–891. [CrossRef]
7. Liu, Z.; Bai, H.; Xu, S.; Sun, D.D. Hierarchical CuO/ZnO "corn-like" architecture for photocatalytic hydrogen generation. *Int. J. Hydrog. Energy* **2011**, *36*, 13473–13480. [CrossRef]
8. Liu, C.; Meng, F.; Zhang, L.; Zhang, D.; Wei, S.; Qi, K.; Fan, J.; Zhang, H.; Cui, X. CuO/ZnO heterojunction nanoarrays for enhanced photoelectrochemical water oxidation. *Appl. Surf. Sci.* **2019**, *469*, 276–282. [CrossRef]
9. Saravanan, R.; Karthikeyan, S.; Gupta, V.K.; Sekaran, G.; Narayanan, V.; Stephen, A.J.M.S. Enhanced photocatalytic activity of ZnO/CuO nanocomposite for the degradation of textile dye on visible light illumination. *Mater. Sci. Eng. C* **2013**, *33*, 91–98. [CrossRef]
10. Khiavi, N.D.; Katal, R.; Eshkalak, S.K.; Masudy-Panah, S.; Ramakrishna, S.; Hu, J.Y. Visible light driven heterojunction photocatalyst of $CuO-Cu_2O$ thin films for photocatalytic degradation of organic pollutants. *Nanomaterials* **2019**, *9*, 1011. [CrossRef]
11. Araújo, E.S.; da Costa, B.P.; Oliveira, R.A.P.; Libardi, J.; Faia, P.M.; de Oliveira, H.P. TiO_2/ZnO hierarchical heteronanostructures: Synthesis, characterization and application as photocatalysts. *J. Environ. Chem. Eng.* **2016**, *4*, 2820–2829. [CrossRef]
12. Pal, S.; Maiti, S.; Maiti, U.N.; Chattopadhyay, K.K. Low temperature solution processed ZnO/CuO heterojunction photocatalyst for visible light induced photo-degradation of organic pollutants. *Cryst. Eng. Comm.* **2015**, *17*, 1464–1476. [CrossRef]
13. Hossain, S.S.; Tarek, M.; Munusamy, T.D.; Karim, K.M.R.; Roopan, S.M.; Sarkar, S.M.; Cheng, C.K.; Khan, M.M.R. Facile synthesis of CuO/CdS heterostructure photocatalyst for the effective degradation of dye under visible light. *Environ. Res.* **2020**, *188*, 109803. [CrossRef] [PubMed]
14. Hassanpour, M.; Safardoust-Hojaghan, H.; Salavati-Niasari, M.; Yeganeh-Faal, A. Nano-sized CuO/ZnO hollow spheres: Synthesis, characterization and photocatalytic performance. *J. Mater. Sci. Mater. Electron.* **2017**, *28*, 14678–14684. [CrossRef]

15. Xu, L.; Zhou, Y.; Wu, Z.; Zheng, G.; He, J.; Zhou, Y. Improved photocatalytic activity of nanocrystalline ZnO by coupling with CuO. *J. Phys. Chem. Solids* **2017**, *106*, 29–36. [CrossRef]
16. Cahino, A.M.; Loureiro, R.G.; Dantas, J.; Madeira, V.S.; Fernandes, P.C.R. Characterization and evaluation of ZnO/CuO catalyst in the degradation of methylene blue using solar radiation. *Ceram. Int.* **2019**, *45*, 13628–13636. [CrossRef]
17. Zhu, W.; Sun, F.; Goei, R.; Zhou, Y. Construction of WO_3–g-C_3N_4 composites as efficient photocatalysts for pharmaceutical degradation under visible light. *Catal. Sci. Technol.* **2017**, *7*, 2591. [CrossRef]
18. Borthakur, S.; Basyach, P.; Kalita, L.; Sonowal, K.; Tiwari, A.; Chetia, P.; Saikia, L. Sunlight assisted degradation of a pollutant dye in water by a WO_3@g-C_3N_4 nanocomposite catalyst. *New J. Chem.* **2020**, *44*, 2947–2960. [CrossRef]
19. Peng, W.; Min, Y.; Sengpei, T.; Feitai, C.; Youji, L. Preparation of cellular C_3N_4/$CoSe_2$/GA composite photocatalyst and its CO_2 reduction activity. *Chem. J. Chin. Univ. -Chin.* **2021**, *42*, 1924–1932.
20. Lin, X.; Du, S.; Li, C.; Li, G.; Li, Y.; Chen, F.; Fang, P. Consciously constructing the robust NiS/g-C_3N_4 hybrids for enhanced photocatalytic hydrogen evolution. *Catal. Lett.* **2020**, *150*, 1898–1908. [CrossRef]
21. Tang, S.; Fu, Z.; Li, Y.; Li, Y. Study on boron and fluorine-doped C_3N_4 as a solid activator for cyclohexane oxidation with H_2O_2 catalyzed by 8-quinolinolato iron^{III} complexes under visible light irradiation. *Appl. Catal. A Gen.* **2020**, *590*, 117342. [CrossRef]
22. Chen, P.; Liu, F.; Ding, H.; Chen, S.; Chen, L.; Li, Y.J.; Au, C.T.; Yin, S.F. Porous double-shell CdS@C_3N_4 octahedron derived by in situ supramolecular self-assembly for enhanced photocatalytic activity. *Appl. Catal. B Environ.* **2019**, *252*, 33–40. [CrossRef]
23. Xiong, S.; Yin, Z.; Zhou, Y.; Peng, X.; Yan, W.; Liu, Z.; Zhang, X. The dual-frequency (20/40 kHz) ultrasound assisted photocatalysis with the active carbon fiber-loaded Fe^{3+}-TiO_2 as photocatalyst for degradation of organic dye. *Bull. Korean Chem. Soc.* **2013**, *34*, 3039–3045. [CrossRef]
24. Singh, J.; Arora, A.; Basu, S. Synthesis of coral like WO_3/g-C_3N_4 nanocomposites for the removal of hazardous dyes under visible light. *J. Alloys Compd.* **2019**, *808*, 151734. [CrossRef]
25. Hanif, M.A.; Akter, J.; Lee, I.; Islam, M.A.; Sapkota, K.P.; Abbas, H.G.; Hahn, J.R. Formation of chemical heterojunctions between ZnO nanoparticles and single walled carbon nanotubes for synergistic enhancement of photocatalytic activity. *J. Photochem. Photobiol. A Chem.* **2021**, *413*, 113260. [CrossRef]
26. Zhang, Q.; Zhang, K.; Xu, D.; Yang, G.; Huang, H.; Nie, F.; Liu, C.; Yang, S. CuO nanostructures: Synthesis, characterization, growth mechanisms, fundamental properties. *Prog. Mater. Sci.* **2014**, *60*, 208–337. [CrossRef]
27. Shi, Y.; Yang, Z.; Liu, Y.; Yu, J.; Wang, F.; Tong, J.; Su, B.; Wang, Q. Fabricating a gC_3N_4/CuO_x heterostructure with tunable valence transition for enhanced photocatalytic activity. *RSC Adv.* **2016**, *6*, 39774. [CrossRef]
28. Meng, J.; Pei, J.; He, Z.; Wu, S.; Lin, Q.; Wei, X.; Li, J.; Zhang, Z. Facile synthesis of g-C_3N_4 nanosheets loaded with WO_3 nanoparticles with enhanced photocatalytic performance under visible light irradiation. *RSC Adv.* **2017**, *7*, 24097–24104. [CrossRef]
29. Hanif, M.A.; Akter, J.; Islam, M.A.; Sapkota, K.P.; Hahn, J.R. Visible-light-driven enhanced photocatalytic performance using cadmium-doping of tungsten (VI) oxide and nanocomposite formation with graphitic carbon nitride disks. *Appl. Surf. Sci.* **2021**, *565*, 150541. [CrossRef]
30. Xiao, T.; Tang, Z.; Yang, Y.; Tang, L.; Zhou, Y.; Zou, Z. In situ construction of hierarchical WO_3/g-C_3N_4 composite hollow microspheres as a Z-scheme photocatalyst for the degradation of antibiotics. *Appl. Catal. B Environ.* **2018**, *220*, 417–428. [CrossRef]
31. Alman, V.; Singh, K.; Bhat, T.; Sheikh, A.; Gokhale, S. Sunlight assisted improved photocatalytic degradation of rhodamine B using Pd-loaded g-C_3N_4/WO_3 nanocomposite. *Appl. Phys. A* **2020**, *126*, 724. [CrossRef]
32. Zhao, D.; Wang, Y.; Dong, C.L.; Huang, Y.C.; Chen, J.; Xue, F.; Shen, S.; Guo, L. Boron-doped nitrogen-deficient carbon nitride-based Z-scheme heterostructures for photocatalytic overall water splitting. *Nat. Energy* **2021**, *6*, 388–397. [CrossRef]
33. Chen, C.; Zheng, Y.; Zhan, Y.; Lin, X.; Zheng, Q.; Wei, K. Reduction of nanostructured CuO bundles: Correlation between microstructure and reduction properties. *Cryst. Growth Des.* **2008**, *8*, 3549–3554. [CrossRef]
34. Akter, J.; Hanif, M.A.; Islam, M.A.; Sapkota, K.P.; Lee, I.; Hahn, J.R. Visible-light-active novel α-Fe_2O_3/Ta_3N_5 photocatalyst designed by band-edge tuning and interfacial charge transfer for effective treatment of hazardous pollutants. *J. Environ. Chem. Eng.* **2021**, *9*, 106831. [CrossRef]
35. Hanif, M.A.; Lee, I.; Akter, J.; Islam, M.A.; Zahid, A.A.S.M.; Sapkota, K.P.; Hahn, J.R. Enhanced photocatalytic and antibacterial performance of ZnO nanoparticles prepared by an efficient thermolysis method. *Catalysts* **2019**, *9*, 608. [CrossRef]
36. Dong, Z.; Wu, Y.; Thirugnanam, N.; Li, G. Double Z-scheme ZnO/ZnS/g-C_3N_4 ternary structure for efficient photocatalytic H_2 production. *Appl. Surf. Sci.* **2018**, *430*, 293–300. [CrossRef]
37. Jourshabani, M.; Shariatinia, Z.; Badiei, A. High efficiency visible-light-driven Fe_2O_3-xS_x/S-doped g-C_3N_4 heterojunction photocatalysts: Direct Z-scheme mechanism. *J. Mater. Sci. Technol.* **2018**, *34*, 1511–1525. [CrossRef]
38. Thomas, M.; Naikoo, G.A.; Sheikh, M.U.D.; Bano, M.; Khan, F. Effective photocatalytic degradation of Congo red dye using alginate/carboxymethyl cellulose/TiO_2 nanocomposite hydrogel under direct sunlight irradiation. *J. Photochem. Photobiol. A Chem.* **2016**, *327*, 33–43. [CrossRef]

Article

Bimetallic PdCo Nanoparticles Loaded in Amine Modified Polyacrylonitrile Hollow Spheres as Efficient Catalysts for Formic Acid Dehydrogenation

Yulin Li [1,†], Ping She [1,†], Rundong Ding [1], Da Li [2], Hongtan Cai [1], Xiufeng Hao [1,*] and Mingjun Jia [1]

1. College of Chemistry, Jilin University, Changchun 130012, China; yulin18@mails.jlu.edu.cn (Y.L.); sheping@jlu.edu.cn (P.S.); dingrd18@mails.jlu.edu.cn (R.D.); caiht21@mails.jlu.edu.cn (H.C.); jiamj@jlu.edu.cn (M.J.)
2. School of Stomatology, Jilin University, Changchun 130022, China; lida19@mails.jlu.edu.cn
* Correspondence: haoxf@jlu.edu.cn; Tel.: +86-431-85155390; Fax: +86-431-85168420
† These authors contributed equally to this work.

Abstract: Polyacrylonitrile hollow nanospheres (HPAN), derived from the polymerization of acrylonitrile in the presence of polystyrene emulsion (as template), were modified by surface amination with ethylenediamine (EDA), and then used as support for loading Pd or PdCo nanoparticles (NPs). The resultant bimetallic catalyst (named PdCo$_{0.2}$/EDA-HPAN) can efficiently catalyze the additive-free dehydrogenation of formic acid with very high activity, selectivity and recyclability, showing turnover frequencies (TOF) of 4990 h^{-1} at 333 K and 915 h^{-1} at 303 K, respectively. The abundant surface amino groups and cyano group as well as the hollow structure of the support offer a suitable environment for achieving high dispersion of the Pd-based NPs on the surface of EDA-HPAN, thus generating ultra-small bimetallic NPs (bellow 1.0 nm) with high stability. The addition of a small portion of Co may adjust the electronic state of Pd species to a certain extent, which can further improve their capability for the dehydrogenation of formic acid. In addition, the surface amino groups may also play an important role in synergistically activating formic acid to generate formate, thus leading to efficient conversion of formic acid to hydrogen at mild conditions.

Keywords: formic acid decomposition; polyacrylonitrile hollow nanospheres; ethylenediamine modification; palladium cobalt nanoparticles

1. Introduction

Hydrogen production and storage are very topical issues in the field of green energy, which is a novel field in engineering with a target to develop idealistic energy systems that have no negative environmental, economic and societal impacts [1,2]. Formic acid (FA) is the simplest and most easily available carboxylic acid, which has the characteristics of relatively high hydrogen content (53 g H$_2$/L), low toxicity and easy storage [3,4]. Since FA can readily release hydrogen (H$_2$) and carbon dioxide (CO$_2$) on demand under mild conditions in presence of suitable catalysts, it has been considered as a safe and convenient hydrogen storage compound [3–9]. Moreover, the concomitant CO$_2$ can also be reutilized to regenerate FA by combining another catalytic process of CO$_2$ hydrogenation, thus being of great significance to achieve the hydrogen storage cycle and CO$_2$ utilization [10–12]. During the past decade, considerable efforts have been devoted to developing highly efficient and stable heterogeneous catalysts for FA dehydrogenation, owing to its great potential in sustainable energy storage and utilization [13–20].

Among the numerous reported catalysts, supported Pd nanoparticles (NPs) catalysts have received increased attention because of their relatively high catalytic activity and dehydrogenation selectivity for FA decomposition [19,20]. In order to enhance the catalytic efficiency and the structure stability of Pd NPs (against aggregation), more recent work has been focused on optimizing the dispersity, electronic state, and coordination environments of Pd species. One of the most effective ways is to addition of a second or even third metal component, which can usually have a positive effect on the physicochemical properties, including catalytic performance of the supported Pd-based NPs [21–26]. For instance, Jiang et al. reported that a carbon supported $Co_{0.30}Au_{0.35}Pd_{0.35}$ nanoalloy could be used as an active and stable catalyst for additive-free FA dehydrogenation with an initial TOF of 80 h^{-1} at room temperature [21]. Wang et al. prepared carbon black-supported PdCo-based nanocatalysts, which can efficiently catalyze FA dehydrogenation at room temperature with sodium formate as a promoter [22]. They proposed that the electron status of the Pd surface could be modified by Co, leading to the decrease in Pd 3d binding energy, which can enhance the CO anti-toxicity ability of Pd [22].

In addition, it was also well known that changing the types of supports or adjusting their surface properties may considerably influence the activity and stability of the supported Pd based NPs catalysts [27–31]. Previous works in the literature demonstrated that introducing some basic groups (such as amino groups) into porous supports could be beneficial for achieving high dispersion of Pd-based NPs and enhancing the structure stability of the metal NPs by building suitable metal-support interaction and space-confined effect [32–40]. For instance, Lu et al. reported that amine-functionalized mesoporous SBA-15 supported bimetallic PdIr NPs exhibited very high catalytic activity and 100% H_2 selectivity for FA dehydrogenation. The amine functional groups of SBA-15-NH_2 can interact strongly with the metal ions, thus leading to the formation of ultrafine and stable PdIr NPs distributed inside the mesopores of the support [32]. Moreover, the surface amino groups may also serve as basic sites to facilitate the O-H bond dissociation in the FA molecule, thus acting as a co-catalyst for achieving FA dehydrogenation at mild condition [32–37]. Currently, it is still a highly attractive subject to modify the surface properties and the morphologies of some easily available supports for constructing more efficient and stable supported Pd-based catalysts.

Previously, the authors and our coworkers also carried out some studies on the preparation of supported Pd-based NPs catalysts for FA dehydrogenation [41–44]. Interestingly, by using low-cost polymer materials such as polyacrylonitrile (PAN) and modified PAN as supports, relatively active supported Pd, Pd-Me (Me = Fe, Co, or Ni) NPs catalysts were obtained [41–43]. It was found that modification of PAN with a suitable amount of ethylenediamine could further improve the dispersity and stability of Pd NPs, thus resulting in formation of a highly active supported Pd catalyst with ultra-small particle size (abound 1.2 nm), which can work well for FA dehydrogenation with a TOF 688 h^{-1} at 303 K [41]. As a continuation of the above work, we here tried to use ethylenediamine modified PAN hollow nanospheres (HPAN) as support to prepare supported Pd and PdCo NPs catalysts. A variety of characterization results demonstrated that the as-synthesized EDA-HPAN could show great advantages in generating and stabilizing the ultrasmall Pd-based NPs with average particle size below 1 nm, thus leading to the formation of more active and stably supported Pd-based NPs catalysts for FA dehydrogenation at ambient temperature.

2. Results and Discussion

2.1. SEM Studies

The HPAN support with hollow sphere structure was synthesized on the basis of a literature work reported for synthesis of polystyrene/polyacrylonitrile composite nanospheres (PS@PAN) [45]. The preparation process of Pd/EDA-HPAN and PdCo$_x$/EDA-HPAN catalysts can be divided into three steps: the preparation of HPAN, the synthesis of amine-functionalized HPAN, and the synthesis of Pd/EDA-HPAN and PdCo$_x$/EDA-HPAN (Scheme 1). As shown in Figure 1, the SEM images of PS, PS@PAN, HPAN and EDA-HPAN reveal that all the samples possess microsphere morphology. The PS microspheres have a smooth surface with particle size around 300–400 nm (Figure 1a). Core-shell PS@PAN formed by polymerization of AN on the surface of PS core shows a larger particle size

(500–600 nm) and a coarse surface (Figure 1b). The support of HPAN hollow sphere, which was obtained by dissolving the PS core of the PS@PAN with THF, shows a much rougher surface with sphere morphology. The hollow structure of HPAN could be confirmed by the appearance of some semi-spherical shell in the SEM image of HPAN (Figure 1c). After modification with EDA, both the particle size and the morphology of EDA-HPAN (Figure 1d) have no obvious change in comparison with the parent of HPAN, indicating that amino modification has little effect on the structure and the morphology of the polymer support.

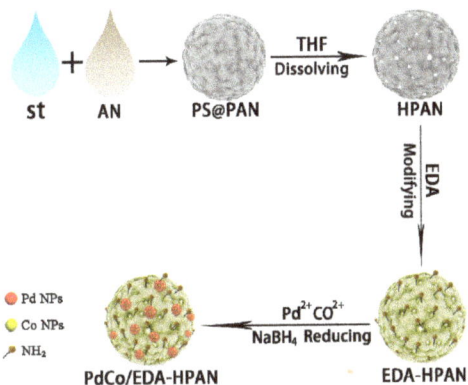

Scheme 1. Preparation route of PdCo$_x$/EDA-HPAN. Step 1: removal of PS template by THF to obtain HPAN. Step 2: EDA-HPAN obtained by EDA modification. Step 3: PdCo/EDA-HPAN obtained by impregnation method.

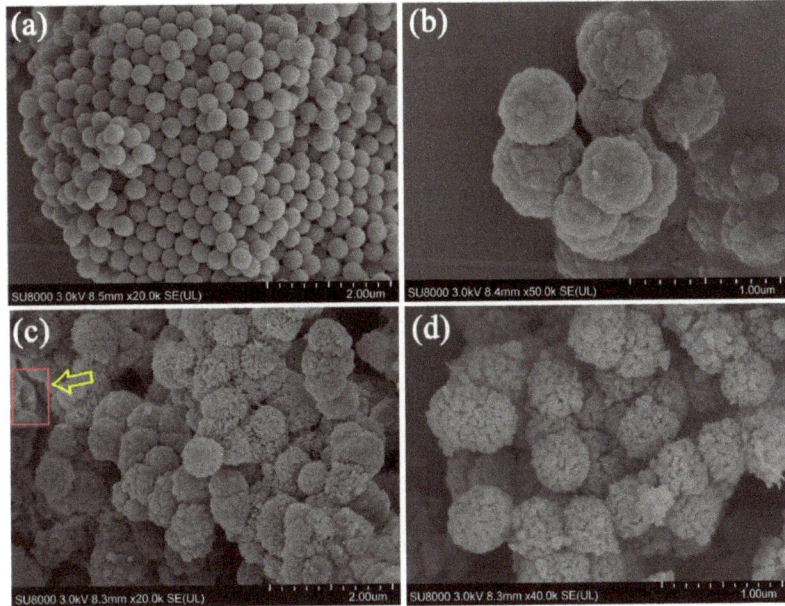

Figure 1. SEM images of the fresh samples. (**a**) PS, (**b**) PS@PAN, (**c**) HPAN, (**d**) EDA-HPAN.

2.2. TEM Studies

The TEM images of Pd/EDA-HPAN and PdCo$_{0.2}$/EDA-HPAN show that a large number of ultra-small Pd-based NPs are uniformly dispersed on the surface of the EDA-

HPAN (Figure 2). The Pd-based NPs observed in the region of the two images have an average diameter of about 0.88 nm and 0.81 nm, respectively, smaller than the previous reported EDA-PAN supported Pd NPs (1.2 nm) [41]. This might be mainly related to the fact that the HPAN support with hollow sphere structure has a rougher spherical surface and a higher specific surface area, as revealed by the following N_2 adsorption-desorption measurements, which can provide more space for the uniform distribution of the amino groups as well as the introduced metal NPs. It is generally believed that the well-distributed amino groups on the solid support have a strong ability to coordinate with the cations (such as Pd^{2+} and Co^{2+}), thus playing an important role in the nucleation and growth of Pd-based NPs, and, meanwhile, inhibiting the aggregation of the generated small NPs during the $NaBH_4$ reduction process. This results in a reduction in the size of Pd NPs. Compared with Pd/EDA-HPAN, the size of metal NPs in the image of $PdCo_{0.2}$/EDA-HPAN is slightly smaller, indicating that the introduction of a small amount of the second metal component may effectively decrease the particle size of Pd-based NPs. This might be mainly caused by the alloy effect, which can adjust the electronic state of Pd-based NPs to a certain extent, and then bring about a positive effect on achieving high dispersion of the bimetallic NPs [42].

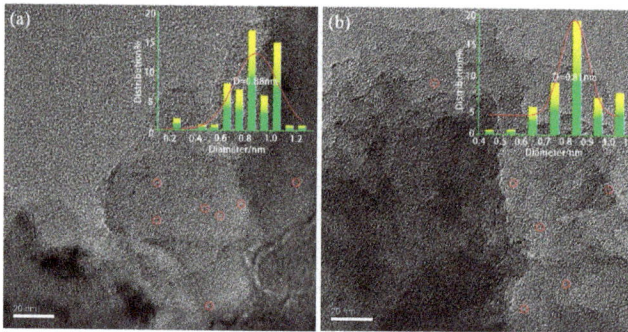

Figure 2. TEM images and histograms of the PdCo nanoparticles size distribution for the fresh catalyst (**a**) Pd/EDA-HPAN and (**b**) $PdCo_{0.2}$/EDA-HPAN.

2.3. IR Spectra Analysis of Samples

Figure 3 shows the infrared spectra of PS, PS@PAN, HPAN, EDA-HPAN and $Pd^{2+}CO^{2+}$/EDA-HPAN, respectively. For the sample of PS sphere (Figure 3a), the stretching vibration peak of C-H on the benzene ring appears at 3060 cm^{-1}, the absorption peak at 2922 cm^{-1} proves the existence of methylene, while the signals appeared at 1452 and 727 cm^{-1} are related to the skeleton vibration peak of the benzene ring and the out of plane bending vibration absorption peak of C-H, respectively [46,47]. As for PAN (Figure 3b), a new absorption peak of 2242 cm^{-1} appears, which belongs to the C≡N functional group in PAN. In the spectrum of HPAN (Figure 3c), the disappearance of the characteristic peak of PS suggests that the PS core has been completely removed after treating the precursor of PS@PAN with THF. Combined with the above characterization results of SEM and FTIR, it could be deduced that the resultant HPAN should possess hollow sphere structure after dissolving the inside core (PS sphere) from PS@PAN. For EDA-HPAN (Figure 3d) and $Pd^{2+}Co^{2+}$/EDA-HPAN (Figure 3e), the signals appear at 1578, 1633 and 1669 cm^{-1} should originate from the stretching vibration peaks of N-H, NH_2 and N-C=N, respectively [48]. Moreover, as shown in Figure S1, the TG curves of HPAN and EDA-HPAN are well consistent with that of PAN reported in the literature [49,50], indicating that the modification of the PAN with EDA did not have an obvious effect on the thermal stability of the polymer support. This might be attributed to the fact that only a small amount of EDA molecules was anchored on the surface of the HPAN support, through a reaction between the -C≡N group in PAN and the amino group in EDA. This point could be further confirmed by the fact that the nitrogen content in EDA-HPAN reached 25.34 wt %, higher than that in HPAN

(23.83 wt %), as given in Table 1. After the introduction of Pd^{2+} and Co^{2+} species, the peak position of N-H bond in the spectrum of $Pd^{2+}Co^{2+}$/EDA-HPAN shifts slightly from 1578 cm^{-1} to 1560 cm^{-1}, which may be due to the coordination action between the metal cations (Pd^{2+} and Co^{2+}) and the amino groups distributed on the surface of EDA-HPAN (Figure 3e).

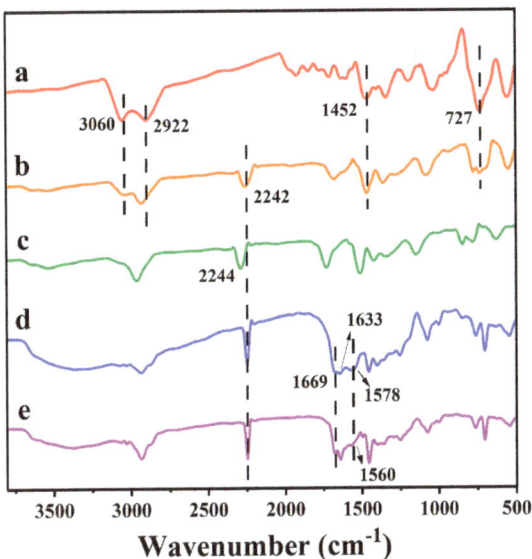

Figure 3. The infrared spectra of (**a**) PS (**b**) PS@PAN (**c**) HPAN (**d**) EDA-HPAN (**e**) $Pd^{2+}Co^{2+}$/EDA-HPAN.

Table 1. The N contents, Pd loading amount and the specific surface areas of various samples.

Sample	N Content (wt %)	Pd Loading (wt %)	BET Surface Area ($m^2 g^{-1}$)
HPAN	23.83	-	45
EDA-HPAN	25.34	-	49
Pd/EDA-HPAN	24.57	3.12	-
$PdCo_{0.2}$/EDA-HPAN	24.43	3.09	-

2.4. Elemental Analyses and Structure Properties of Samples

The ICP-AES results show that the loading amounts of Pd species in Pd/EDA-HPAN and $PdCo_{0.2}$/EDA-HPAN samples are 3.12 wt % and 3.09 wt %, respectively (Table 1), which are very close to the theoretical value of 3.19 wt %. It can be inferred from the TEM image in Figure 2 that the amine groups introduced by amination should be uniformly distributed on the surface of EDA-HPAN. Combined with results of the N_2 adsorption isotherms of HPAN and EDA-HPAN, the specific surface areas of the HPAN and EDA-HPAN supports calculated by the Brunauer-Emmett-Teller method are 45 and 48 m^2/g, respectively (Table 1), which are higher than those of coral-like PAN and EDA-PAN reported in our previous work [41].

The XRD patterns of HPAN, EDA-HPAN, Pd/EDA-HPAN and $PdCo_{0.2}$/EDA-HPAN samples are shown in Figure 4. All the samples present a strong signal at 17°, which is the characteristic peak of PAN appearing at 17° [41,51]. Compared with the support of HPAN and EDA-HPAN, the characteristic peaks in the supported Pd-based catalysts became weaker due to the incorporation of a certain amount of metal NPs. It is worth noting that the characteristic peaks of the metal NPs are not detected in the patterns, confirming further the high dispersion of the small metal NPs on the surface of the polymer support.

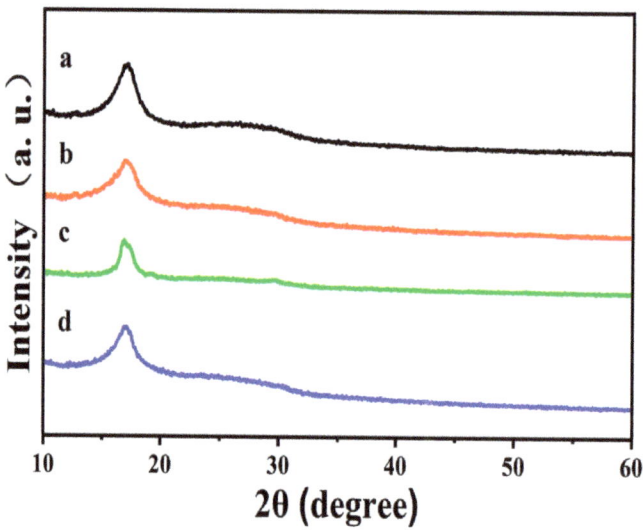

Figure 4. XRD patterns of (**a**) HPAN, (**b**) EDA-HPAN, (**c**) Pd/EDA-HPAN and (**d**) PdCo$_{0.2}$/EDA-HPAN.

2.5. X-ray Photoelectron Spectroscopy Analyses

The Pd 3d XPS spectra of the Pd/EDA-HPAN and PdCo$_{0.2}$/EDA-HPAN catalysts were shown in Figure 5. There are two groups of symmetrical characteristic peaks in the energy spectra, corresponding to the existence of two kind of Pd species in both catalysts. For Pd/EDA-HPAN (Figure 5a), the peaks at 342.8 and 337.8 eV are related to the Pd 3d$_{3/2}$ and Pd 3d$_{5/2}$ characteristic peaks of Pd^{2+}, while the peaks at 341.1 and 335.9 eV are associated to the Pd 3d$_{3/2}$ and Pd 3d$_{5/2}$ characteristic peaks of Pd^0. For PdCo$_{0.2}$/EDA-HPAN (Figure 5b), both kinds of Pd species are still present, while the signals belonging to Pd^0 shift slightly toward lower binding energies compared with the single metal catalyst, possibly related to the alloy effect of Pd and Co, i.e., some electrons transferred from Co to Pd [21]. Combined with the results of the above TEM images (Figure 2), it can be deduced that the Pd NPs with ultra-small particle size should be easily oxidized, thus resulting in the appearance of a number of oxidized Pd^{2+} species in the supported Pd-based catalyst. It can be seen from Figure 5c that the characteristic signals of Co 2p in Co/EDA-HPAN appear at 780.9 and 796.4 eV, while the signals of Co 2p in PdCo$_{0.2}$/EDA-HPAN move upward, appearing at 781.1 and 797.3 eV, respectively. Combined with the above results of Pd 3d spectra, it can be proposed that there is indeed a strong electron interaction between Pd and Co in PdCo$_{0.2}$/EDA-HPAN. Such strong interaction could lead to the change of the electronic state of palladium, which may considerably influence the catalytic activity and stability of the supported Pd-based NPs.

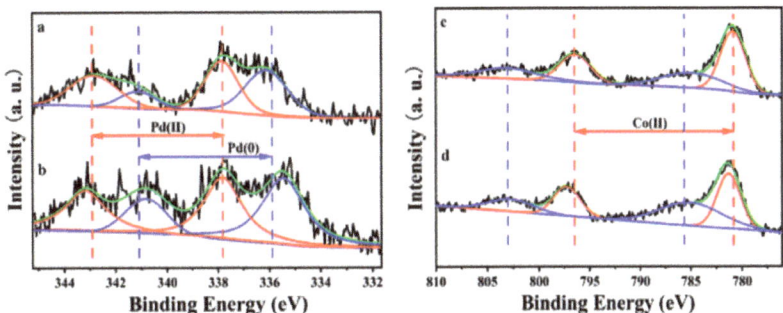

Figure 5. XPS spectra of (**a**) Pd/ EDA-HPAN, (**b**) PdCo$_{0.2}$/EDA-HPAN, (**c**) Co/EDA-HPAN, (**d**) PdCo$_{0.2}$/EDA-HPAN.

2.6. Evaluation of Catalytic Activity

The catalytic properties of Pd/EDA-HPAN and PdCo$_{0.2}$/EDA-HPAN catalysts were measured for the dehydrogenation of FA. As shown in Figure 6a,b, both Pd/EDA-HPAN and PdCo$_{0.2}$/EDA-HPAN exhibit good catalytic activity in the temperature range of 303–353 K, and the dehydrogenation rate of PdCo$_{0.2}$/EDA-HPAN is higher than that of Pd/EDA-HPAN. No CO signal was detected by GC analysis during the reaction term (Figure S2). In addition, after CO$_2$ was absorbed by NaOH solution trap (10 mol/L), the total volume of gas product was reduced by half (Figure S3), indicating that the catalyst had excellent dehydrogenation selectivity. It should be pointed out that both of HPAN and EDA-HPAN are nearly inactive for FA dehydrogenation (Figure S4). The dehydrogenation rate increases with the increase of reaction temperature. The activation energies of FA dehydrogenation on Pd/EDA-HPAN and PdCo$_{0.2}$/EDA-HPAN are 39.30 and 37.25 kJ/mol, respectively, which are comparable with the most active supported Pd-based catalysts reported in studies in the literature [33–37]. At 333 K, the TOF value of PdCo$_{0.2}$/EDA-HPAN catalyst is 4990 h^{-1}, which is higher than that of Pd/EDA-HPAN catalyst (4330 h^{-1}). It is worth noting that the aminated PdCo$_{0.2}$/EDA-HPAN catalyst could also efficiently convert FA to H$_2$ and CO$_2$ at ambient temperature (303 K) with a TOF of 915 h^{-1}, higher than Pd/EDA-HPAN and the previously reported PAN supported Pd-based catalysts [41–43]. The catalytic activity of PdCo$_{0.2}$/EDA-HPAN is comparable or even better than that of the recently reported MOF- or carbon-supported Pd-based catalysts (Table S1), suggesting that HPAN with hollow sphere structure could be used as a suitable support to prepare supported Pd-based NPs catalysts with excellent catalytic activity for FA dehydrogenation.

In addition, the stability and recyclability of Pd/EDA-HPAN and PdCo$_{0.2}$/EDA-HPAN catalysts were also studied for FA dehydrogenation at 333 K. As shown in Figure 7, both Pd/EDA-HPAN and PdCo$_{0.2}$/EDA-HPAN catalysts still show very high activity after five cycles, suggesting the excellent stability of these catalysts. SEM images shown in Figure S5 demonstrate that the morphologies of the spent catalysts of Pd/EDA-HPAN and PdCo$_{0.2}$/EDA-HPAN are kept well in comparison with the fresh catalysts. The TEM results revealed that the average particle sizes of the Pd-based NPs in the spent catalysts of Pd/EDA-HPAN and PdCo$_{0.2}$/EDA-HPAN are still quite small (around 1.0 nm and 0.9 nm) after five catalytic cycles (Figure S6), suggesting that these two supported Pd-based catalysts have strong capability against aggregation.

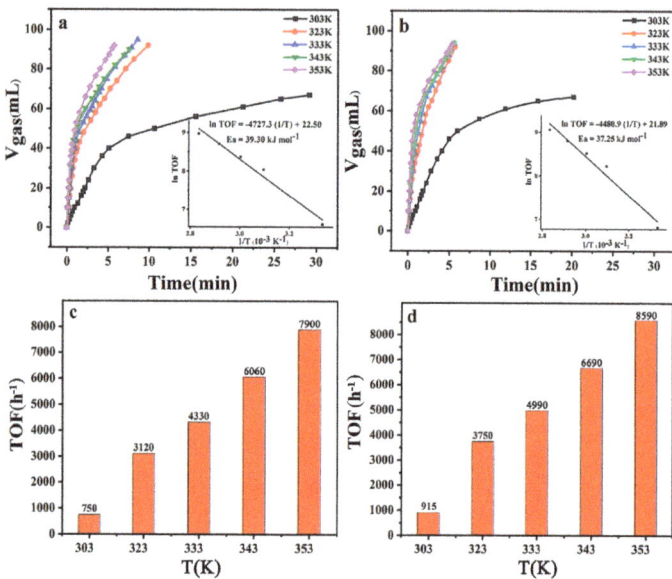

Figure 6. (**a**,**b**) Gas evolution volume ($CO_2 + H_2$) versus time for the dehydrogenation of FA over Pd/EDA-HPAN and $PdCo_{0.2}$/EDA-HPAN catalysts and (inset) the corresponding Arrhenius plot; (**c**,**d**) TOF values at different temperatures.

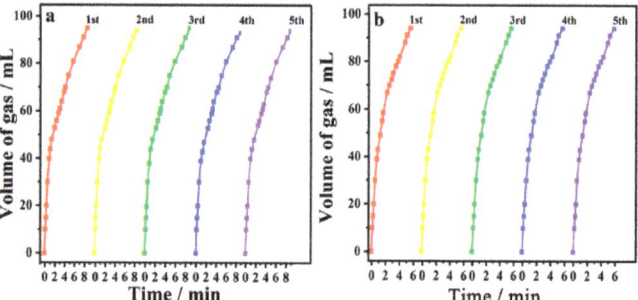

Figure 7. Durability tests for the dehydrogenation of FA solution over (**a**) Pd/EDA-HPAN and (**b**) $PdCo_{0.2}$/EDA-HPAN catalysts at 333 K.

The content of Co in the supported Pd-based NPs should be also a key factor in influencing the catalytic performance of the catalysts. To check this point, a series of supported $PdCo_x$/EDA-HPAN catalysts with different Co/Pd molar ratios were also prepared by using the same procedure as described in the experimental section. As shown in Figure S7, the catalytic activities of the supported bimetallic $PdCo_x$ catalysts are all higher than that of the single metal Pd based catalyst. Among them, the catalyst with Co/Pd ratio of 0.2 exhibits the highest TOF, suggesting that optimizing the Pd/Co ratio can adjust the catalytic performance to a certain extent, which might be due to the fact that the addition of different Co contents could generate different effects on the electron density of Pd as well as the dispersion state of Pd-based NPs, thus changing the activation ability for FA molecules. Further work is still required to clarify this point.

2.7. Catalytic Mechanism

On this basis of the above results and the related literature reports [15,37], a possible mechanism of FA dehydrogenation on PdCo$_x$/EDA-HPAN catalyst can be proposed. As shown in Scheme 2, the surface functional groups such as cyanide and amino can act as basic sites to promote deprotonation of FA to generate [HCOO]$^-$ (Step 1). Then the adjacent PdCo species interacted with [HCOO]$^-$ to form an intermediate PdCo-[HCOO]$^-$ (Step 2). The C–H bond of the intermediate was then dissociated to produce CO$_2$ and [H]$^-$ (Step 3). Finally, [H]$^-$ combined with the initial H$^+$ to produce H$_2$ to complete the whole cycle.

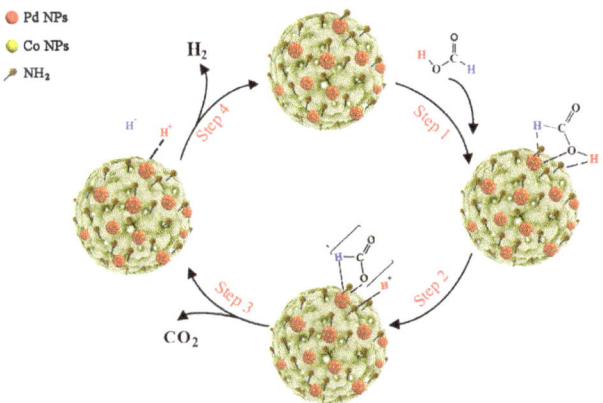

Scheme 2. Possible mechanism of dehydrogenation of FA.

3. Materials and Methods

3.1. Chemicals and Materials

Acrylonitrile (AN), styrene (St) and polyvinyl pyrrolidone (PVP) were supplied by Sinopharm Chemical Reagent Co., Ltd. (Sinopharm Chemical Reagent Co., Ltd., Shanghai, China). Azodiisobutyronitrile (AIBN) was purchased from Tianjin Institute of Fine Chemicals retrocession (Tianjin Institute of Fine Chemicals retrocession, Tianjin, China). FA and EDA were procured from Aladdin Chemistry Co., Ltd. (Aladdin Chemistry Co., Ltd., Los Angeles, SC, USA). Ethanol and tetrahydrofuran (THF) were bought from Beijing Chemical Factory (Beijing Chemical Factory, Beijing, China). PdCl$_2$ was supplied by Sinopharm Chemical Reagent Co., Ltd. (Sinopharm Chemical Reagent Co., Ltd., Shanghai, China). NaBH$_4$ and Co(NO$_3$)$_2$ were purchased from Shandong Xiya Chemical Industry Co., Ltd. (Xiya Chemical Industry Co., Ltd., Chengdu, China).

3.2. Preparation of Polyacrylonitrile Hollow Nanospheres (HPAN)

HPAN was produced by emulsifier-free emulsion polymerization and dissolution method [45]. To begin, 200 mg of PVP and 10 mL of St were added to a three necked flask containing 50 mL of deionized water and 50 mL of ethanol. After stirring for 5 min, 50 mg of AIBN was added. The resulting mixture was then heated at 358 K for 4 h. Subsequently, heating was stopped and the polystyrene (PS) emulsion was cooled to room temperature. Next, 55 mL of PS emulsion was mixed with 5 mL of AN for 2 h. Under the magnetic stirring, 20 mg of AIBN, 10 mL of deionized water and ethanol were added. Subsequently, the mixture was heated for 5 h at 360 K, then heating was stopped, and the emulsion was cooled to room temperature. The centrifuged powder product of PS@PAN was obtained by grinding. Finally, 700 mg of the powder and 10 mL of THF were mixed and dissolved by magnetic stirring for 12 h, then centrifuged and washed. The above operation was repeated twice and the product was lyophilized to obtain a white powder product of HPAN.

3.3. Synthesis of Amine-Functionalized HPAN

First, 400 mg of HPAN powder was mixed with 32 mL of H_2O and treated with ultrasound for 3 min. Then, 4 mL of EDA was added by stirring and the mixture was heated to 368 K for 3 h. The sample was collected by centrifugation and washed three times with water. Finally, the product was dried overnight at 333 K to obtain EDA-HPAN.

3.4. Synthesis of Pd/EDA-HPAN and PdCo$_x$/EDA-HPAN

To begin, 100 mg of EDA-HPAN powder was dispersed in 20 mL water by ultrasonic treatment, and then 1.5 mL H_2PdCl_4 (0.02 mol/L) and 0.3 mL $Co(NO_3)_2$ (0.02 mol/L) were added after ultrasonic treatment. After stirring for 8 h, the mixture was centrifuged, and the separated solid samples of $Pd^{2+}Co^{2+}$/EDA-HPAN were washed several times with ethanol and deionized water and dried in an oven at 333 K. Subsequently, the solid samples of $Pd^{2+}Co^{2+}$/EDA-HPAN were ultrasonically dispersed in 20 mL deionized water, and the newly prepared $NaBH_4$ solution (30 mg $NaBH_4$ dissolved in 2 mL H_2O) was rapidly injected under strong stirring for 6 h. The reduced samples of $PdCo_{0.2}$/EDA-HPAN were collected by centrifugation and washed with water three times. Finally, the product was dried overnight in an oven at 333 K. For comparison, $PdCo_x$/EDA-HPAN samples (where x represents the ratio of Co to Pd, which is 0, 0.4 and 0.6, respectively) were prepared according to the same procedure as above.

3.5. Catalyst Characterisation

FTIR: The samples were characterized by a Nicolet iS5 Spectrometer in the region of from 500 to 4000 cm^{-1}.

Elemental analysis: The content of N was analyzed by using a Vario EL cube.

SEM: Patterns of the catalysts powder was collected by a SU8020 microscope (HI-TACHI Company, Tokyo, Japan) scanning electron microscopy (SEM) with an acceleration voltage of 10.0 kV and 100 kV.

TEM: Transmission electron microcopy (TEM) measurement was performed on FEI Tecnai F20 instrument operated at 200 kV.

ICP-AES: The contents of Pd and Co were analyzed by Thermo iCAP Qc inductively coupled plasma atomic emission spectrometry (ICP-AES).

TG: The pyrolysis experiments were performed on Bruker V70, the sample was heated with a higher heating rate of 10 °C/min from room temperature to 600 °C.

BET: The surface area of samples was calculated using the Brunauer-Emmett-Teller (BET) method, which performed on ASAP2020PlusHD88 at 77 K. All samples were degassed at 110 °C for 12 h prior to analyses.

XRD: The patterns of the catalysts powder were performed on a PANalytical B.V. Empyrean diffractometer with a CuKα radiation (λ = 1.5406 Å) at a voltage of 40 kV and a current of 40 mA.

XPS: X-ray photoelectron spectroscopy (XPS) measurements were performed on a Thermo ESCA LAB 250 spectrometer equipped with an MgKα source (1254.6 eV).

Gas Analysis: The gas composition analysis by a Shimadzu GC-8A gas chromatograph (GC) equipped with a TCD (detection limit: 10 ppm for CO).

3.6. Catalytic Test

An FA dehydrogenation experiment at different reaction temperatures using FA aqueous solution as raw material was conducted. In the test, 0.5 mL of FA (4 mol/L) aqueous solution was injected into a double-necked flask containing 9.5 mL of water and 50 mg of the catalyst powder without any additives under agitation. After each reaction, the catalyst was centrifuged and washed. Finally, it was dried in an oven at 333 K. The gas released from FA aqueous solution was collected and measured by the gasometric method, and real-time mapping.

In order to check the recoverability of the catalysts, Pd/EDA-HPAN and PdCo$_x$/EDA-HPAN samples were collected after the catalytic test by centrifugation. The separated solid catalyst was washed several times and finally dried in an oven at 333 K before circulation experiment.

The catalytic activity of the catalysts is calculated as follows:

$$\text{TOF} = P_0 V / (2RT n_{Pd} t) \qquad (1)$$

In which P_0 represents the atmospheric pressure (101,325 Pa), V is the volume of generated H_2 when the conversion rate of FA reaches 20% (m^3), and R is the universal gas constant. T is the ambient temperature (298 K), n_{Pd} is the total mole number of the catalyst (mol), and t is the reaction time (h) when the conversion rate of FA reaches 20%.

4. Conclusions

In summary, PdCo NPs supported on surface aminated polyacrylonitrile hollow nanospheres (EDA-HPAN) were prepared and could be used as a highly active and stable catalyst for dehydrogenation of FA. PdCo NPs with ultra-small particle size (below 1 nm) could be uniformly dispersed on the surface of EDA-HPAN support due to the special hollow sphere structure, rough surface and the abundant basic groups of the polymer support. The existence of a relatively strong interaction between Pd and Co, as well as the metal-support interaction could modify the electronic structure of Pd to a certain extent, and lead to the formation of a stable structure and catalytically active Pd-based NPs for FA dehydrogenation. The abundant surface basic groups such as cyanide and amino groups could enhance the dispersity of the Pd-based NPs, and may also participate in the activation of FA molecules, thus resulting in the formation of a highly efficient FA dehydrogenation catalyst that can work well under ambient conditions. We believe that more efficient Pd-based catalysts will be developed for the dehydrogenation of FA under mild conditions by further tuning the morphologies and surface properties of the polymer support of polyacrylonitrile, which can certainly provide novel opportunities for promoting the development of green-energy engineering in the near future.

Supplementary Materials: The following are available online at https://www.mdpi.com/article/10.3390/catal12010033/s1, Figure S1: The pyrolysis TG curves of HPAN and EDA-HPAN, Figure S2: Gas chromatography analysis results of pure H_2 (b), pure CO (d) and generated gas for PdCo/EDA-HPAN catalyzed additive-free dehydrogenation of FA at 333 K (a) and (c), Figure S3: The volume of generated gas versus time for PdCo/EDA-HPAN catalyzed additive-free dehydrogenation of FA at 298 K in the absence (a) or presence (b) of NaOH trap, Figure S4: The volume of generated gas versus time for HPAN and EDA-HPAN catalyzed additive-free dehydrogenation of FA at 333 K. Figure S5: SEM images of the spent catalysts of (a) Pd/EDA-HPAN and (b) PdCo/EDA-HPAN after five times reaction, Figure S6: The TEM images and the corresponding Pd particle size distributions of the spent catalysts of (a) Pd/EDA-HPAN and (b) PdCo/EDA-HPAN after five times reaction, Figure S7: The TOF values of $PdCo_x$/EDA-HPAN catalysts with different Co/Pd molar ratios. Table S1: Comparison of catalytic activities with literature reported heterogeneous catalysts.

Author Contributions: Conceptualization, X.H. and M.J.; Methodology and writing, Y.L. and P.S.; Investigation, H.C.; Data curation, D.L. and R.D. The manuscript was amended and supplemented by all authors. All authors have read and agreed to the published version of the manuscript.

Funding: The authors greatly appreciate the financial support of the Jilin Institute of Chemical Technology School-level Major Project (No. 2018006) and General Project (No. 2017022), the National Natural Science Foundation of China (No. 22172058), and the Shanghai Engineering Research Center of Green Energy Chemical Engineering (No. 18DZ2254200).

Conflicts of Interest: The authors declare no conflict of interest.

References

1. Li, X. Green energy for sustainability and energy security. In *Green Energy*; British Library: London, UK, 2011; pp. 1–16. ISBN 978-1-84882-646-5.
2. Oncel, S.S. Green energy engineering: Opening a green way for the future. *J. Clean Prod.* **2017**, *142*, 3095–3100. [CrossRef]

3. Centi, G.; Perathoner, S. Opportunities and prospects in the chemical recycling of carbon dioxide to fuels. *Catal. Today* **2009**, *148*, 191–205. [CrossRef]
4. Dörthe, M.; Peter, S.; Henrik, J.; Beller, M. Formic acid as a hydrogen storage material—Development of homogeneous catalysts for selective hydrogen release. *Chem. Soc. Rev.* **2016**, *45*, 3954–3988.
5. Jiang, Y.Q.; Fan, X.L.; Xiao, X.Z.; Qin, T.; Zhang, L.T.; Jiang, F.L.; Li, M.; Li, S.Q.; Ge, H.W.; Chen, L.X. Novel AgPd hollow spheres anchored on graphene as an efficient catalyst for dehydrogenation of formic acid at room temperature. *J. Mater. Chem. A* **2016**, *4*, 657–666. [CrossRef]
6. Zhao, X.; Dai, P.; Xu, D.Y.; Tao, X.M.; Liu, X.E.; Ge, Q.J. Ultrafine PdAg alloy nanoparticles anchored on NH_2-functionalized 2D/2D TiO_2 nanosheet/rGO composite as efficient and reusable catalyst for hydrogen release from additive-free formic acid at room temperature. *J. Energy Chem.* **2021**, *59*, 455–464. [CrossRef]
7. Singh, A.K.; Singh, S.; Kumar, A. Hydrogen energy future with formic acid: A renewable chemical hydrogen storage system. *Catal. Sci. Technol.* **2016**, *6*, 12–40. [CrossRef]
8. Sordakis, K.; Tang, C.; Vogt, L.K.; Junge, H.; Dyson, P.J.; Beller, M.; Laurenczy, G. Homogeneous catalysis for sustainable hydrogen storage in formic acid and alcohols. *Chem. Rev.* **2018**, *118*, 372–433. [CrossRef]
9. Johnson, T.C.; Morris, D.J.; Wills, M. Hydrogen generation from formic acid and alcohols using homogeneous catalysts. *Chem. Soc. Rev.* **2010**, *39*, 81–88. [CrossRef]
10. Aresta, M.; Dibenedetto, A.; Angelini, A. Catalysis for the valorization of exhaust carbon: From CO_2 to chemicals, materials, and fuels. Technological use of CO_2. *Chem. Rev.* **2014**, *114*, 1709–1742. [CrossRef]
11. Cokoja, M.; Bruckmeier, C.; Rieger, B.; Herrmann, W.A.; Kühn, F.E. Transformation of carbon dioxide with homogeneous transition-metal catalysts: A molecular solution to a global challenge? *Angew. Chem. Int. Ed.* **2011**, *50*, 8510–8537. [CrossRef] [PubMed]
12. Mura, M.G.; Luca, L.D.; Giacomelli, G.; Porcheddu, A. Formic acid: A promising bio-renewable feedstock for fine chemicals. *Adv. Synth. Catal.* **2012**, *354*, 3180–3186. [CrossRef]
13. Yao, M.Q.; Ye, Y.; Chen, H.L.; Zhang, X.M. Porous carbon supported Pd as catalysts for boosting formic acid dehydrogenation. *Int. J. Hydrogen Energy* **2020**, *45*, 17398–17409. [CrossRef]
14. Qi, Y.Y.; Li, J.J.; Zhang, D.J.; Liu, C.B. Reexamination of formic acid decomposition on the Pt (111) surface both in the absence and in the presence of water, from periodic DFT calculations. *Catal. Sci. Technol.* **2015**, *5*, 3322–3332. [CrossRef]
15. Ojeda, M.; Iglesia, E. Formic acid dehydrogenation on Au-based catalysts at near ambient temperatures. *Angew. Chem. Int. Ed.* **2009**, *48*, 4800–4803. [CrossRef]
16. Morris, D.J.; Clarkson, G.J.; Wills, M. Insights into hydrogen generation from formic acid using ruthenium complexe. *Organometallics* **2009**, *28*, 4133–4140. [CrossRef]
17. Coffey, R.S. The decomposition of formic acid catalysed by soluble metal complexes. *Chem. Commun.* **1967**, 923–924. [CrossRef]
18. Fukuzumi, S.; Kobayashi, T.; Suenobu, T. Efficient catalytic decomposition of formic acid for the selective generation of H_2 and H/D exchange with a water-soluble rhodium complex in aqueous solution. *ChemSusChem* **2008**, *1*, 827–834. [CrossRef]
19. Shao, M.H.; Odell, J.; Humbert, M.; Yu, T.; Xia, Y.N. Electrocatalysis on shape-controlled palladium nanocrystals: Oxygen reduction reaction and formic acid oxidation. *J. Phys. Chem. C* **2013**, *117*, 4172–4180. [CrossRef]
20. Sousa-Castillo, A.; Li, F.; Carbo'-Argibay, E.; Correa-Duarteb, M.A.; Klinkova, A. Pd-CNT-SiO_2 nanoskein: Composite structure design for formic acid dehydrogenation. *Chem. Commun.* **2019**, *55*, 10733–10736. [CrossRef] [PubMed]
21. Wang, Z.L.; Yan, J.M.; Ping, Y.; Wang, H.L.; Zheng, W.T.; Jiang, Q. An efficient CoAuPd/C catalyst for hydrogen generation from formic acid at room temperature. *Angew. Chem. Int. Ed.* **2013**, *52*, 4406–4409. [CrossRef] [PubMed]
22. Qin, Y.L.; Liu, Y.C.; Liang, F.; Wang, L.M. Preparation of Pd-Co-based nanocatalysts and their superior applications in formic acid decomposition and methanol oxidation. *ChemSusChem* **2015**, *8*, 260–263. [CrossRef] [PubMed]
23. Song, F.Z.; Zhu, Q.L.; Yang, X.C.; Zhan, W.W.; Pachfule, P.; Tsumori, N.; Xu, Q. Metal-organic framework templated porous carbon-metal oxide/reduced graphene oxide as superior support of bimetallic nanoparticles for efficient hydrogen generation from formic acid. *Adv. Energy Mater.* **2018**, *8*, 1701416. [CrossRef]
24. Tedsree, K.; Li, T.; Jones, S.; Chan, C.W.A.; Yu, K.M.K.; Bagot, P.A.J.; Marquis, E.A.; Smith, G.D.W.; Tsang, S.C.E. Hydrogen production from formic acid decomposition at room temperature using a Ag-Pd core-shell nanocatalyst. *Nat. Nanotechnol.* **2011**, *6*, 302–307. [CrossRef] [PubMed]
25. Xu, L.X.; Yao, F.; Luo, J.L.; Wan, C.; Ye, M.F.; Cui, P.; An, Y. Facile synthesis of amine-functionalized SBA-15-supported bimetallic Au-Pd nanoparticles as an efficient catalyst for hydrogen generation from formic acid. *RSC Adv.* **2017**, *7*, 4746–4752. [CrossRef]
26. Hosseini, H.; Mahyari, M.; Bagheri, A.; Shaabani, A. Pd and PdCo alloy nanoparticles supported on polypropylenimine dendrimer-grafted graphene: A highly efficient anodic catalyst for direct formic acid fuel cells. *J. Power Sources* **2014**, *247*, 70–77. [CrossRef]
27. Karatas, Y.; Bulut, A.; Yurderi, M.; Ertas, I.E.; Alal, O.; Gulcan, M.; Celebi, M.; Kivrak, H.; Kaya, M.; Zahmakiran, M. PdAu-MnO_x nanoparticles supported on amine-functionalized SiO_2 for the room temperature dehydrogenation of formic acid in the absence of additives. *Appl. Catal. B-Environ.* **2016**, *180*, 586–595. [CrossRef]
28. Liu, Q.G.; Yang, X.F.; Huang, Y.Q.; Xu, S.T.; Su, X.; Pan, X.L.; Xu, J.M.; Wang, A.Q.; Liang, C.H.; Wang, X.K.; et al. A Schiff base modified gold catalyst for green and efficient H_2 production from formic acid. *Energy Environ. Sci.* **2015**, *8*, 3204–3207. [CrossRef]

29. Gao, S.T.; Liu, W.H.; Feng, C.; Shang, N.Z.; Wang, C. A Ag-Pd alloy supported on an amine functionalized UiO-66 as an efficient synergetic catalyst for the dehydrogenation of formic acid at room temperature. *Catal. Sci. Technol.* **2016**, *6*, 869–874. [CrossRef]
30. Bulut, A.; Yurderi, M.; Karatas, Y.; Zahmakiran, M.; Kivrak, H.; Gulcan, M.; Kaya, M. Pd-MnO$_x$ nanoparticles dispersed on amine-grafted silica: Highly efficient nanocatalyst for hydrogen production from additive-free dehydrogenation of formic acid under mild conditions. *Appl. Catal. B-Environ.* **2015**, *164*, 324–333. [CrossRef]
31. Zhu, Q.L.; Tsumoriab, N.; Xu, Q. Sodium hydroxide-assisted growth of uniform Pd nanoparticles on nanoporous carbon MSC-30 for efficient and complete dehydrogenation of formic acid under ambient conditions. *Chem. Sci.* **2014**, *5*, 195–199. [CrossRef]
32. Nie, W.D.; Luo, Y.X.; Yang, Q.F.; Feng, G.; Yao, Q.L.; Lu, Z.H. An amine-functionalized mesoporous silica-supported PdIr catalyst: Boosting room-temperature hydrogen generation from formic acid. *Inorg. Chem. Front.* **2020**, *7*, 709–717. [CrossRef]
33. Yurderi, M.; Bulut, A.; Caner, N.; Celebi, M.; Kayab, M.; Zahmakiran, M. Amine grafted silica supported CrAuPd alloy nanoparticles: Superb heterogeneous catalysts for the room temperature dehydrogenation of formic acid. *Chem. Commun.* **2015**, *51*, 11417–11420. [CrossRef] [PubMed]
34. Koh, K.; Jeon, M.; Yoon, C.W.; Asefa, T. Formic acid dehydrogenation over Pd NPs supported on amine-functionalized SBA-15 catalysts: Structure-activity relationships. *J. Mater. Chem. A* **2017**, *5*, 16150–16161. [CrossRef]
35. Mori, K.; Dojo, M.; Yamashita, H. Pd and Pd−Ag nanoparticles within a macroreticular basic resin: An efficient catalyst for hydrogen production from formic acid decomposition. *ACS Catal.* **2013**, *3*, 1114–1119. [CrossRef]
36. Yadav, M.; Akita, T.; Tsumoriab, N.; Xu, Q. Strong metal-molecular support interaction (SMMSI): Amine-functionalized gold nanoparticles encapsulated in silica nanospheres highly active for catalytic decomposition of formic acid. *J. Mater. Chem.* **2012**, *22*, 12582–12586. [CrossRef]
37. Song, F.Z.; Zhu, Q.L.; Tsumori, N.; Xu, Q. Diamine-alkalized reduced graphene oxide: Immobilization of sub-2 nm palladium nanoparticles and optimization of catalytic activity for dehydrogenation of formic acid. *ACS Catal.* **2015**, *5*, 5141–5144. [CrossRef]
38. Yadav, M.; Singh, A.K.; Tsumoriab, N.; Xu, Q. Palladium silica nanosphere-catalyzed decomposition of formic acid for chemical hydrogen storage. *J. Mater. Chem.* **2012**, *22*, 19146–19150. [CrossRef]
39. Qiu, X.Y.; Wu, P.; Xu, L.; Tang, Y.W.; Lee, J.M. 3D graphene hollow nanospheres@palladium-networks as an efficient electrocatalyst for formic acid oxidation. *Adv. Mater. Interfaces* **2015**, *2*, 1500321. [CrossRef]
40. Fang, B.Z.; Kim, M.S.; Yu, J.S. Hollow core/mesoporous shell carbon as a highly efficient catalyst support in direct formic acid fuel cell. *Appl. Catal. B-Environ.* **2008**, *84*, 100–105. [CrossRef]
41. Li, Y.L.; Chen, L.L.; Jia, Y.H.; Li, D.; Hao, X.F.; Jia, M.J. The enhanced role of surface amination on the catalytic performance of polyacrylonitrile supported palladium nanoparticles in hydrogen generation from formic acid. *J. Appl. Polym. Sci.* **2021**, *138*, e50456. [CrossRef]
42. Wang, Z.Z.; Zhang, H.Y.; Li, L.; Miao, S.S.; Wu, S.J.; Hao, X.F.; Zhang, W.X.; Jia, M.J. Polyacrylinitrile beads supported Pd-based nanoparticles as superior catalysts for dehydrogenation of formic acid and reduction of organic dyes. *Catal. Commun.* **2018**, *114*, 51–53. [CrossRef]
43. Chen, L.L.; Hao, D.F.; Wang, Z.Z.; Li, Y.L.; Gao, G.; Hao, X.F.; Zhang, W.X.; Jia, M.J. Use of amidoxime polyacrylonitrile bead-supported Pd-based nanoparticles as high efficiency catalysts for dehydrogenation of formic acid. *J. Nanosci. Nanotechnol.* **2020**, *20*, 2389–2394. [CrossRef] [PubMed]
44. Ding, R.D.; Li, Y.L.; Leng, F.; Jia, M.J.; Yu, J.H.; Hao, X.F.; Xu, J.Q. PdAu nanoparticles supported by diamine-containing UiO-66 for formic acid dehydrogenation. *ACS Appl. Nano Mater.* **2021**, *4*, 9790–9798. [CrossRef]
45. Chen, A.B.; Xia, K.C.; Yu, Y.F.; Sun, H.X.; Liu, L.; Ren, S.F.; Li, Y.T. "Dissolution-Capture" strategy to monodispersed nitrogen-doped hollow mesoporous carbon spheres. *J. Electrochem. Soc.* **2016**, *163*, A3063–A3068. [CrossRef]
46. Monsores, K.G.D.C.; da Silva, A.O.; Oliveira, S.D.S.A.; Weber, R.P.; Filho, P.F.; Monteiro, S.N. Influence of ultraviolet radiation on polystyrene. *J. Mater. Res. Technol.* **2021**, *13*, 359–365. [CrossRef]
47. Botan, R.; Nogueira, T.R.; Lona, L.M.F. Síntese e caracterização de nanocompósitos esfoliados de poliestireno-hidróxido duplo lamelar via polimerização in situ. *Polímeros* **2011**, *21*, 34–38. [CrossRef]
48. Ko, Y.G.; Choi, U.S. Observation of metal ions adsorption on novel polymeric chelating fiber and activated carbon fiber. *Sep. Purif. Technol.* **2007**, *57*, 338–347. [CrossRef]
49. Kuo, H.H.; Chen, W.C.; Wen, T.C.; Gopalan, A. A novel composite gel polymer electrolyte for rechargeable lithium batteries. *J. Power Sources* **2002**, *110*, 27–33. [CrossRef]
50. Park, K.O.; Lee, S.H.; Joh, H.I.; Kim, J.K.; Kang, P.H.; Lee, J.H.; Ku, B.C. Effect of functional groups of carbon nanotubes on the cyclization mechanism of polyacrylonitrile (PAN). *Polymer* **2012**, *53*, 2168–2174. [CrossRef]
51. Wang, M.L.; Jiang, T.T.; Lu, Y.; Liu, H.J.; Chen, Y. Gold nanoparticles immobilized in hyperbranched polyethylenimine modified polyacrylonitrile fiber as highly efficient and recyclable heterogeneous catalysts for the reduction of 4-nitrophenol. *J. Mater. Chem. A* **2013**, *1*, 5923–5933. [CrossRef]

Article

Promotion Effect of the Keggin Structure on the Sulfur and Water Resistance of Pt/CeTi Catalysts for CO Oxidation

Tong Zhang [1], Wenge Qiu [1,*], Hongtai Zhu [1], Xinlei Ding [1], Rui Wu [2] and Hong He [1]

[1] Beijing Key Laboratory for Green Catalysis and Separation, Faculty of Environmental and Life, Beijing University of Technology, Beijing 100124, China; wahxbjut@163.com (T.Z.); zhuhongtai624@163.com (H.Z.); dxliu0516@163.com (X.D.); hehong@bjut.edu.cn (H.H.)

[2] Beijing Fangxinlihua Science and Technology Ltd., Beijing 100025, China; wuruicoco@sina.com

* Correspondence: qiuwenge@bjut.edu.cn

Abstract: Developing a catalyst with high SO_2 and H_2O resistance to achieve high-performance CO oxidation for specific industrial applications is highly desirable. Here, three catalysts were prepared using cerium titanium composite oxide (CeTi), molybdophosphate with Keggin structure-modified CeTi (Keg-CeTi), and molybdophosphate without Keggin structure-modified CeTi (MoP-CeTi) as supports, and their sulfur and water resistance in CO oxidation were tested. The characterization of XRD, BET, SO_2/H_2O-DRIFTS, XPS, TEM, SEM, NH_3/SO_2-TPD, H_2-TPR, and ICP techniques revealed that the high SO_2 and H_2O resistance of Pt/Keg-CeTi in CO oxidation was related to its stronger surface acidity, better reduction of surface cerium and molybdenum species, and lower SO_2 adsorption and transformation compared to Pt/CeTi and Pt/MoP-CeTi.

Keywords: CO oxidation; sulfur and water resistance; platinum-based catalyst; Keggin structure

Citation: Zhang, T.; Qiu, W.; Zhu, H.; Ding, X.; Wu, R.; He, H. Promotion Effect of the Keggin Structure on the Sulfur and Water Resistance of Pt/CeTi Catalysts for CO Oxidation. *Catalysts* **2022**, *12*, 4. https://doi.org/10.3390/catal12010004

Academic Editors: Florica Papa, Anca Vasile and Gianina Dobrescu

Received: 25 November 2021
Accepted: 16 December 2021
Published: 22 December 2021

Publisher's Note: MDPI stays neutral with regard to jurisdictional claims in published maps and institutional affiliations.

Copyright: © 2021 by the authors. Licensee MDPI, Basel, Switzerland. This article is an open access article distributed under the terms and conditions of the Creative Commons Attribution (CC BY) license (https://creativecommons.org/licenses/by/4.0/).

1. Introduction

Carbon monoxide (CO) is one of the major air pollutants mainly originating from the incomplete combustion of fossil fuels (e.g., coal and oil) or from industrial production. It not only affects the atmospheric chemistry and climate but also the health of human beings and animals [1]. Different approaches have been developed to reduce CO emissions, including adsorption [2], CO methanation [3,4], and catalytic oxidation [5–9]. Among them, the catalytic oxidation of CO has proven to be one of the most effective techniques for removing this pollutant. In this process, the catalyst preparation is a key step. In the past century, much research has been devoted to the development of CO oxidation catalysts with high activity and high selectivity, including Pt-group-metal (PGMs: Pt, Pd, Ir, Rh, Ru) catalysts [9–11], gold catalysts [12–14], transition metal oxides (Fe_2O_3 [15], MnO_x [16,17], CuO [18], V_2O_5 [19], CeO_2 [20], Co_3O_4 [21], etc.), metal coordination polymers [2], and metal composite oxides [8] with spinel or perovskite structure catalysts. The gold catalysts and a few transition metal oxide catalysts (such as Co_3O_4) commonly show a higher performance in CO oxidation at low temperature compared to the PGM ones, but the two former types of catalysts suffer from the easy deactivation in the presence of sulfur and moisture. The PGM catalysts, particularly platinum, have been investigated for nearly a century since Langmuir's pioneering work [22], which showed high activities for CO oxidation in the temperature range of 150 to 250 °C with high resistance to sintering and water tolerance [23,24]. In the past several decades, three-way catalysts (TWCs) have been demonstrated as an effective approach to purify vehicle exhaust, which can remove 99% of CO emissions [25,26]. However, the pollutants emitted from separable means of transport are generally low compared to the image of the industrial chimney. Coke oven smoke is a big source of CO emission, in which the CO concentration is even more than 2000 mg/m^3. However, the coexistence of SO_2 and H_2O in the flue gas with CO may reduce the active sites and change the catalyst chemical structure, leading to the diminishing of the catalytic activity of CO oxidation due to the absorption and transformation of the SO_2 and H_2O molecules on the catalyst surface. Therefore, developing a catalyst with high sulfur and water resistance to achieve high-performance CO oxidation, rendering it useful for specific industrial applications, is highly desirable. It was demonstrated that the introduction of a helper component [27] and protective shell [28] in the catalyst, as well as using an optimal catalyst preparation method [29], were beneficial to the improvement of CO oxidation activity and sulfur resistance. Very recently, Jiang [30] investigated the CO oxidation on a Pt single-atom catalyst supported by graphene with a single carbon vacancy (Pt-SG) and double carbon vacancy (Pt-DG) by using first-principles calculations, and they found that Pt-DG possessed a higher sulfur and water resistance due to the fact that carbon divacancy makes Pt less attractive toward SO_2 and H_2O molecules compared to Pt-SG, revealing the effect of the support structure to catalyst performance.

Heteropolyacids (HPAs) are useful molecular metal oxide acids, which have served as catalysts for a number of processes due to their good acidity and high-performance oxidation–reduction [31]. Early research results of Golodov [32] and Zhizhina [33,34] showed that HPAs could oxidize CO with H_2O in the presence of O_2 and Pd or Pt. The presence of HPAs with a Keggin structure could also significantly improve the CO tolerance of a Pt/C catalyst in PEMFCs [35]. Recently, Yoshida [36] reported that the Keggin-type polyoxometalate-supported gold catalyst exhibited a high catalytic activity for CO oxidation at low temperature and extremely high stability. In previous work, we found that the V_2O_5-MoO_3/TiO_2 catalyst with the Keggin structure had more surface Brönsted and Lewis acid sites, which was beneficial to improve significantly the activity and SO_2 resistance in the NH_3-SCR reaction [37]. The above results inspired us to investigate the promotion of the Keggin structure to the sulfur and water resistance of Pt/CeTi catalysts for CO oxidation. Herein, three Pt catalysts using cerium titanium composite oxide (CeTi), ammonium molybdophosphate with Keggin structure-modified CeTi (Keg-CeTi), and molybdophosphate without Keggin structure-modified CeTi (MoP-CeTi) as supports were prepared by an impregnation method under similar conditions. We observed that the Pt/Keg-CeTi catalyst showed a higher SO_2 and H_2O tolerance in CO oxidation compared to the Pt/CeTi and Pt/MoP-CeTi catalysts, revealing the remarkable promotion effect of the Keggin structure.

2. Results and Discussion

2.1. Catalytic Activity and SO_2/H_2O Durability

The catalytic performances of the Pt/CeTi, Pt/Keg-CeTi, and Pt/MoP-CeTi catalysts in the absence and presence of H_2O or SO_2, or both, are displayed in Figure 1. It was found that Pt/CeTi showed a much higher CO oxidation activity in the absence of H_2O and SO_2, and the CO conversion was up to 100% at 150 °C, which is consistent with the literature result [38]. Meanwhile, the intrinsic activities of Pt/Keg-CeTi and Pt/MoP-CeTi were much lower compared to Pt/CeTi (Figure 1a), and their CO complete conversion temperatures were 210 °C and 220 °C, respectively. It is known that the presence of ceria in the support can improve the catalytic performance of noble metal catalysts for CO oxidation by storing oxygen during oxidation and releasing it during reduction [39]. The introduction of molybdophosphate led to the partial covering of the CeTi surface, which restrained the synergism between Pt and cerium species. When a small amount of SO_2 was introduced into the feed mixture, obvious decreases in the activities of the three samples were observed, particularly for Pt/CeTi, due to the significantly detrimental effect of SO_2. In the presence of SO_2 and O_2, Ce(IV) ions could transform into Ce(III) ions according to the following reaction [40,41]: $2CeO_2 + 3SO_2 + O_2 \rightarrow Ce_2(SO_4)_3$, which caused the redox cycle between Ce (IV) and Ce (III) to be terminated. The Pt/Keg-CeTi showed the highest activity in the presence of SO_2 compared to the two others (Figure 1b). On the other hand, the introduction of H_2O had, to a certain extent, a positive effect on the activities of the Pt/CeTi and Pt/Keg-CeTi catalysts (Figure 1c), and their CO complete conversion temperatures decreased to 130 °C and 180 °C, respectively. However, the activity of Pt/MoP-CeTi declined significantly, and the CO could not convert completely even under high temperature. This might be explained by the adsorption of H_2O molecules on the active centers, inhibiting the catalytic process due to the structure change of the support used. In the presence of both H_2O and SO_2, the activity of Pt/Keg-CeTi was also slightly higher than that of Pt/CeTi and was much better than that of Pt/MoP-CeTi (Figure 1d), revealing the better sulfur and water resistance.

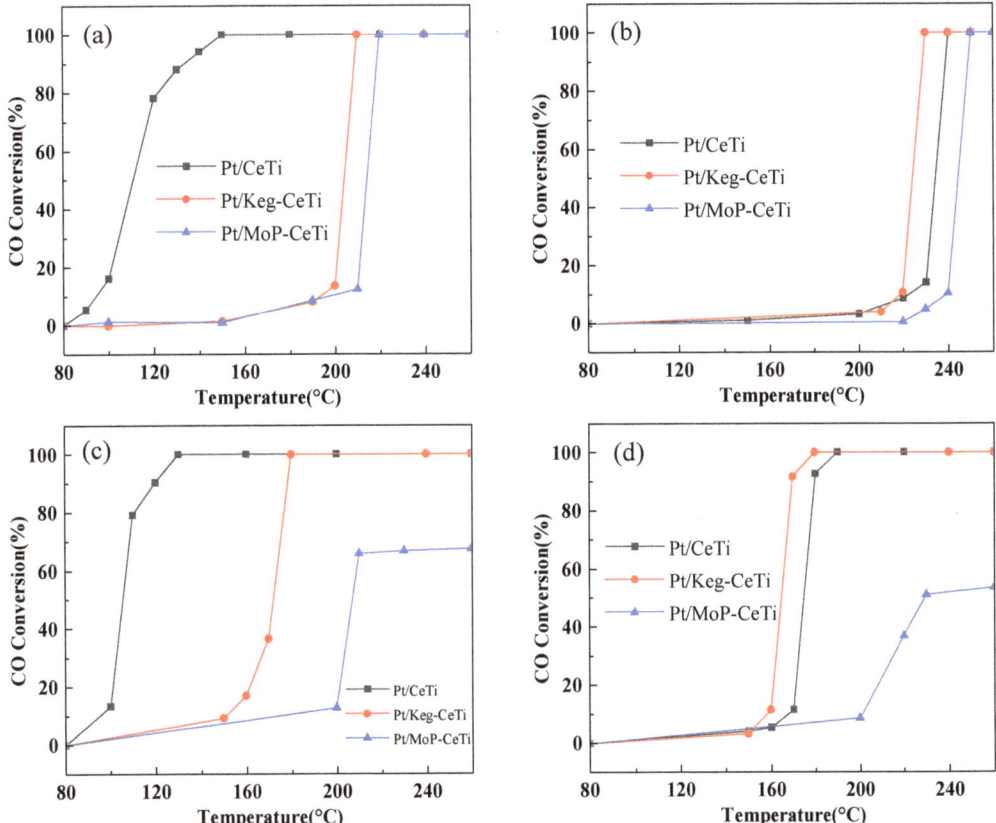

Figure 1. CO conversions as a function of reaction temperature over the Pt/CeTi, Pt/Keg-CeTi, and Pt/MoP-CeTi catalysts in the absence of H_2O and SO_2 (**a**), and in the presence of SO_2 (**b**), H_2O, (**c**) or both H_2O and SO_2 (**d**), respectively. Reaction conditions: [CO] = 1%, [O_2] = 6 vol%, [H_2O] = 10% (when used), [SO_2] = 100 ppm (when used), balance He, total flow rate = 667 mL/min, GHSV = 4×10^5 h^{-1}.

The SO_2/H_2O durability of the three catalysts was also tested (Figure 2). The activities of the Pt/CeTi and Pt/MoP-CeTi catalysts at 250 °C in the presence of SO_2 decreased rapidly from 100% to 95% within the first two hours, and then decreased gradually (Figure 2a). Moreover, the deactivation rate of Pt/MoP-CeTi was faster than that of Pt/CeTi. The CO conversions over Pt/CeTi and Pt/MoP-CeTi after continuing the test on the reaction stream for 30 h were 87% and 74%, respectively. However, the durability of Pt/Keg-CeTi was much better than those of the two others, and its activity data after 30 h were still maintained at above 95%. In the presence of 10% H_2O only, the Pt/CeTi and Pt/Keg-CeTi catalysts showed a high stability, and their CO oxidation activities at 200 °C only showed a slightly decrease when the reaction time extended to 30 h. Nevertheless, the Pt/MoP-CeTi catalyst deactivated much faster under similar conditions, and its CO conversion decreased quickly from 67% to 31% in the first 5 h on stream (Figure 2b). When both H_2O and SO_2 were added, a similar deactivation phenomenon for Pt/MoP-CeTi was also observed. The CO conversion dropped rapidly to less than 30% in the first 5 h (Figure 2c). The Pt/CeTi catalyst showed a middle stability. Its CO conversion was maintained at above 95% within the first 5 h, and then decreased gradually to 61%. However, for Pt/Keg-CeTi, no obvious deactivation was observed within the test time, and the activity data were maintained at above 93% after 30 h. The results further showed that Pt/Keg-CeTi had a higher resistance to SO_2 and H_2O poisoning than the two others.

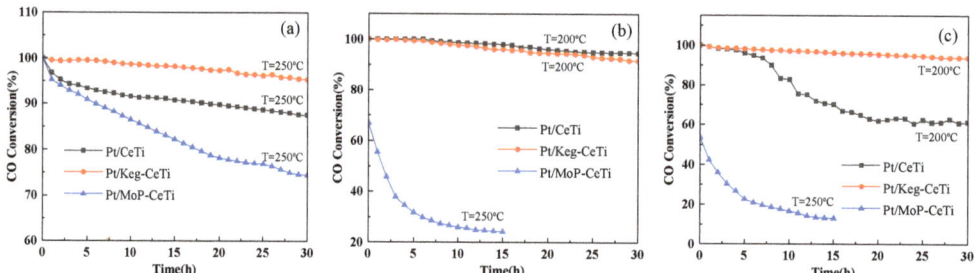

Figure 2. Durability of the Pt/CeTi, Pt/Keg-CeTi, and Pt/MoP-CeTi catalysts at constant temperatures in the presence of SO_2 (**a**), H_2O (**b**), and both of them (**c**). Reaction conditions: [CO] = 1%, [O_2] = 6 vol%, [H_2O] = 10% (when used), [SO_2] = 100 ppm (when used), balance He, total flow rate = 667 mL/min, GHSV = 4×10^5 h^{-1}.

2.2. Structure and Morphology

The XRD patterns of the three samples are illustrated in Figure 3. The diffraction peaks at 2θ = 25.2, 37.7, 47.8, 53.7, 55.9, and 62.5° were attributed to anatase TiO_2, and the peaks at 28.5, 33.0, and 56.3° were assigned to the cubic fluorite-type CeO_2. Moreover, the intensities of the corresponding peaks for the three catalysts were similar, implying that the addition of molybdophosphate with the Keggin structure or molybdophosphate without the Keggin structure had no obvious effect on the structure of cerium and titanium composite oxide. For the Pt/Keg-CeTi sample, several additional sharp peaks at 2θ = 10.5, 15.5, 26.5, 30.5, 36.1, and 40.0° were also detected, demonstrating the formation of a Keggin structure. Meanwhile no similar diffraction peaks were observed on the Pt/MoP-CeTi catalyst, indicating the destruction of the Keggin structure after the high-temperature treatment. In addition, no peak that could be assigned to metallic platinum or its oxides was observed in the XRD curves of the three samples, implying the high dispersion of Pt. This was also corroborated by the TEM data (Figure 4).

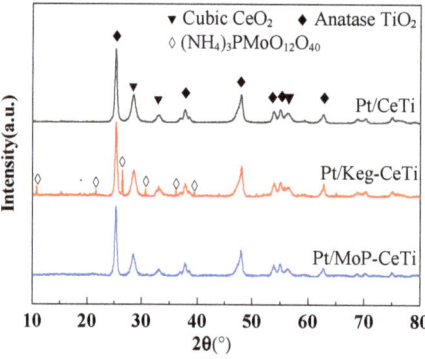

Figure 3. X-ray diffraction patterns of the Pt/CeTi, Pt/Keg-CeTi, and Pt/MoP-CeTi catalysts.

The porosities of the Pt/CeTi, Pt/Keg-CeTi, and Pt/MoP-CeTi catalysts were characterized by N_2 adsorption-desorption isotherms (Figure S1). The three samples all showed typical type II isotherms and type H3 hysteresis loops in a high relative pressure range of $P/P_0 > 0.5$, indicating the existence of stacking channels and the sorption behavior of mesopores. Meanwhile, there was no obvious saturated adsorption platform in the low-pressure range, revealing that there were few micropores on the catalyst surface (Figure S1). This was in agreement with their relatively low specific surface area (41–63 m^2/g) (Table 1). The pore volumes of Pt/Keg-CeTi and Pt/MoP-CeTi were comparatively smaller than that of Pt/CeTi possibly due to the blocking of partial pores of CeTi by the molybdophosphate, and the pore diameter of Pt/Keg-CeTi was slightly larger than those of the two others, which might be related to the difference in particles' size.

Table 1. Textural and structural properties of the catalyst samples.

Samples	S_{BET} (m^2/g)	S_{mic} (m^2/g)	V_{tot} (cm^3/g)	Pore Size (nm)
Pt/CeTi	63	0.0	0.210	9.6
Pt/Keg-CeTi	53	7.4	0.152	12.3
Pt/MoP-CeTi	41	3.5	0.152	9.6

The SEM images clearly showed that the three catalysts had a similar morphology, which were comprised of an irregular conglomeration of particles that were formed by many fine particles with different diameters (Figure 4d–f). TEM images of the Pt/CeTi and Pt/MoP-CeTi catalysts showed that the platinum species were well dispersed on the surface of the support (Figure 4a,c), and most of the Pt particle sizes were in the ranges of <5 nm. Meanwhile, on the Pt/Keg-CeTi sample, few clear Pt particles could be observed, indicating a better dispersion of platinum species on the Keggin structure-modified CeTi composite oxide.

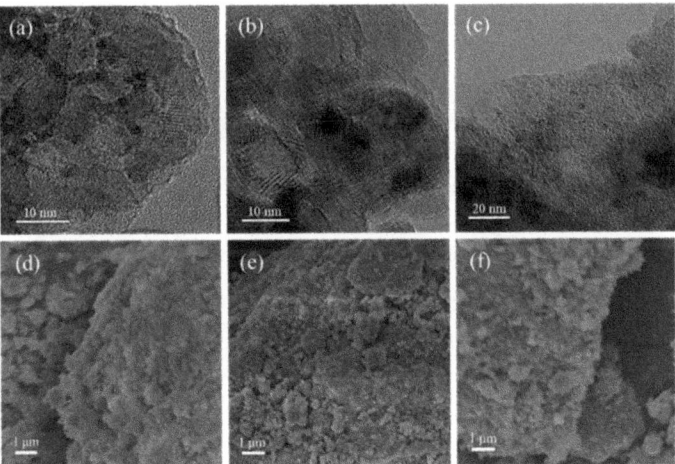

Figure 4. TEM and SEM images of the Pt/CeTi (**a,d**), Pt/Keg-CeTi (**b,e**), and Pt/MoP-CeTi (**c,f**) catalysts.

2.3. Adsorption and Desorption of SO_2 on the Catalysts

In order to understand the difference in sulfur resistance of the Pt/CeTi, Pt/Keg-CeTi, and Pt/MoP-CeTi catalysts, the adsorption of SO_2 on the three catalysts under the same conditions was investigated in detail by in situ IR spectroscopy and temperature-programmed desorption experiments. The DRIFTS results of SO_2 adsorption experiments in the presence of 150 ppm of SO_2 and a large excess of O_2 at 250 °C are shown in Figure 5. For the

Pt/CeTi sample, after the introduction of SO_2, four adsorption bands at 1451, 1350, 1294, and 1145 cm^{-1} were observed and their intensities increased with time. According to other studies [42–46], the peaks at 1451, 1350, and 1294 cm^{-1} were attributed to the surface sulfate species. The band centered at 1145 cm^{-1} was assigned to the bulk-like sulfate, revealing that the surface and bulk sulfate species were formed nearly simultaneously on the Pt/CeTi surface. For the Pt/Keg-CeTi sample, a peak at 1065 cm^{-1} was detected in the first 5 min, which was attributed to symmetrical oscillations of sulfites [47], which could be detected in the whole experimental time. The signals attributed to the surface and bulk sulfates were much weaker, implying that the accumulation of sulfate species on the catalyst was inhibited significantly due to the modification of the Keggin structure. For the Pt/MoP-CeTi sample, two wide and weak peaks at 1400 and 1168 cm^{-1} were detected, which were assigned to surface sulfate and bulk sulfate species [48], respectively. It can be seen that both surface sulfate and bulk sulfate species were detected on the three catalysts, but the corresponding peak intensities were quite different. Obviously, the amount of sulfate species on Pt/Keg-CeTi was much lower than those of Pt/CeTi and Pt/MoP-CeTi.

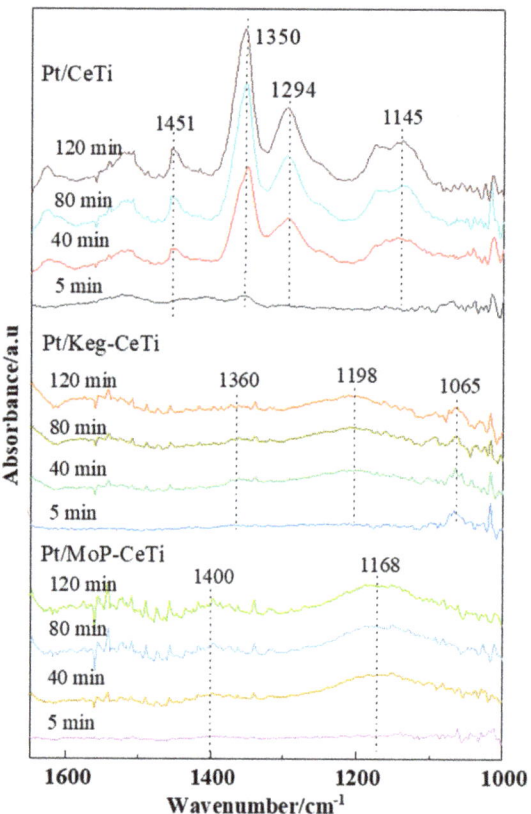

Figure 5. Changes in DRIFTS spectra of Pt/CeTi, Pt/Keg-CeTi, and Pt/MoP-CeTi with time. Conditions: $[O_2]$ = 16 vol%, $[SO_2]$ = 150 ppm, balance N_2, total flow rate = 50 mL/min, at 250 °C.

To further understand the nature and content of sulfate species formed on the poisoned samples better, TPD analysis was examined (Figure 6). It clearly showed that two peaks at 675 and 765 °C were observed for Pt/CeTi, which were attributed to the decomposition of $Ce(SO_4)_2$ and $Ce_2(SO_4)_3$ [49], respectively. There was one main SO_2 desorption peak at 705 °C for Pt/MoP-CeTi, while the peak that appeared at 695 °C for Pt/Keg-CeTi was much weaker compared to the two others. It was clear that the amount of sulfate species accumulated on Pt/Keg-CeTi was much lower than that on Pt/MoP-CeTi and Pt/CeTi, which was also in agreement with the ICP data (Table S1) and the DRIFTS spectra (Figure 5). The results revealed that the SO_2 adsorption and transformation on Pt/Keg-CeTi were highly inhibited, which might be one main reason why it showed good sulfur resistance. However, the low activity and sulfur resistance of Pt/MoP-CeTi may also relate to the low surface Pt concentrations (see Table 2) except its moderate SO_2 adsorption and transformation.

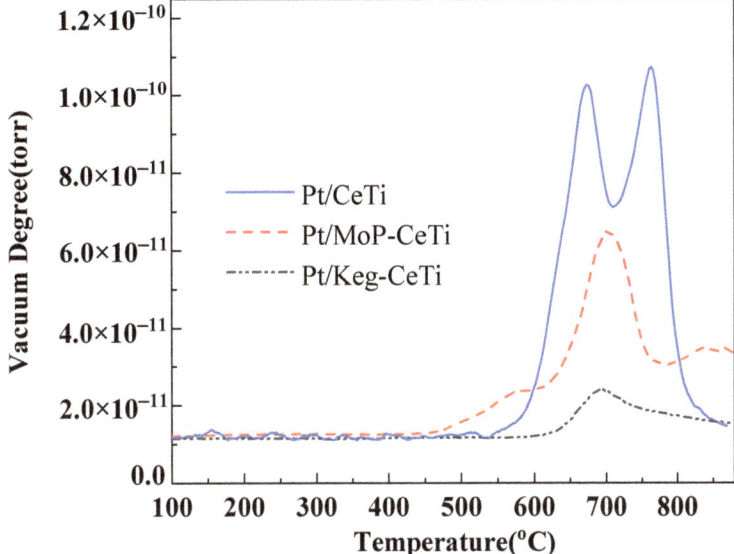

Figure 6. SO_2-TPD of Pt/MoP-CeTi, Pt/Keg-CeTi, and Pt/CeTi after exposure to 1000 ppm of SO_2 with 16% O_2 in Ar atmosphere for 1 h at 250 °C.

Table 2. Surface atomic concentration and atomic ratio of Pt/CeTi, Pt/Keg-CeTi, and Pt/MoP-CeTi catalysts.

Samples	Surface Atomic Concentration (Atom%)			Atomic Ratio (%)		
	Pt	Ce	O	$Pt^{4+}/(Pt^{2+} + Pt^{4+})$	$Ce^{3+}/(Ce^{3+} + Ce^{4+})$	$O_{ads}/(O_{ads} + O_{latt})$
Pt/CeTi	0.23	1.78	54.39	23.3	18.0	10.1
Pt/Keg-CeTi	0.16	1.33	45.86	24.7	31.0	22.2
Pt/MoP-CeTi	0.11	1.61	44.54	55.7	42.3	20.1

2.4. Surface Properties and Redox Property

The Pt 4f and Ce 3d XPS spectra of the three samples are illustrated in Figure 7, and the surface atomic compositions, $Pt^{4+}/(Pt^{2+} + Pt^{4+})$, $Ce^{3+}/(Ce^{3+} + Ce^{4+})$, and $O_{ads}/(O_{ads} + O_{latt})$ molar ratios, and binding energies are summarized in Table 2 and Table S2.

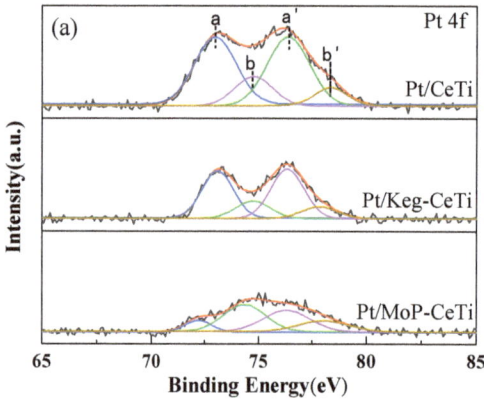
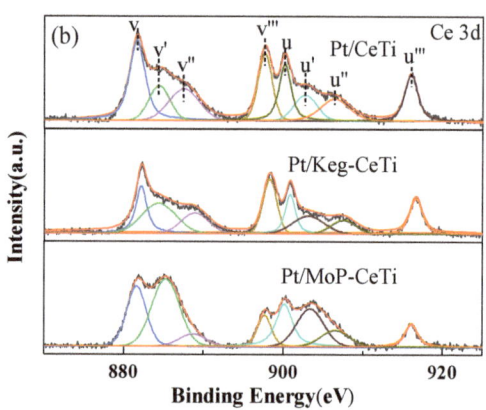

Figure 7. XPS spectra of Pt 4f (**a**) and Ce 3d (**b**) in the Pt/CeTi, Pt/Keg-CeTi, and Pt/MoP-CeTi catalysts.

It was observed that the binding energy values of Pt 4f in the three samples were different (Figure 7a). The peaks were labeled as "a" and "b," representing Pt^{2+} and Pt^{4+} [50], respectively. This suggested that Pt^{2+} and Pt^{4+} coexisted on the three catalysts' surface with different surface atomic concentrations and atomic ratios, and no Pt^0 species were detected. The surface atomic concentrations of Pt for Pt/CeTi, Pt/Keg-CeTi, and Pt/MoP-CeTi were 0.23%, 0.16%, and 0.11% (Table 2), respectively, revealing that the addition of molybdophosphate with or without the Keggin structure resulted in a decrease in the surface active sites. This might be one reason why Pt/CeTi has the best catalytic activity in the absence of H_2O and SO_2. The atomic ratios of $Pt^{4+}/(Pt^{2+} + Pt^{4+})$ for Pt/CeTi, Pt/Keg-CeTi, and Pt/MoP-CeTi were 23.3%, 24.7%, and 55.7% (Table 2), indicating that the introduction of molybdophosphate with the Keggin structure had no obvious effect on the Pt valence state. However, the Pt valence state of Pt/MoP-CeTi changed a lot. During the destruction of the Keggin structure under high temperature, more Ce^{3+} species were generated, resulting in more adsorbed oxygen species on the catalyst surface compared to Pt/CeTi, which might oxidize partial Pt^{2+} to Pt^{4+}. The XPS spectra of Pt 4f in the recovered Pt/CeTi, Pt/Keg-CeTi, and Pt/MoP-CeTi catalysts indicated that the valence state distribution of Pt species changed significantly during the reaction (Figure S2), and the Pt^{4+} species in the used Pt/Keg-CeTi and Pt/MoP-CeTi samples disappeared, unlike the corresponding Pt/CeTi. The photoelectron spectrum of Ce 3d is also given in Figure 7b, in which the peaks labeled "u" represented Ce $3d_{3/2}$ contribution, and those labeled "v" were assigned to $3d_{5/2}$ [51]. The spectral lines denoted as v, v″, v‴ and u, u″, u‴ were characteristic of the Ce^{4+}, while v′ and u′ were assigned to the Ce^{3+}, suggesting the coexistence of Ce^{3+} and Ce^{4+} on the three catalyst samples. It is known that oxygen vacancies and unsaturated chemical bonds are related to the presence of Ce^{3+} and benefit the formation of chemisorbed oxygen on the surface [52]. The atomic ratios of $Ce^{3+}/(Ce^{3+} + Ce^{4+})$ for Pt/CeTi, Pt/Keg-CeTi, and Pt/MoP-CeTi were 18.0%, 31.0%, and 42.3%, respectively (Table 2). The relatively high Ce^{3+} ratio of Pt/Keg-CeTi might be attributed to the interaction between CeO_2 and molybdophosphate, while the highest Ce^{3+} ratio of Pt/MoP-CeTi was possibly also related to the high-temperature treatment, except the interaction of CeO_2 and molybdophosphate. The O 1s XPS (Figure S3) signals of the three catalysts could be fitted into two groups referred to the lattice oxygen (O_{latt}) at around 529.8 eV and the surface chemisorbed oxygen (O_{ads}) in the range of 530.9–531.8 eV [53]. Compared to Pt/CeTi, the concentrations of chemisorbed oxygen species in Pt/Keg-CeTi and Pt/MoP-CeTi were higher due to the existence of more oxygen vacancies on their surface (Table 2). The difference in binding energy of oxygen species also revealed the diversity of the surface microenvironment of the catalysts.

The surface acidity of the catalysts was characterized using the NH$_3$-TPD technique. From the TPD profile (Figure 8), two broad NH$_3$ desorption peaks centered at 98 and 273 °C, 98 and 246 °C, and 289 and 419 °C could be observed, respectively, for the Pt/MoP-CeTi, Pt/CeTi, and Pt/Keg-CeTi samples between 50 °C and 600 °C. The signal in the low-temperature range (50–200 °C) was assigned to the desorption of physiosorbed NH$_3$ on the weak acid sites, and the signal in the high-temperature range (200–600 °C) was ascribed to the desorption of chemisorbed NH$_3$ on the strong acid sites [54]. It was known that the high surface acidity could inhibit the adsorption of SO$_2$ on the catalyst surface [55]. It could be seen that there were many more acid sites, particularly for strong acid sites on the surface of Pt/Keg-CeTi. The calculated data showed that the number of total acid sites was in the order of Pt/Keg-CeTi > Pt/CeTi > Pt/MoP-CeTi (Table S3), which was consistent with their sulfur resistance, implying that the surface acidity of the catalyst might be one factor influencing its sulfur resistance in CO oxidation.

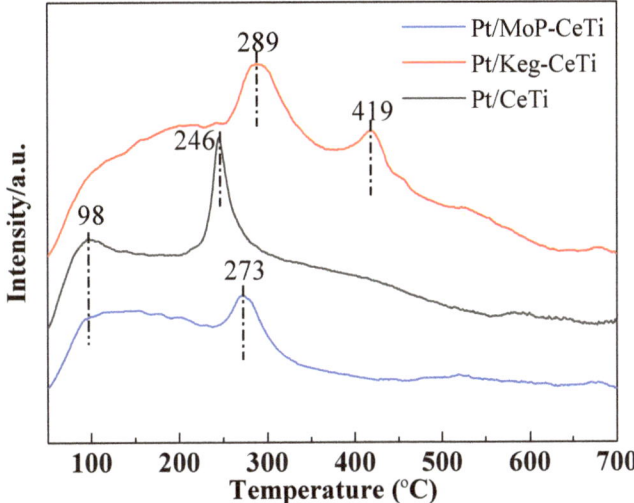

Figure 8. NH$_3$-TPD profiles of the Pt/CeTi, Pt/Keg-CeTi, and Pt/MoP-CeTi catalysts.

The H$_2$-TPR technique was employed to investigate the reduction property of the catalysts. Figure 9 illustrates the H$_2$-TPR profiles of the three catalysts performed under an Ar atmosphere (Table S4). One could see that there were five reduction peaks at 84, 224, 334, 585, and 640 °C for Pt/Keg-CeTi, as well as five peaks at 76, 193, 366, 613, and 657 °C for Pt/MoP-CeTi. For Pt/CeTi, only three reduction peaks at 83, 379, and 645 °C could be observed. As a result of the hard reduction of titanium oxides below 700 °C, all the signals of the various catalysts were attributed to the reduction of corresponding platinum, cerium, and molybdenum species. The reduction peak below 100 °C could be assigned to the reduction of Pt oxides. The Pt reduction temperatures of Pt/CeTi (83 °C), Pt/Keg-CeTi (84 °C), and Pt/MoP-CeTi (76 °C) are comparably higher than the data of bulk Pt oxides' reduction, which is normally below room temperature, possibly due to the strong interaction between Pt and CeO$_2$ and the formation of Pt−O−Ce species [56]. The reduction peaks in the ranges of 150–500 °C and 550–700 °C originated from the reduction of the surface and bulk of CeO$_2$, respectively [57,58]. For the Pt/Keg-CeTi and Pt/MoP-CeTi catalysts, there were two reduction peaks of surface cerium species between 150 and 500 °C, and their position moved to a lower temperature range compared to the data of Pt/CeTi, possibly due to the modification of molybdophosphate and the interaction between Pt and CeO$_2$. In addition, the shoulder peaks at 585 and 613 °C were attributed to the reduction of molybdenum species in Pt/Keg-CeTi and Pt/MoP-CeTi [59], respectively, implying that it was easier to reduce the molybdenum species within the Keggin structure compared to the MoO_3 in Pt/MoP-CeTi. The better reduction of surface cerium and molybdenum species for Pt/Keg-CeTi was beneficial to improve its catalytic performance for CO oxidation.

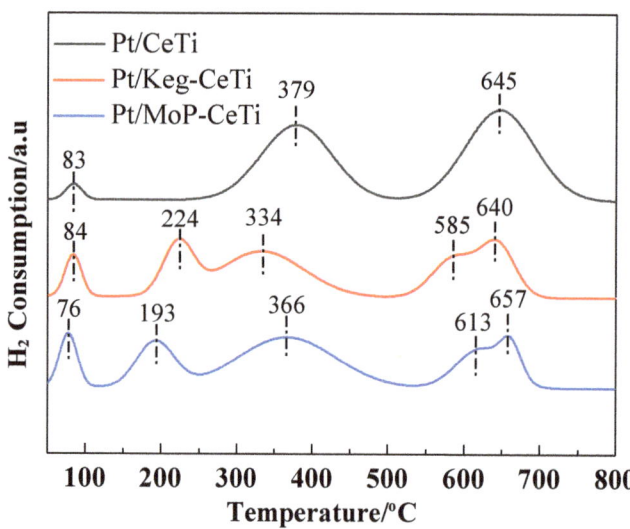

Figure 9. H_2-TPR profiles of the Pt/CeTi, Pt/Keg-CeTi, and Pt/MoP-CeTi catalysts.

2.5. Water Adsorption on the Catalysts

According to Feng's study [60], H_2O could enhance the catalytic oxidation of CO, and the promoting effect of H_2O was greater than the inhibiting effect of SO_2 when H_2O and SO_2 coexisted in the atmosphere. Similar phenomena were observed over Pt/CeTi and Pt/Keg-CeTi, but it was different for Pt/MoP-CeTi (Figures 1 and 2). In order to explain the phenomena, H_2O-DRIFTS experiments over Pt/Keg-CeTi and Pt/MoP-CeTi were conducted at 250 °C with 3% H_2O (Figure 10). During the experiment, 3% H_2O was supplied in the first 60 min, and then it was cut off. For the Pt/Keg-CeTi sample, with H_2O exposure, two peaks at 1620 and 3440 cm^{-1} grew in intensity with time (Figure 10a). Moreover, their intensities remained after H_2O was cut off, indicating a stable adsorption of H_2O molecules on the Pt/Keg-CeTi surface at 250 °C. For the Pt/MoP-CeTi sample, two peaks 1650 and 3200 cm^{-1} were detected in the first 35 min. However, after that, the H_2O adsorption peaks almost disappeared and two new peaks centered at 1843 and 2650 cm^{-1} appeared, which remained even after cutting off the water. The bands at 1843 and 2650 cm^{-1} could be attributed to the PO-H stretching vibration of hydrogen phosphate and dihydrogen phosphate, revealing the change in the support chemical structure. The formation of hydrogen phosphate and dihydrogen phosphate on the Pt/MoP-CeTi surface in the presence of moisture at 250 °C could lead to the inhibition of H_2O adsorption and inaccessibility of active sites, which might be the main reason for the deactivation of Pt/MoP-CeTi in the presence of H_2O.

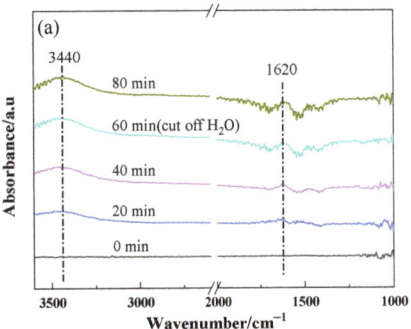 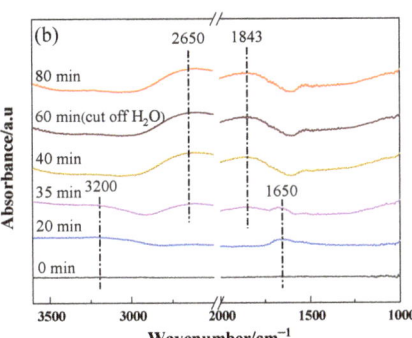

Figure 10. Changes in H_2O-DRIFTS spectra of Pt/Keg-CeTi (**a**) and Pt/MoP-CeTi (**b**) with time. Conditions: [H_2O] = 3 vol%, balance N_2, total flow rate = 50 mL/min, at 250 °C.

3. Experimental

3.1. Chemicals

All chemical reagents were from commercial sources and were used directly without any further purification. The CeO_2, $(NH_4)_6Mo_7O_{24}$ (99%), and $NH_4H_2PO_4$ (99%) were of analytical grade and were purchased from Fuchen (Tianjin, China). Platinum nitrate solution (14.99%) was from Helishi, (Shanghai, China). TiO_2 was an industrial product from Xinhua, (Chongqing, China).

3.2. Catalyst Preparation

The CeTi composite oxide support was prepared by the ball milling method. Commercial anatase TiO_2 (16.0 g) and CeO_2 (4.0 g) were placed into a 500 cm^3 sintered zirconium oxide grinding jar with agate balls (20, 15, and 10 mm in diameter). The ball-to-powder mass ratio was 10:1, and the rotation speed and time were 500 rpm and 1 h, respectively. The received powder was calcined at 500 °C for 2 h, giving the CeTi support. The Pt/CeTi catalyst was fabricated by an impregnation method. In a typical procedure, a 4 g CeTi support was mixed with a calculated amount of platinum nitrate solution of 0.02 mol/L according to the Pt loading (1 wt%), stirred at 80 °C for 2 h, and then dried at 80 °C for 6 h. The resulting solid was ground into powder and calcined at 350 °C for 2 h, giving the Pt/CeTi catalyst. Certain amounts of CeTi support, $(NH_4)_6Mo_7O_{24}$, and $NH_4H_2PO_4$ were added to 200 mL of distilled water, stirred, and then nitric acid solution was added dropwise to adjust the system pH = 1 at room temperature, leading to the formation of $(NH_4)_3PMo_{12}O_{40}$ with the Keggin structure. After 2 h, the mixture was dried at 80 °C for 6 h and calcined at 350 °C for 2 h, giving the molybdophosphate with Keggin structure-modified CeTi (Keg-CeTi), in which the $(NH_4)_3PMo_{12}O_{40}$ loading was 10 wt%. The corresponding catalyst, denoted as Pt/Keg-CeTi, was prepared using a similar impregnation method as above. In addition, the Keg-CeTi support was calcinated at 500 °C for 2 h to fully destroy the Keggin structure, and then the active Pt species was loaded by a similar impregnation method. The received catalyst was denoted as Pt/MoP-CeTi.

3.3. Catalyst Characterization

The X-ray diffraction (XRD) instrument was a Bruker D8 ADVANCE (Karlsruhe, Germany). The X-ray radiation source was Cu Kα (λ = 1.54 Å), and the voltage between the cathode and anode and the current were 50 kV and 35 mA, respectively. The 2θ angle was in the range of 10~80°, and the scanning speed was 8 s/step with a step of 0.02°. The N_2 adsorption was detected by using Autosorb iQ automatic physical adsorption made by Quantachrome Instruments (Boynton Beach, FL, USA). The sample was pretreated at 200 °C for 4 h under vacuum conditions, and the N_2 adsorption isotherms were performed at 77 K. The BET (Brunauer–Emmett–Teller) method was used to calculate the specific surface area of the catalyst, and the Barrett–Joyner–Halenda (BJH) method was used to calculate the pore size distribution and pore volume of the catalyst. The profiles of H_2-temperature programmed reduction (H_2-TPR) was performed on an Autochem II 2920, Micromeritics (Norcross, PA, USA) chemisorption apparatus. Before experiments, 50 mg of catalyst powder was pretreated in pure N_2 at 200 °C for 60 min. After cooling to room temperature, a 10% H_2/Ar mixture was introduced to purge the sample. When the baseline was stable, the temperature was programmed to 900 °C with a heating rate of 10 °C/min; meanwhile, the H_2 signal was analyzed with a TCD detector. Temperature-programmed desorption of ammonia (NH_3-TPD) was used to investigate the surface acidities of the catalysts by a ChemBET Pulsar TPR/TPD, Quantachrome company (Boynton Beach, FL, USA). First, 100 mg of catalyst powder was pretreated at 200 °C for 1 h in a helium atmosphere. After that, when it was cooled to 30 °C, 2% NH_3/He gas was switched on for purging for 1 h, and then it was purged with helium gas for 1 h. After the baseline was stable, the desorbed NH_3 signal was detected by a thermal conductivity detector (TCD) under a heating rate of 10 °C/min. The total gas flow was 20 mL/min in each step. The temperature-programmed desorption of sulfur dioxide (SO_2-TPD) was performed on the same apparatus as for NH_3-TPD. An amount of 50 mg of catalyst powder was pretreated at 200 °C for 1 h in an Ar atmosphere. After that, 1000 ppm SO_2 + 16 vol% O_2 was switched on for 1 h at 250 °C. When it was cooled to 30 °C, Ar gas was used to purge for 1 h. After the baseline was stable, the desorbed SO_2 signal was detected under a heating rate of 10 °C/min. The total gas flow was 50 mL/min in each step. Transmission electron microscopy (TEM) images were recorded over a JEM 2100, JEOL (Tokyo, Japan) microscope and operated at an acceleration voltage of 200 kV and an electric current of 20 mA. Scanning electron microscopy (SEM) images were collected with a JEOL JSM-35C (Tokyo, Japan) instrument and operated at 20 kV acceleration voltages. X-ray photoelectron spectra (XPS) were carried out on an X-ray photoelectron spectrometer (Thermo Scientific K-Alpha, Waltham, MA, USA), using monochromatic Al Kα radiation (1486.6 eV). Inductively coupled plasma–atomic emission spectrometry (ICP–AES) was used to accurately determine the accumulation of sulfur on the catalysts by ICP-AES: Aglient 7800 (Palo Alto, CA, USA). Certain amounts of Pt/CeTi, Pt/Keg-CeTi, and Pt/MoP-CeTi were treated with SO_2 (1000 ppm) + O_2 (16%) at 250 °C for 1 h. After cooling, one part of the samples was used to measure the content of sulfur, and another part of the samples was treated at 900 °C for 1 h before the sulfur measurement. The adsorption and transformation of sulfur species and water on the catalysts under different conditions were investigated by in situ DRIFTS experiments using a FTIR spectrometer (Nicolet 6700, Madison WI, USA) with a diffuse reflectance chamber and a KBr window. The high-sensitivity mercury-cadmium-telluride (MCT) detector was cooled by liquid nitrogen. The sample (about 120 mg) was pretreated in an N_2 flow (50 mL/min) at 300 °C for 1 h. All the IR spectra were recorded in 32 accumulative scans with a resolution of 4 cm^{-1} in the range of 4000–400 cm^{-1}. The background spectra were collected at corresponding temperatures after pretreatment. For SO_2-DRIFTS, 150 ppm SO_2 and 16 vol% O_2 were introduced, and the balance gas was N_2. For H_2O-DRIFTS, 3% H_2O was introduced and balanced with N_2 too. The total flow rate was 50 mL/min.

3.4. Catalytic Activity Test

CO oxidation activity was measured in a fixed-bed quartz tube reactor (10 mm internal diameter) containing 1 mL of catalyst (40–60 mesh). The FTIR data of the three catalysts at 250 °C in the presence of CO and oxygen revealed that CO_2 was the only product of CO oxidation, and there were few or no carbonate species on the catalysts' surface (Figure S4). In order to demonstrate the reproducibility of the preparation method, three parallel samples were prepared and tested (Figure S5). The activity was detected from 80 °C to 300 °C under a heating rate of 10 °C/min. The typical composition of reactant gas was as follows: [CO] = 1%, [O_2] = 6%, [SO_2] = 100 ppm (when used), [H_2O] = 10% (when used), and He as balance. The total flow rate was 667 mL/min, which corresponded to an hourly space velocity (GHSV) of approximately 4×10^5 h^{-1}. The SO_2/H_2O durability experiment was evaluated at 200 or 250 °C under similar conditions. The reaction was carried out under atmospheric pressure, and the CO conversion was calculated as follows:

$$X = \frac{[CO]_{in} - [CO]_{out}}{[CO]_{in}} \times 100\%$$

4. Conclusions

In this work, three catalysts, Pt/CeTi, Pt/Keg-CeTi, and Pt/MoP-CeTi, were prepared by similar impregnation methods and used for CO oxidation. It was found that the Pt/Keg-CeTi catalyst showed a higher resistance to SO_2 and H_2O compared to Pt/CeTi and Pt/MoP-CeTi, which could be associated with its stronger surface acidity, better reduction of surface cerium and molybdenum species, and much lower SO_2 adsorption and transformation than the two others due to the modification of molybdophosphate with the Keggin structure. However, the Pt/MoP-CeTi catalyst displayed a much lower resistance to SO_2 and H_2O, which might be attributed to the low stability of molybdophosphate without the Keggin structure as a result of the formation of hydrogen phosphate and dihydrogen phosphate in the presence of H_2O under the reaction temperature, as well as the low surface Pt concentrations and moderate SO_2 adsorption and transformation. This work may offer a simple strategy to improve the catalyst performance for CO oxidation.

Supplementary Materials: The following are available online at https://www.mdpi.com/article/10.3390/catal12010004/s1, Figure S1: Nitrogen adsorption/desorption isotherms of the Pt/CeTi, Pt/Keg-CeTi, and Pt/MoP-CeTi catalysts, Figure S2: Results of XPS of O 1s, Ti 2p, and Mo 3d in the Pt/CeTi, Pt/Keg-CeTi, and Pt/MoP-CeTi catalysts, Figure S3: Results of XPS of O 1s, Ti 2p and Mo 3d in the Pt/CeTi, Pt/Keg-CeTi and Pt/MoP-CeTi catalysts, Figure S4: Changes of FTIR spectra of Pt/CeTi, Pt/Keg-CeTi and Pt/MoP-CeTi with time under the following conditions: [CO] = 2%, [O_2] = 10 vol %, balance N_2, total flow rate = 50mL/min, T=250 °C, Figure S5: CO conversions as a function of reaction temperature over the three parallel samples of Pt/CeTi (a), Pt/Keg-CeTi (b) and Pt/MoP-CeTi (c), respectively. Reaction conditions: [CO]= 1%, [O_2] = 6 vol %, balance He, total flow rate = 667 ml/min, GHSV = 4×10^5 h^{-1}, Table S1: ICP results of Pt/CeTi, Pt/Keg-CeTi, and Pt/MoP-CeTi after being poisoned by SO_2 and the poisoned samples treated at 900 °C, Table S2: XPS binding energies (eV) of Pt/CeTi, Pt/Keg-CeTi, and Pt/MoP-CeTi, Table S3: Relative area of NH_3-TPD desorption peak of Pt/CeTi, Pt/Keg-CeTi, and Pt/MoP-CeTi, Table S4: H_2 consumption of Pt/CeTi, Pt/Keg-CeTi, and Pt/MoP-CeTi calculated from the H_2-TPR curves.

Author Contributions: Conceptualization, T.Z., H.H. and W.Q.; methodology, T.Z. and W.Q.; validation, T.Z. and H.Z.; formal analysis, T.Z., H.Z. and R.W.; investigation, T.Z., H.Z. and W.Q.; resources, W.Q.; data curation, T.Z. and W.Q.; writing—original draft preparation, T.Z.; writing—review and editing, W.Q.; visualization, T.Z. and W.Q.; supervision, W.Q. and H.H.; project administration, W.Q.; funding acquisition, W.Q. All authors have read and agreed to the published version of the manuscript.

Funding: This research was funded by the National Natural Science Foundation of China (21577005; 22075005).

Data Availability Statement: Not applicable.

Conflicts of Interest: The authors declare no conflict of interest.

References

1. Kumar, G.M.; Sampath, S.; Jeena, V.S.; Anjali, R. Carbon monoxide pollution levels at environmentally different sites. *J. Ind. Geophys. Union.* **2008**, *12*, 31–40.
2. DeCoste, J.B.; Peterson, G.W. Metal-organic frameworks for air purification of toxic chemicals. *Chem. Rev.* **2014**, *114*, 5695–5727. [CrossRef]
3. Wang, W.; Gong, J.L. Methanation of carbon dioxide: An overview. *Front. Chem. Sci. Eng.* **2011**, *5*, 2–10.
4. Park, E.D.; Lee, D.; Lee, H.C. Recent progress in selective CO removal in a H_2-rich stream. *Catal. Today* **2009**, *139*, 280–290. [CrossRef]
5. Patel, D.M.; Kodgire, P.; Dwivedi, A.H. Low temperature oxidation of carbon monoxide for heat recuperation: A green approach for energy production and a catalytic review. *J. Clean Prod.* **2020**, *245*, 118838. [CrossRef]

6. Soubaihi, R.M.A.; Saoud, K.M.; Dutta, J. Critical review of low-temperature CO oxidation and hysteresis phenomenon on heterogeneous catalysts. *Catalysts* **2018**, *8*, 660. [CrossRef]
7. Dobrosz-Gomez, I.; Gomez-Garcia, M.-A.; Rynkowski, J.M. The origin of $Au/Ce_{1-x}Zr_xO_2$ catalyst's active sites in low-temperature CO oxidation. *Catalysts* **2020**, *10*, 1312. [CrossRef]
8. Prasad, R.; Singh, P. A review on CO oxidation over copper chromite catalyst. *Catal. Rev. Sci. Eng.* **2012**, *54*, 224–279. [CrossRef]
9. Lin, J.; Wang, X.D.; Zhang, T. Recent progress in CO oxidation over Pt-group-metal catalysts at low temperatures. *Chin. J. Catal.* **2016**, *37*, 1805–1813. [CrossRef]
10. Dey, S.; Dhal, G.C. Property and structure of various platinum catalysts for low temperature carbon monoxide oxidations. *Mater. Today Chem.* **2020**, *16*, 100228. [CrossRef]
11. Beniya, A.; Higashi, S.; Ohba, N.; Jinnouchi, R.; Hirata, H.; Watanabe, Y. CO oxidation activity of non-reducible oxide-supported mass-selected few-atom Pt single-clusters. *Nat. Commun.* **2020**, *11*, 1888. [CrossRef]
12. Haruta, M.; Kobayashi, T.; Sano, H.; Yamada, N. Novel gold catalysts for the oxidation of carbon monoxide at a temperature far below 0 °C. *Chem. Lett.* **1987**, *16*, 405–408. [CrossRef]
13. Haruta, M.; Yamada, N.; Kobayahsi, T.; Iijima, S. Gold catalysts prepared by coprecipitation for low-temperature oxidation of hydrogen and of Carbon Monoxide. *J. Catal.* **1989**, *115*, 301–309. [CrossRef]
14. Haruta, M. Gold rush. *Nature* **2005**, *437*, 1098–1099. [CrossRef]
15. Liu, G.; Walsh, A.G.; Zhang, P. Synergism of iron and platinum species for low-temperature CO oxidation: From two-dimensional surface to nanoparticle and single-atom catalysts. *J. Phys. Chem. Lett.* **2020**, *11*, 2219–2229. [CrossRef]
16. Lamb, A.B.; Bray, W.C.; Frazer, J. The removal of carbon monoxide from air. *Ind. Eng. Chem.* **1920**, *12*, 13–221. [CrossRef]
17. Dey, S.; Dhal, G.C. Deactivation and regeneration of hopcalite catalyst for carbon monoxide oxidation: A review. *Mater. Today Chem.* **2019**, *14*, 100180. [CrossRef]
18. Pillai, U.R.; Deevi, S. Room temperature oxidation of carbon monoxide over copper oxide catalyst. *Appl. Catal. B Environ.* **2006**, *64*, 146–151. [CrossRef]
19. Abdul-Kareem, H.K.; Hudgins, R.R.; Silveston, P.L. Forced cycling of the catalytic oxidation of CO over a V_2O_5 catalyst II Temperature cycling. *Chem. Eng. Sci.* **1980**, *35*, 2085–2088. [CrossRef]
20. Dey, S.; Dhal, G.C. Cerium catalysts applications in carbon monoxide oxidations. *Mat. Sci. Energy Technol.* **2020**, *3*, 6–24. [CrossRef]
21. Xie, X.W.; Li, Y.; Liu, Z.Q.; Haruta, M.; Shen, W.J. Low-temperature oxidation of CO catalysed by Co_3O_4 nanorods. *Nature* **2009**, *458*, 746–749. [CrossRef] [PubMed]
22. Langmuir, I. The mechanism of the catalytic action of platinum in the reactions $2CO + O_2 = 2CO_2$ and $2H_2 + O_2 = 2H_2O$. *Trans. Faraday Soc.* **1922**, *17*, 621–654. [CrossRef]
23. Kummer, J. Use of noble metals in automobile exhaust catalysts. *J. Phys. Chem.* **1986**, *90*, 4747–4752. [CrossRef]
24. Gandh, H.S.; Graham, G.W.; McCabe, R.W. Automotive exhaust catalysis. *J. Catal.* **2003**, *216*, 433–442. [CrossRef]
25. Bartholomew, C.H.; Farrauto, R.J. *Fundamentals of Industrial Catalytic Processes*, 2nd ed.; John Wiley & Sons, Inc.: Hoboken, NJ, USA, 2005.
26. Ramalingam, S.; Rajendran, S.; Ganesan, P. Performance improvement and exhaust emissions reduction in biodiesel operated diesel engine through the use of operating parameters and catalytic converter: A review. *Renew. Sustain. Energy Rev.* **2018**, *81*, 3215–3222. [CrossRef]
27. Shin, H.; Baek, M.; Kim, D.H. Sulfur resistance of Ca-substituted $LaCoO_3$ catalysts in CO oxidation. *Mol. Catal.* **2019**, *468*, 148–153. [CrossRef]
28. Yan, D.; Li, Q.; Zhang, H.; Zhou, X.; Chen, H. A highly dispersed mesoporous zeolite@TiO_2-supported Pt for enhanced sulfur-resistance catalytic CO oxidation. *Catal. Commun.* **2020**, *142*, 106042. [CrossRef]
29. Shin, H.; Baek, M.; Ro, Y.; Song, C.; Lee, K.-Y.; Song, I.K. Improvement of sulfur resistance of Pd/Ce-Zr-Al-O catalysts for CO oxidation. *Appl. Surf. Sci.* **2018**, *429*, 102–107. [CrossRef]
30. Jiang, Q.; Huang, M.; Qian, Y.; Miao, Y.; Ao, Z. Excellent sulfur and water resistance for CO oxidation on Pt single-atom-catalyst supported by defective graphene: The effect of vacancy type. *Appl. Surf. Sci.* **2021**, *566*, 150624. [CrossRef]
31. Hu, J.; Burns, R.C. Homogeneous-phase catalytic H_2O_2 oxidation of isobutyraldehyde using Keggin, Dawson and transition metal-substituted lacunary heteropolyanions. *J. Mol. Catal. A Chem.* **2002**, *184*, 451–464. [CrossRef]
32. Golodov, V.A.; Jumakaeva, B.S. Catalytic oxidation of CO by heteropolyacids (HPA) and dioxygen in the presence of Pd(II) salt-HPA-H_2O system. *J. Mol. Catal.* **1986**, *35*, 309–315. [CrossRef]
33. Zhizhina, E.G.; Kuznetsova, L.I.; Maksimovskaya, R.I.; Pavlova, S.N.; Maweev, K.I. Oxidation of CO to CO_2 by heteropolyacids in the presence of Palladium. *J. Mol. Catal.* **1986**, *38*, 345–353. [CrossRef]
34. Zhizhina, E.G.; Matveev, K.I. Low-temperature oxidation of CO to CO_2 in solutions of halide complexes of Pt and heteropolyacids (HPA). *React. Kinet. Catal. Lett.* **1992**, *47*, 255–262. [CrossRef]
35. Stanis, R.J.; Kuo, M.-C.; Turner, J.A.; Herring, A.M. Use of W, Mo, and V substituted heteropolyacids for CO mitigation in PEMFCs. *J. Electrochem. Soc.* **2008**, *155*, B155–B162. [CrossRef]
36. Yoshida, T.; Murayama, T.; Sakaguchi, N.; Okumura, M.; Ishida, T.; Haruta, M. Carbon monoxide oxidation by polyoxometalate-supported gold nanoparticulate catalysts: Activity, stability, and temperature dependent activation properties. *Angew. Chem. Int. Ed.* **2018**, *130*, 1539–1543. [CrossRef]

37. Wu, R.; Zhang, N.Q.; Liu, X.J.; Li, L.C.; Song, L.Y.; Qiu, W.G.; He, H. The Keggin structure: An important factor in governing NH_3-SCR activity over the V_2O_5-MoO_3/TiO_2 catalyst. *Catal. Lett.* **2018**, *148*, 1228–1235. [CrossRef]
38. Simsek, E.; Ozkara, S.; Aksoylu, A.E.; Onsan, Z.I. Preferential CO oxidation over activated carbon supported catalysts in H_2-rich gas streams containing CO_2 and H_2O. *Appl. Catal. A* **2007**, *316*, 169–174. [CrossRef]
39. Trimm, D.L. Minimisation of carbon monoxide in a hydrogen stream for fuel cell application. *Appl. Catal. A* **2005**, *296*, 1–11. [CrossRef]
40. Gao, S.; Wang, P.L.; Yu, F.X.; Wang, H.Q.; Wu, Z.B. Dual resistance to alkali metals and SO_2: Vanadium and cerium supported on sulfated zirconia as an efficient catalyst for NH_3-SCR. *Catal. Sci. Technol.* **2016**, *6*, 8148–8157. [CrossRef]
41. Smirnov, M.Y.; Kalinkin, A.V.; Pashis, A.V.; Sorokin, A.M.; Noskov, A.S.; Bukhtiyarov, V.I.; Kharas, K.C.; Rodkin, M.A. Comparative XPS study of Al_2O_3 and CeO_2 sulfation in reactions with SO_2, $SO_2 + O_2$, $SO_2 + H_2O$, and $SO_2 + O_2 + H_2O$. *Kinet. Catal.* **2003**, *44*, 575–583. [CrossRef]
42. Luo, T.; Gorte, R. Characterization of SO_2-poisoned ceria-zirconia mixed oxides. *Appl. Catal. B* **2004**, *53*, 77–85. [CrossRef]
43. Waqif, M.; Bazin, P.; Saur, O.; Lavalley, J.C.; Blanchard, G.; Touret, O. Study of ceria sulfation. *Appl. Catal. B* **1997**, *11*, 193–205. [CrossRef]
44. Kwon, D.W.; Park, K.H.; Hong, S.C. Enhancement of SCR activity and SO_2 resistance on VO_x/TiO_2 catalyst by addition of molybdenum. *Chem. Eng. J.* **2016**, *284*, 315–324. [CrossRef]
45. Tumuluri, U.; Li, M.J.; Cook, B.G.; Sumpter, B.; Dai, S.; Wu, Z.L. Surface structure dependence of SO_2 interaction with ceria nanocrystals with well-defined surface facets. *J. Phys. Chem. C* **2015**, *119*, 28895–28905. [CrossRef]
46. Goodman, A.L.; Bernard, E.T.; Grassian, V.H. Spectroscopic study of nitric acid and water adsorption on oxide particles: Enhanced nitric acid uptake kinetics in the presence of adsorbed water. *J. Phys. Chem. A* **2001**, *105*, 6443–6457. [CrossRef]
47. Tan, W.; Wang, J.M.; Yu, S.H.; Liu, A.N.; Li, L.L.; Guo, K.; Luo, Y.D.; Xie, S.H.; Gao, F.; Liu, F.D.; et al. Morphology-Sensitive sulfation effect on ceria catalysts for NH_3-SCR. *Top. Catal.* **2020**, *63*, 932–943. [CrossRef]
48. Bazin, P.; Sauf, O.; Lavalley, J.C.; Blanchard, G.; Visciglio, V.; Touret, O. Influence of platinum on ceria sulfation. *Appl. Catal. B* **1997**, *13*, 265–274. [CrossRef]
49. Xu, W.Q.; He, H.; Yu, Y.B. Deactivation of a Ce/TiO_2 Catalyst by SO_2 in the Selective Catalytic Reduction of NO by NH_3. *J. Phys. Chem. C* **2009**, *113*, 4426–4432. [CrossRef]
50. Pei, W.B.; Liu, Y.X.; Deng, J.G.; Zhang, K.F.; Hou, Z.Q.; Zhao, X.T.; Dai, H.X. Partially embedding Pt nanoparticles in the skeleton of 3DOM Mn_2O_3: An effective strategy for enhancing catalytic stability in toluene combustion. *Appl. Catal. B* **2019**, *256*, 117814. [CrossRef]
51. Tsunekawa, S.; Fukuda, T.; Kasuya, A. X-ray photoelectron spectroscopy of monodisperse CeO_{2-x} nanoparticles. *Surf. Sci.* **2000**, *457*, L437–L440. [CrossRef]
52. Liu, C.; Chen, L.; Li, J.; Ma, L.; Arandiyan, H.; Du, Y.; Xu, J.; Hao, J. Enhancement of activity and sulfur resistance of CeO_2 supported on TiO_2-SiO_2 for the selective catalytic reduction of NO by NH_3. *Environ. Sci. Technol.* **2012**, *46*, 6182–6189. [CrossRef] [PubMed]
53. Sellick, D.R.; Aranda, A.; Garcia, T.; Lopez, J.M.; Solsona, B.; Mastral, A.M.; Morgan, D.J.; Carley, A.F.; Taylor, S.H. Influence of the preparation method on the activity of ceria zirconia mixed oxides for naphthalene total oxidation. *Appl. Catal. B* **2013**, *132–133*, 98–106. [CrossRef]
54. Chmielarz, L.; Dziembaj, R.; Grzybek, T.; Klinik, J.; Lojewski, T.; Olszewska, D.; Wegrzyn, A. Pillared smectite modified with carbon and manganese as catalyst for SCR of NO_x with NH_3. *Catal. Lett.* **2000**, *70*, 51–56. [CrossRef]
55. Wu, J.; Jin, S.; Wei, X.; Gu, F.; Han, Q.; Lan, Y.; Qian, C.; Li, J.; Wang, X.; Zhang, R.; et al. Enhanced sulfur resistance of $H_3PW_{12}O_{40}$-modifed Fe_2O_3 catalyst for NH_3-SCR: Synergistic effect of surface acidity and oxidation ability. *Chem. Eng. J.* **2021**, *412*, 128712. [CrossRef]
56. Lee, J.; Ryou, Y.S.; Chan, X.J.; Kim, T.J.; Kim, D.H. How Pt interacts with CeO_2 under the reducing and oxidizing environments at elevated temperature: The origin of improved thermal stability of Pt/CeO_2 compared to CeO_2. *J. Phys. Chem. C* **2016**, *120*, 25870–25879. [CrossRef]
57. Damyanova, S.; Perez, C.A.; Schmal, M.; Bueno, J.M.C. Characterization of ceria-coated alumina carrier. *Appl. Catal. A* **2002**, *234*, 271–282. [CrossRef]
58. Cheng, K.; Song, W.Y.; Cheng, Y.; Liu, J.; Zhao, Z.; Wei, Y.C. Selective catalytic reduction over size-tunable rutile TiO_2 nanorod microsphere-supported CeO_2 catalysts. *Catal. Sci. Technol.* **2016**, *6*, 4478–4490. [CrossRef]
59. North, J.; Poole, O.; Alotaibi, A.; Bayahia, H.; Kozhevnikova, E.F.; Alsalme, A.; Siddiqui, M.R.H.; Kozhevnikov, I.V. Efficient hydrodesulfurization catalysts based on Keggin polyoxometalates. *Appl. Catal. A* **2015**, *508*, 16–24. [CrossRef]
60. Feng, C.L.; Liu, X.L.; Zhu, T.Y.; Hu, Y.T.; Tian, M.K. Catalytic oxidation of CO over Pt/TiO_2 with low Pt loading: The effect of H_2O and SO_2. *Appl. Catal. A* **2021**, *622*, 118218. [CrossRef]

Review

Modified Catalysts and Their Fractal Properties

Gianina Dobrescu [1,*], Florica Papa [1,*], Razvan State [1], Monica Raciulete [1], Daniela Berger [2], Ioan Balint [1] and Niculae I. Ionescu [1]

[1] Surface Chemistry and Catalysis Laboratory, "Ilie Murgulescu" Institute of Physical-Chemistry of the Romanian Academy, 202 Spl. Independentei, 060021 Bucharest, Romania; rstate@icf.ro (R.S.); mpavel@icf.ro (M.R.); ibalint@icf.ro (I.B.); ionescu@icf.ro (N.I.I.)

[2] Faculty of Applied Chemistry and Materials Science, University Politehnica of Bucharest, 1-7 Gheorghe Polizu St., 011061 Bucharest, Romania; daniela.berger@upb.ro

* Correspondence: gdobrescu@icf.ro (G.D.); frusu@icf.ro (F.P.)

Abstract: Obtaining high-area catalysts is in demand in heterogeneous catalysis as it influences the ratio between the number of active surface sites and the number of total surface sites of the catalysts. From this point of view, fractal theory seems to be a suitable instrument to characterize catalysts' surfaces. Moreover, catalysts with higher fractal dimensions will perform better in catalytic reactions. Modifying catalysts to increase their fractal dimension is a constant concern in heterogeneous catalysis. In this paper, scientific results related to oxide catalysts, such as lanthanum cobaltites and ferrites with perovskite structure, and nanoparticle catalysts (such as Pt, Rh, Pt-Cu, etc.) will be reviewed, emphasizing their fractal properties and the influence of their modification on both fractal and catalytic properties. Some of the methods used to compute the fractal dimension of the catalysts (micrograph fractal analysis and the adsorption isotherm method) and the computed fractal dimensions will be presented and discussed.

Keywords: fractal dimension; modified catalysts; fractal analysis; perovskite; nanoparticles

1. Introduction

The power of self-similarity as a fractal property was first emphasized in 1975 by B.B Mandelbort [1,2]. Following this finding, many processes and phenomena were analyzed as fractal behavior: light scattering on rough surfaces [3], fractal antennae [4], diffusion-limited aggregation [5], fractures [6], reaction kinetics [7], tumor diagnosis and cancer therapy [8,9] and, recently, mechanical responses of cell membranes [10].

In 1984, David Avnir, Dina Farin and Peter Pfeifer [11] reported that, at the molecular scale, the surfaces of most materials are fractal. This property leads to scaling laws of great interest in the description of various processes specific to heterogeneous chemistry: physical adsorption, chemisorption, and catalytic processes. Lately, a series of articles regarding the fractal analysis of surfaces of some catalysts and catalytic reactions have appeared in literature [12–26].

Tailoring catalysts with high activity in specific reactions is a challenging field of interest. Strategies implying the influence of particle size on catalytic properties [27] or metal-support interaction [28] or morphological controlling, metal deposition and chemical treatment [29] are largely seen in the literature. In the following, we shall focus only on the influence of fractal behavior self-similarity on catalytic properties.

Briefly, from a geometric point of view, catalytic reactions are favored by the existence of a large number of active centers arranged on the irregular surfaces of the catalysts, surfaces that have large specific surface areas (BET). Therefore, the surface of a catalyst can't be described as a flat surface, but rather as a sum of convoluted flat surfaces. Thus, fractal geometry can describe the surface of a catalyst better than classic, Euclidean geometry. Fractal geometry deals with the description of certain properties and characteristics of catalysts as scale sizes, not as their sums for small entities, keeping the same mathematical properties at different scales.

Not every random structure is fractal [1,2]: it is necessary to verify the existence of the self-similarity property on a sufficiently large scaling area to be able to conclude that the object itself is fractal.

The characterization of a fractal object is related to the measurement of two properties: the fractal dimension and the scaling domain. The fractal dimension, which is often a fractional number between 2 and 3 for surfaces, and, respectively, between 1 and 2 for powders, measures the degree of space occupancy, irregularity and roughness, and becomes close to three for surfaces that tend to "fill" the entire volume and two for surfaces that tend towards the plane, while the domain of self-similarity is the scaling area where the fractal properties are manifested [1,2]. The greater the field of self-similarity, the closer the fractal gets to an ideal mathematical fractal.

This work's focus is our results regarding the fractal characterization of oxide catalysts with perovskite structure (cobaltite and ferrite), as well as of some supported and unsupported metallic nanoparticles (NP) of Pd, Pt, Rh and bimetallic Pd-Cu, Pd-Ag and Pt-Cu. The fractal characteristics of these catalysts will be correlated with their catalytic properties. Fractal analysis will be performed both by SEM or TEM image analysis of micrographs and by the nitrogen adsorption isotherms method.

Therefore, based on the analyzed catalytic materials, the present paper aims to highlight the fractal character of various investigated catalysts and the means in which the fractal characteristics are tailored based on the modified catalysts. The impact on the catalytic properties will also be emphasized.

2. Results

2.1. Influences of Synthesis Parameters on the Fractal Dimension

2.1.1. Precursor Type Influences the Fractal Dimension of Perovskite

Obtaining catalysts with high fractal dimension, hoping that higher fractal dimensions will lead to higher catalytic activities, is a challenging objective.

To achieve this purpose, systematic studies must be performed to analyze the relation between synthesis parameters—in every case—and catalyst fractal dimensions. From this point of view, it is a large field of research and a vast domain.

Oxides with perovskite structure of $LaCoO_3$ and $LaFeO_3$ were obtained in different preparation conditions by thermal decomposition of the precursors with maleic acid, alpha-alanine, urea and sorbitol [30,31].

Analyzing the nitrogen adsorption isotherms for lanthanum cobaltites ($LaCoO_3$) [30] and lanthanum ferrites ($LaFeO_3$) [31] obtained from different precursors, only Dubinin–Radushkevitch isotherm can be used to fit the experimental data.

The computed fractal dimension of oxide cobaltites are presented in Table 1. Results show the fractal behavior of the analyzed samples related to the BET surface; as expected, a higher fractal dimension corresponds to a higher BET surface.

Table 1. Fractal dimensions of perovskite oxides $LaCoO_3$ obtained by the direct fit of the DR isotherm; fractal dimension obtained by the Avnir–Jaroniec method; BET specific surface area (m^2/g).

Precursor	Fractal Dimension DR Isotherm	Fractal Dimension AJ Method	BET Surface (m^2/g)
Maleic acid	2.3±0.06	2.34 ± 0.06 Fractional filling 0.40–0.80	20.42
Alpha-alanine	2.62±0.06	2.62 ± 0.06 Fractional filling 0.68–0.88	32.50
Urea	2.43±0.03	2.43 ± 0.03 Fractional filling 0.45–0.82	22.62

In the case of Lanthanum ferrites, the calculation of the fractal dimension by direct fitting of the isotherm and by using the Avnir–Jaroniec method leads to low and medium values for the fractal dimension (Table 2). This indicates that J(x), the pore distribution, is not very steep, so there are more and more wide pores in the microporosity regime when the fractal dimension decreases. In this case, multilayer filling of the micropores requires higher pressure values compared to samples with larger fractal dimensions. This behavior is described by type II isotherms. Ferrites obtained using alpha-alanine or sorbitol have similar values of the fractal dimension, indicating a similar microstructure [31].

Table 2. Fractal dimensions of lanthanum ferrites (LaFeO$_3$) obtained under various preparation conditions. Reprinted from ref. [31], Copyright (2003), with permission from Elsevier.

Precursor	Fractal Dimension DR Isotherm Fitting	Fractal Dimension Avnir–Jaroniec Method
Maleic acid	2.11 ± 0.04	2.07 ± 0.01 0.4–0.6 (fractional filling range)
Alpha-alanine	2.42 ± 0.02	2.44 ± 0.05 0.46–0.93 (fractional filling range)
Urea	2.40 ± 0.02	2.49 ± 0.01 0.77–0.86 (fractional filling range)

Both Tables 1 and 2 show good concordance between the values of the fractal dimension obtained by the two methods (i.e., direct fitting by fractal adsorption isotherms and the Avnir–Jaroniec Method) and, also, that the fractal dimension value is strongly influenced by the preparation method. Larger fractal dimensions are obtained when alanine or sorbitol precursors are used.

2.1.2. The Dopant (Sr) Influences the Fractal Dimension. The Limits of Self-Similarity

Strontium-doped lanthanum cobaltites with concentrations of 0.1–0.3 have a strong effect on the fractal dimension. Our results indicate that the fractal dimension of the doped samples is larger than the fractal dimension of the pure sample (Tables 3 and 4). Good agreement between the fractal dimension computed by SEM micrographs analysis and by fitting the adsorption isotherms was found (Table 3).

Table 3. Fractal Dimension dependence on Sr concentration x for doped lanthanum cobaltites (La$_{1-x}$Sr$_x$CoO$_3$); samples were obtained by thermal decomposition of the complex precursor molar ratio La:Sr:Co:acid maleic = 1 − x:x:1:8.6. Reprinted from ref. [32], Copyright (2003), with permission from JOAM.

x	SEM Analysis	Self-Similarity Limits (nm) (SEM Analysis)	Fractal Dimension (DR Adsorption Isotherm)
0	2.32 ± 0.01	250–1110	2.34 ± 0.06
0.1	2.51 ± 0.02	100–440	2.58 ± 0.02
0.2	2.43 ± 0.01	30–330	2.48 ± 0.01

For a more accurate study, Wojsz and Terzyk [33] showed that the detailed isotherm can be fitted with a general Dubinin–Astakhov type isotherm [16], computing the limits of the micropore range as the minimum and the upper limit of the pores size at which fractal behavior is emphasized.

The same conclusions were taken from the analysis of the La$_{1-x}$Sr$_x$CoO$_3$ (x = 0–0.3) samples obtained by the thermal decomposition of alpha-alanine precursors. The results were presented in detail [34] and summarized in Table 4.

Table 4. Fractal dimension obtained by direct fitting of the nitrogen adsorption data with DR adsorption isotherms. Reprinted from ref. [34], Copyright (2018), with permission from RRC.

Sample	Fractal Dimension	Determination Coefficient
$LaCoO_3$	2.39 ± 0.03	0.987
$La_{0.9}Sr_{0.1}CoO_3$	2.45 ± 0.01	0.989
$La_{0.8}Sr_{0.2}CoO_3$	2.48 ± 0.01	0.999
$La_{0.7}Sr_{0.3}CoO_3$	2.62 ± 0.03	0.970

Extending the research on other oxide catalysts with perovskite structure such as $LaMnO_3$, both pure and doped with Sr, a different behavior was emphasized. Adding Sr into the Mn perovskite structure seems to have an inverse influence—it decreases the fractal dimension. Although the effect is not very strong, it is an unexpected observation (Table 5).

Table 5. Fractal dimensions of $La_{1-x}Sr_xMnO_3$ samples with perovskite structure. The fractal dimension was obtained by analysis of SEM micrographs and by direct fitting of the adsorption isotherms using DR fractal isotherm.

Samples	Method	Fractal Dimension	Determination Coefficient	Self-Similarity Limit (nm)
$LaMnO_3$-alanine	SEM—correlation function method	2.49 ± 0.01	0.999	100–282
		2.70 ± 0.01	0.983	282–2326
	SEM—variable scale method	2.53 ± 0.02	0.998	1000–4000
		2.74 ± 0.01	0.992	4000–12,000
	DR fractal isotherm	2.54 ± 0.04	0.972	-
$La_{0.9}Sr_{0.1}MnO_3$	SEM—correlation function method	2.20 ± 0.01	0.994	20–116
		2.62 ± 0.01	0.984	116–820
	DR fractal isotherm	2.19 ± 0.02	0.975	-
$La_{0.8}Sr_{0.2}MnO_3$	SEM—correlation function method	2.18 ± 0.01	0.996	20–136
		2.43 ± 0.01	0.988	136–700
	SEM—variable scale method	-	-	-
	DR fractal isotherm	2.20 ± 0.02	0.989	-

It can be observed that regardless of the type of catalyst ($LaCoO_3$ or $LaMnO_3$), the introduction of Sr as a dopant will reduce the self-similarity limits of the samples. This means that irrespective of whether the fractal dimension increases or decreases, the fractal properties of the catalyst manifest themselves on a smaller scaling width. The destruction of the fractal character of perovskite materials with the addition of Sr as a dopant seems to be more accentuated in the case of the $LaMnO_3$ sample than in the case of $LaCoO_3$ sample.

Our results show that there are differences between the values of the fractal dimensions obtained by the analysis of SEM micrographs and those obtained from direct fitting of the adsorption isotherms. This behavior can be explained by the fact that TEM micrographs "expose" the surface as seen by the microscope, while the adsorption isotherm "describes" the surface as "seen" by adsorbed nitrogen molecules. In other words, there will be various hidden areas in the case of TEM images not "seen" by nitrogen molecules in the case of adsorption isotherms and, thus, not counted.

To compare the results regarding the catalytic activity of perovskites, the use of the fractal dimensions calculated from the fit of the adsorption isotherms or by the Avnir–Jaroniec method is indicated. These are methods that give a more accurate description of the specific surface, which is responsible for the existence of active centers.

2.2. The Dependence of the Fractal Dimension on Catalytic Activity

Our previous results indicate that both pure and doped $LaCoO_3$ samples [34] exhibit catalytic activity in the hydrogen peroxide decomposition reaction following the sequence $LaCoO_3 < La_{0.9}Sr_{0.1}CoO_3 < La_{0.8}Sr_{0.2}CoO_3 < La_{0.7}Sr_{0.3}CoO_3$. Therefore, as the fractal dimension increases, the catalytic activity also increases (Table 4). This correlation can be explained by the increase in catalytic activity due to the number of vacancies generated by partial substitution of La by the Sr ions. [35,36].

At the same time, there is a linear dependence of the logarithm of the pre-exponential factor lnA and the apparent activation energy for the hydrogen peroxide decomposition reaction, indicating a compensation effect (Figure 1). This effect can be explained by the existence of a non-uniform energetic surface and/or by the dependence of the active centers number on the presence of the Sr ions on the catalyst surface.

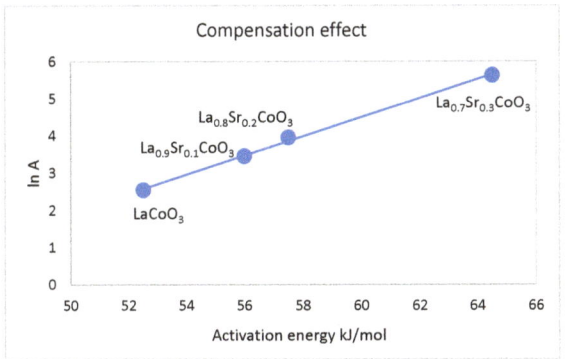

Figure 1. Compensation effect in the H_2O_2 decomposition reaction. Reprinted from ref. [34], Copyright (2018), with permission from RRC.

Trypolskyi and al. [37] showed that the activation energy in catalytic processes depends on the fractal surface dimension. Our results emphasize the same behavior as the reported paper (Figures 2 and 3).

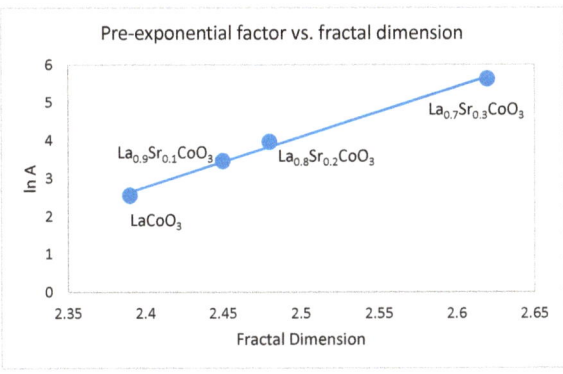

Figure 2. Pre-exponential factor vs. fractal dimension. Reprinted from ref. [34], Copyright (2018), with permission from RRC.

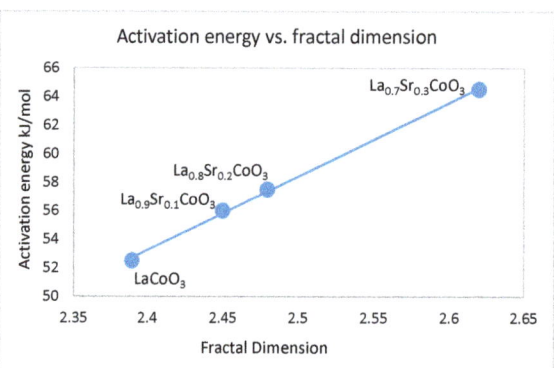

Figure 3. Activation energy versus fractal dimension. Reprinted from ref. [34], Copyright (2018), with permission from RRC.

Sr-doped $LaMnO_3$ catalyst with various concentrations (x = 0–0.2) were also investigated in the methane combustion reaction (Figure 4).

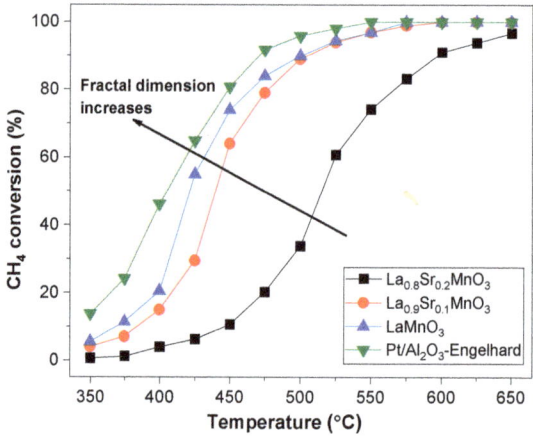

Figure 4. The catalytic activity for methane combustion.

It is observed that the conversion of methane vs. temperature in the total oxidation of methane depends on the used catalyst. Better results were obtained for the pure catalyst compared to those doped with strontium cation. For comparison, the results obtained using Pt/Al_2O_3 1 wt.% commercial catalyst (Engelhard) are also presented. The reactivity of the three samples decreases following the sequence $La_{0.8}Sr_{0.2}MnO_3 < La_{0.9}Sr_{0.1}MnO_3 < LaMnO_3$. This is in good agreement with the decrease of the computed fractal dimension according to the same sequence. Moreover, the activation energy of the methane oxidation reaction decreases with the decreases of fractal dimension (Figure 5).

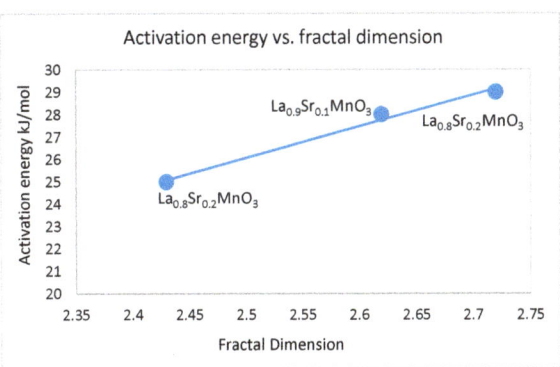

Figure 5. The dependence of the activation energy on the fractal dimension in the methane oxidation reaction when perovskite catalysts of the $La_{1-x}Sr_xMnO_3$ type are used.

2.3. The Fractal Structure of Mono- and Bi-Metallic Nanoparticles

Mono- and bi-metallic nanoparticles of Pd, Pd-Cu and Pd-Ag were obtained under different conditions, as is described in detail in [38]. Fractal behavior was investigated by TEM micrographs analysis. The D_0 fractal dimensions and the lacunarities [20] of the nanoparticles were computed using the "box counting" method and the modified black and white TEM images, considering nanoparticles as 2D black disks (Tables 6 and 7).

Table 6. The «box-counting» fractal dimension and the lacunarity of monometallic and alloy nanoparticles. Reprinted from ref. [38], Copyright (2018), with permission from Elsevier.

Sample	Fractal Dimension D_0	Lacunarity A_0
Pd	1.427 ± 0.457	8.39×10^4
Pd-Cu 4:1 Alloy	1.912 ± 0.014	8.24×10^5
Pd-Cu 1:1 Alloy	1.681 ± 0.015	2.57×10^5
Pd-Cu 1:4 Alloy	1.799 ± 0.035	2.36×10^5

Table 7. The "box-counting" fractal dimensions D_0, and the fractal dimensions $D(\lambda_m)$, $D(\lambda_1)$ and $D(\lambda_2)$ obtained by gray level analysis, where $D(\lambda_1)$ and $D(\lambda_2)$ are the fractal dimensions characteristic for the nucleation processes, the "core" fractal dimension and the "shell" fractal dimension. Meanwhile, $D(\lambda_m)$ is the fractal dimension of the whole structure. Reprinted from ref. [38], Copyright (2018), with permission from Elsevier.

Sample	D_0	λ_m	$D(\lambda_m)$	λ_1	$D(\lambda_1)$	λ_2	$D(\lambda_2)$
Pd-Cu Core-shell	1.841 ± 0.009	140	1.773 ± 0.048	170	1.632 ± 0.074	130	1.826 ± 0.037
Cu-Pd Inverse core-shell	1.855 ± 0.010	110	1.796 ± 0.026	130	1.827 ± 0.029	100	1.813 ± 0.010
Pd-Ag Core-shell	1.854 ± 0.018	130	1.819 ± 0.107	160	1.669 ± 0.061	120	1.863 ± 0.079
Ag-Pd Inverse Core-shell	1.875 ± 0.032	100	1.836 ± 0.023	110	1.776 ± 0.018	90	1.884 ± 0.021

The results presented in (Tables 6 and 7) indicate self-similarity and fractal properties for all the studied catalysts. It should be noted that the grey-level fractal analysis of TEM images leads to the identification of the structure of nanoparticles (alloy or core-shell type), obtaining the fractal dimension both for the core and for the shell [38].

Another bimetallic nanoparticle system characterized by fractal properties is Pt-Cu prepared in two synthesis variants: with low molar ratio of $PVP/Pt^{4+} = 5$ (Pt-Cu)$_L$, and high molar ratio (Pt-Cu)s, $PVP/Pt^{4+} = 10$, respectively, as described in the literature [39]. Fractal properties are summarized in (Table 8).

Table 8. Results of the fractal analysis performed on Pt-Cu nanoparticles. Reprinted from ref. [39], Copyright (2015), with permission from RSC.

Sample	Calculation Method	Fractal Dimension	Correlation Coefficient	Self-Similarity Domain/nm
(Pt-Cu)$_L$	Correlation function	2.50 ± 0.01	0.991	1.4–2.5
		2.70 ± 0.01	0.981	2.5–5.0
	Variable length scale	2.68 ± 0.01	0.984	5.0–27.5
(Pt-Cu)$_S$	Correlation function	2.40 ± 0.01	0.998	1.4–2.2
		2.73 ± 0.01	0.996	7.5–12.5
	Variable length scale	2.88 ± 0.01	0.992	12.5–22.5

As was described in detail in article [39], the PVP/Pt^{4+} low molar ratio leads to the formation of larger nanoparticles (between 2 nm and 5 nm), while a PVP/Pt^{4+} high molar ratio leads to nanoparticles of diameters between 1 nm and 2 nm. This observation can be seen in Table 8 from the analysis of self-similarity limits. The fractal behavior of (Pt-Cu)$_L$ nanoparticles indicates a mixture of small particles 1.4 nm–2.5 nm and, respectively, large 2.5 nm–5 nm ones, between which there are long-distance correlations characterized by a fractal size of 2.68. On the other hand, (Pt-Cu)$_S$ sample has an homogeneous structure composed of particles of 1.4 nm–2.2 nm, with medium and long-correlated fractal dimensions of 2.73 and 2.88, respectively.

One of the conclusions of the cited article is that unsupported (Pt-Cu)$_S$ nanoparticles have a much better catalytic performance than (Pt-Cu)$_L$ nanoparticles in terms of the catalytic reduction of NO_3^- ions. Beyond the explanations provided in detail by the cited article, it is observed that catalytic performance can be correlated with long-distance fractal dimension: the larger fractal dimension (Pt-Cu)$_S$ nanoparticles, D = 2.88, will favor catalytic activity compared to the lower fractal dimension (Pt-Cu)$_L$ nanoparticles, D = 2.68.

The catalytic applications of bimetallic nanoparticles are diverse: oxidative conversion of methane [40], oxidation of CO [41], reduction of nitrates [42], oxidation of methanol [43], etc. In most cases, bimetallic nanoparticles have shown better catalytic activity than their constituent metals.

We will further refer to alumina supported Pt-Cu nanoparticles studied in the context of total oxidation of methane. The catalytic behavior of the above mentioned nanoparticles was compared with an available catalyst Pt/Al$_2$O$_3$ 1 wt.% (Engelhard) with the same metal loading level (1wt. %) (Figure 6) [44].

Fractal analysis of TEM images (images for which the background was removed in order to improve image quality) shows a bimodal fractal behavior characterized by two fractal dimensions: one for low scales and another for large scales. The inflection point (contact between the two self-similarity domains) is located at 1.2 nm and it is an indication of the average radius of Pt-Cu particles. For wide scaling domains (greater than 4.5 nm) the structure does not show fractal behavior (Table 9).

Table 9. The fractal dimension of Pt-Cu/Al$_2$O$_3$ obtained using the correlation function method. Reprinted from ref. [44], Copyright (2011), with permission from Elsevier.

Fractal Dimension	Linear Correlation Coefficient	Self-Similarity Domain
2.39 ± 0.01	0.996	0.18–1.24
2.81 ± 0.01	0.975	1.24–4.50

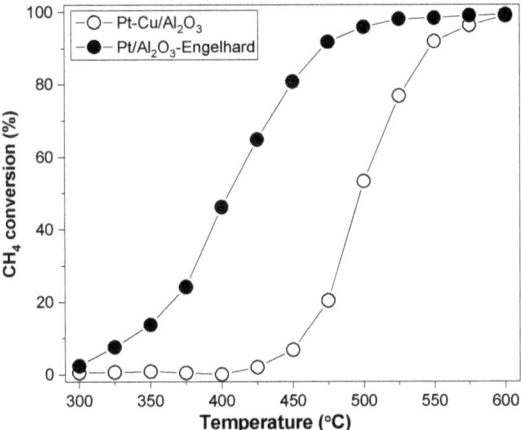

Figure 6. Methane conversion versus temperature using two different catalysts: Pt-Cu/Al$_2$O$_3$ and Pt/Al$_2$O$_3$. Reprinted from ref. [44], Copyright (2011), with permission from Elsevier.

Although, the fractal size of the Pt-Cu/Al$_2$O$_3$ bimetallic system is quite large on a wide scale (2.81) compared to a commercial Pt/Al$_2$O$_3$ catalyst, from the obtained experimental results we observed that there is a decrease in the catalytic activity; an explanation lies in the nanoparticles' surface chemical composition due to copper enrichment.

2.4. The Influence of the Fractal Dimension of Supported Nanoparticles on Surface Basicity

In order to deepen how fractal character influences the physico-chemical properties of the catalysts, we further analyze a series of Rh nanoparticles on various supports: Al$_2$O$_3$, TiO$_2$ and WO$_3$ [45].

The studied based-nanoparticle catalysts Rh/Al$_2$O$_3$, Rh/TiO$_2$, Rh/WO$_3$ showed fractal behavior (Tables 10 and 11), both before and after CO$_2$-TPD experiments, on a wide self-similarity domain. There is a strong correlation between the fractal dimension and the basicity of the studied catalysts.

Table 10. Fractal dimensions before CO$_2$-TPD from the analysis of TEM micrographs, using two methods: "C" meaning the correlation function method, and "S" the variable length scale method [45].

Sample	Fractal Dimension	Correlation Coefficient	Self-Similarity Domain (nm)
Rh/Al$_2$O$_3$	2.872 ± 0.001 (C) 2.784 ± 0.051 (S)	0.9910 0.8186	4.4–14.4 7.2–11.8
Al$_2$O$_3$	2.952 ± 0.001 (C) 2.962 ± 0.004 (S)	0.9375 0.9549	0.7–8.1 1.4–2.8
Rh/TiO$_2$	2.733 ± 0.001 (C) 2.832 ± 0.009 (S)	0.9884 0.9649	4.9–14.4 5.4–17.3
TiO$_2$	2.831 ± 0.001 (C) 2.911 ± 0.017 (S)	0.857 0.925	6.5–11 2.4–3.00
Rh/WO$_3$	2.490 ± 0.001 (C) 2.330 ± 0.001 (C) 2.469 ± 0.035 (S) 2.226 ± 0.047 (S)	0.9975 0.9991 0.9747 0.9334	0.2–2.7 2.7–13.8 4.5–11 11–29.3
WO$_3$	2.660 ± 0.004 (C) 2.293 ± 0.002 (C) 2.652 ± 0.042 (S)	0.990 0.986 0.949	0.2–1.2 1.8–5.0 4.1–5.9

Table 11. Fractal dimensions (before and after CO$_2$-TPD) obtained by direct fitting of adsorption data with DR adsorption isotherms [45].

Sample	Fractal Dimension	Correlation Coefficient	Self-Similarity Domain (p/p$_0$)	CO$_2$-TPD
Rh/Al$_2$O$_3$	2.607 ± 0.004	0.9986	0.033–0.850	Before
	2.534 ± 0.009	0.9973	0.011–0.350	After
	2.623 ± 0.002	0.9998	0.350–0.800	After
Rh/TiO$_2$	2.604 ± 0.003	0.9989	0.011–0.750	Before
	2.292 ± 0.011	0.9980	0.005–0.350	After
	2.548 ± 0.005	0.9987	0.350–0.850	After
Rh/WO$_3$	2.448 ± 0.012	0.9977	0.005–0.200	Before
	2.589 ± 0.014	0.9889	0.200–0.750	Before
	2.595 ± 0.008	0.9970	0.005–0.350	After
	2.003 ± 0.003	0.9993	0.350–0.875	After

Results showed (Tables 10 and 11) that adding Rh nanoparticles on the corresponding supports leads to a decrease in the fractal dimension of the Rh/support system, compared to the fractal dimension of the support itself. This decrease in fractal dimension can be explained by blocking the surface pores on the support and/or by encapsulating Rh via strong metal support-interaction (SMSI) [28]. The bimodal character of the WO$_3$ substrate is preserved even when Rh nanoparticles are added. The fractal dimensions obtained by fitting the adsorption isotherms with the DR isotherms (in the capillary condensation regime) are smaller than the fractal dimensions obtained by the image analysis of the TEM micrographs for Rh/Al$_2$O$_3$ and Rh/TiO$_2$ samples. This behavior can be explained by the fact that fractal dimension obtained from the analysis of the adsorption isotherm is an expression of the pore filling capacity on the surface of the adsorbate, while the fractal dimension obtained from the TEM image analysis measures correlations and similarities of all points (visible in TEM) on the studied surface. The last sample (Rh/WO$_3$) has the same bimodal characteristic, both in terms of the DR fractal dimension and in terms of the TEM fractal dimension.

Figures 7–10 present the number of basic centers dependence on the fractal dimension of the system/support.

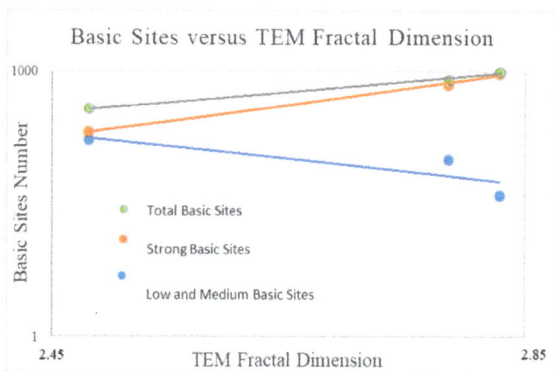

Figure 7. Basic sites: total basic sites, strong basic sites and low and medium basic sites dependencies on TEM fractal dimension of NP/support (double-logarithmic scale).

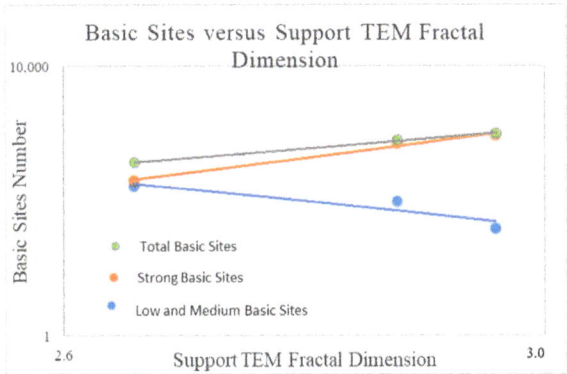

Figure 8. Basic sites: total basic sites, strong basic sites and low and medium basic sites dependencies on TEM fractal dimension of the support (double-logarithmic scale).

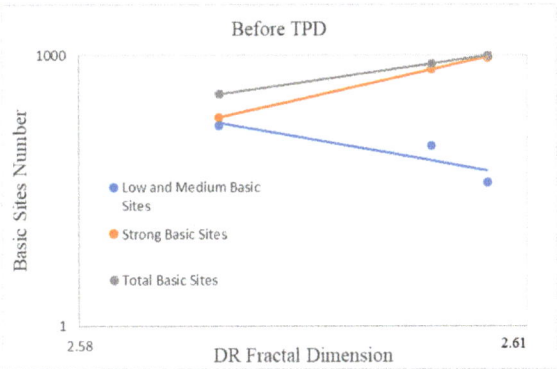

Figure 9. Basic sites versus DR Fractal Dimension before CO_2-TPD (double-logarithmic scale).

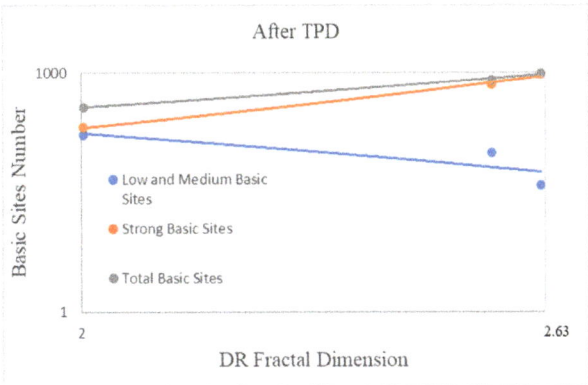

Figure 10. Basic sites vs. DR Fractal Dimension after CO_2-TPD measurements (double-logarithmic scale).

It is observed that the number of low and medium basic centers decreases slightly with the TEM fractal dimension of the supported nanoparticles, but also with the TEM fractal dimension of the support. On the other hand, the total number of strong basic centers increases with the same fractal dimensions. The same behavior is observed in the case of the fractal dimension obtained by analyzing the adsorption isotherm, both before and after CO_2-TPD. This behavior leads to the idea that large fractal dimensions favor strong basic centers, while small fractal dimensions favor weak basic centers.

A system with a large fractal dimension will favor the strong metal support-interaction (SMSI) leading to the formation of monodentate carbonate (Rh/Al_2O_3 and Rh/TiO_2) and implicitly of the strong basic centers. Weak and medium centers, usually attributed to the HO group, forming bicarbonate species with CO_2 or bidentate carbonate, are favored by surfaces with fewer defects, pores, fewer irregularities and lower fractal dimension.

In order to improve the DRM (Dry Reforming of Methane), catalytic activity of Rh/Al_2O_3, $Rh/TiO2$ and Rh/WO_3 and a large number of active centers are needed [46,47]. One way to achieve this is to use supports with a large fractal dimension; supports that will lead to NP systems/supports with a large fractal dimension and, therefore, to a large number of basic centers.

3. Discussion Regarding the Fractal Properties of Catalysts

The novelty element of this study is its correlation of the fractal behavior (fractal dimension) of catalysts with their catalytic activity as well as their specific chemical properties. We started from the general observation that a "perfect" catalyst should have an "ideal" fractal structure [12]. This idea involves structures of catalysts with very large specific surfaces and therefore many active centers capable of favoring chemical reactions. In reality, catalysts cannot be ideal mathematical fractals. No real structure can be described as an ideal mathematical fractal. It would require that it could be characterized by an infinite field of self-similarity and a deterministic self-similar structure. In the real world, systems are characterized by fractal properties only on restricted domains of self-similarity, which is always of necessary mention in the characterization of systems. Obviously, if we study a phenomenon on the order of nanometers, we are interested in self-similarity properties at this scale and not at scale lengths on the order of meters.

Self-similarity domains, especially the intersection areas of the domains with different fractal dimensions, give information regarding particle size, bi-modality of the sample, etc.

One of the methods used in determining fractal size is the direct fitting of the adsorption data with DR adsorption isotherms. The usual way to construct DR fractal isotherms is to consider the object as a fractal pore system. However, are real catalysts actually real fractal pore systems? In addition, if they are not, can we talk about their fractal dimension? Alternatively, can fractal adsorption isotherms, which are based on the idea of a pore fractal system [16], be used to determine the fractal dimension?

There are studies [48] showing that an optimal catalyst cannot be described as a fractal pore system realistically, though the fractal dimension could be an indication of the roughness of the analyzed samples.

Rudzinschi and. al. [49] show that fractal adsorption isotherms can be deduced without assuming the existence of a fractal pore system. The actual nature of the partially correlated solid surfaces is sufficient to deduce the fractal isotherms used. Surface energy heterogeneities in relation to fractal geometric non-uniformities lead to the deduction of the Dubinin–Radushkevitch isotherm without resorting to the questionable fractal distribution of pores [49]. In conclusion, the fractal dimension calculated by direct fitting of the adsorption isotherms with the DR equation reflects a real characteristic of self-similarity of the catalysts and is not a mathematical artefact.

4. Materials and Methods

4.1. Synthesis of Materials

Mixed perovskite-type oxides (cobaltite and ferrite) were prepared by calcination in air of isolated complex precursors (urea-base precursor, alpha-alanine-base precursor, sorbitol-base precursor, acid maleic-base precursor). Details on catalyst synthesis are presented in extenso in [30–32,34,50,51].

Bimetallic nanoparticle samples were prepared using the alkaline polyol method, an easy and versatile synthesis method described in detail in the literature [38,39,44,45,52].

4.2. Methods

4.2.1. Fractal Dimension Determination Using Image Analysis

The mathematical determination of the fractal dimension has its basis in the fundamental fractal property of self-similarity [2]. Self-similarity is the property of an object to appear the same when seen from near or far. The mathematical description of this property is given by the following formula:

$$N(r/R) \sim (r/R)^{-D} \tag{1}$$

where D is the fractal dimension, N(r/R) is the number of r size boxes that can cover an object of size R.

Starting from this definition and from its mathematical expression, various methods for determining both the fractal dimension and the scaling domain can be imagined, i.e., the values for r for which Equation (1) is valid.

SEM, TEM and AFM micrographs can be analyzed using various methods, such as the Fourier transform method [53], the box-counting method [2], the mass-radius dependence method [54], the correlation function method [53,55] and the variable scaling method [56].

4.2.2. Fractal Dimension Determination Using Adsorption Isotherms

The fractal size of the catalysts can be determined by adsorption experiments, either by direct fitting the adsorption isotherms or by the Avnir–Jaroniec method. Experimental adsorption isotherms have been shown to be efficiently fitted with the Dubinin–Radushkevitch isotherm [16]:

$$\theta = K[\ln(p_o/p)]^{D-3} \tag{2}$$

where θ is the relative adsorption, K is a characteristic constant, p_0 and p are saturation and equilibrium pressures and D is the fractal dimension.

Other isotherms (such as those presented below—Equations (3) and (4)) did not lead to viable results regarding the fractal dimension.

- the BET fractal isotherm obtained by Fripiat [57]:

$$\theta = N/N_m = \frac{c}{1+(c-1)x} \sum_{n=1}^{\infty} n^{2-D} x^n \tag{3}$$

- the Frenkel-Halsey-Hill [58] isotherm:

$$N/N_m = (z/a)^{3-D} = [\gamma/(-\ln x)]^{(3-D)/3}$$
$$\gamma \equiv \alpha/(kTa^3),\ 2 \leq D < 3, \tag{4}$$
$$x \equiv P/P_o, z = [\alpha/(kT\ln(P_o/P))]^{1/3}$$

where N_m, is the monolayer volume, a is the monolayer thickness and α is the difference of the interaction constants.

It is noteworthy that although the Frenkel–Halsey–Hill isotherm shape resembles the Dubinin–Radushkevitch fractal isotherm, the exponent is different for the two isotherms. The use of the Frenkel–Halsey–Hill isotherm can lead, in some cases, to values of fractal size of over 4, which is obviously unrealistic.

4.2.3. Avnir–Jaroniec Method

This method is described in detail in the literature [24]. It is based on the micropores volume calculation and, thus, the dependence of the fractional filling on pressure is computed, leading to the local determination of the Avnir–Jaroniec fractal dimension. The zone where the fractal dimension is constant gives the value of the fractal dimension, but also the domain of self-similarity.

5. Conclusions

The present paper depicted some catalysts that have been shown to have fractal properties, with a large domain of self-similarity.

The experimental results revealed that the synthesis method of catalysts influences the fractal dimension. Both the nature of the precursor as well as the introduction of dopants changes the fractal dimension for mixed oxide perovskite-type catalysts. Moreover, fractal analysis was used to obtain information regarding the morphology/geometry of the samples.

The fractal dimension is, as well, an indicator of the number of basic centers on the surface of the supported nanoparticles studied, basic centers that are directly related to the catalytic activity of these catalysts.

Moreover, very importantly, both catalytic activity in various reactions and the activation energy strongly depend on the fractal dimension of the catalysts. Catalytic activity is undoubtedly favored by catalysts with large fractal dimensions.

Author Contributions: Conceptualization, G.D. and F.P.; methodology, G.D.; investigation, D.B., F.P., R.S. and M.R.; data curation, I.B. and F.P.; writing—original draft preparation, G.D. and N.I.I.; writing—review and editing, G.D. and F.P.; supervision, I.B. and N.I.I.; funding acquisition, F.P. All authors have read and agreed to the published version of the manuscript.

Funding: This research was funded by Unitatea Executiva pentru Finantarea Invatamantului Superior, a Cercetarii, Dezvoltarii si Inovarii (UEFISCDI), grant number PN III—26PTE/2020 DENOX.

Conflicts of Interest: The authors declare no conflict of interest.

References

1. Mandelbrot, B.B. *Fractals: Form, Chance and Dimension*; Freeman, W.H., Ed.; MANDELBROT. WH Freeman and Co.: San Francisco, CA, USA, 1977.
2. Mandelbrot, B.B. *The Fractal Geometry of Nature*; Freeman, W.H., Ed.; MANDELBROT. WH Freeman and Co.: San Francisco, CA, USA, 1982.
3. Sorensen, C.M. Light scattering by fractal aggregates: A review. *Aerosol Sci. Technol.* **2001**, *35*, 648–687. [CrossRef]
4. Abed, A.T.; Abu-AlShaer, M.J.; Jawad, A.M. Fractal antennas for wireless communications. In *Modern Printed-Circuit Antennas*; Al-Rizzo, H., Ed.; IntechOpen: London, UK, 2020.
5. Witten, T.A.; Sander, L.M. Diffusion-limited aggregation. *Phys. Rev. B* **1983**, *27*, 5686–5697. [CrossRef]
6. Zhou, H.W.; Xie, H. Direct estimation of the fractal dimensions of a fracture surface of rock. *Surf. Rev. Lett.* **2003**, *10*, 751–762. [CrossRef]
7. Kopelman, R. Fractal Reaction Kinetics. *Science* **1988**, *241*, 1620–1626. [CrossRef] [PubMed]
8. Mattfeldt, H.W.; Gottfried, H.W.; Schmidt, V.; Kestler, H.A. Classification of spatial textures in benign and cancerous glandular tissues by stereology and stochastic geometry using artificial neural networks. *J. Microsc.* **1999**, *198*, 143–158.
9. Socoteanu, R.; Anastasescu, M.; Dobrescu, G.; Boscencu, R.; Vasiliu, G.; Constantin, C. AFM imaging, fractal analysis and in vitro cytotoxicity evaluation of Zn (II) vs. Cu (II) porphyrins. *Chaos Solit. Fractals* **2015**, *77*, 304–309. [CrossRef]
10. Hang, J.T.; Kang, Y.; Xu, G.-K.; Gao, H. A hierarchical cellular structural model to unravel the universal power-law rheological behavior of living cells. *Nat. Commun.* **2021**, *12*, 6067. [CrossRef] [PubMed]
11. Avnir, D.; Farin, D.; Pfeifer, P. Surface geometric irregularity of particulate materials: The fractal approach. *J. Colloid Interface Sci.* **1985**, *103*, 112–123. [CrossRef]
12. Rothschild, W.G. Fractals in heterogeneous catalysis. *Catal. Rev. Sci. Eng.* **1991**, *33*, 71–107. [CrossRef]
13. Pfeifer, P.; Obert, M.; Cole, M.W. Fractal BET and FHH theories of adsorption: A comparative study. *Proc. R. Soc. Lond. A* **1989**, *423*, 169–188.
14. Pfeifer, P.; Avnir, D.; Farin, D. *Complex Surface Geometry in Nano-Structure Solids: Fractal Versus Bernal-Type Models. Large Scale Molecular Systems—Quantum and Stochastic Aspects*; NATO ASI, Series B; Gans, W., Blumen, A., Amann, A., Eds.; Plenum: New York, NY, USA, 1991; pp. 215–229.

15. Meakin, P. Fractals and Reactions on Fractals. In *Reactions in Compartmentalized Liquids*; Knoche, W., Schomäcker, R., Eds.; Springer: Berlin/Heidelberg, Germany, 1989; pp. 173–198.
16. Avnir, D.; Jaroniec, M. An isotherm equation for adsorption on fractal surfaces of heterogeneous porous materials. *Langmuir* **1989**, *5*, 1431–1433. [CrossRef]
17. Pfeifer, P.; Kenntner, J.; Cole, M.W. Detecting capillary condensation in the absence of adsorption/desorption hysteresis. In *Fundamentals of Adsorption*; American Institute of Chemical Engineers: New York, NY, USA, 1991; pp. 689–700.
18. Ludlow, D.K.; Moberg, T.P. Technique for determination of surface fractal dimension using a dynamic flow adsorption instrument. *Instrum. Sci. Technol.* **1990**, *19*, 113–123. [CrossRef]
19. Farin, D.; Avnir, D.; Pfeifer, P. Fractal dimensions of surfaces. The use of adsorption data for the quantitative evaluation of geometric irregularity. *Particul. Sci. Technol.* **1984**, *2*, 27–35. [CrossRef]
20. Pfeifer, P.; Stella, A.L.; Toigo, F.; Cole, M.W. Scaling of the dynamic structure factor of an adsorbate on a fractal surface. *Europhys. Lett.* **1987**, *3*, 717–722. [CrossRef]
21. Cheng, E.; Cole, M.W.; Pfeifer, P. Defractalization of films adsorbed on fractal surfaces. *Phys. Rev. B* **1989**, *39*, 12962–12965. [CrossRef]
22. Avnir, D.; Farin, D. Fractal scaling laws in heterogeneous chemistry: Part I: Adsorptions, chemisorptions and interactions between adsorbates. *N. J. Chem.* **1990**, *14*, 197–206.
23. Pfeifer, P.; Johnston, G.P.; Deshpande, R.; Smith, D.M.; Hurd, A.J. Structure analysis of porous solids from preadsorbed films. *Langmuir* **1991**, *7*, 2833–2843. [CrossRef]
24. Kaneko, K.; Sato, M.; Suzuki, T.; Fujiwara, Y.; Nishikawa, K.; Jaroniec, M. Surface fractal dimension of microporous carbon fibres by nitrogen adsorption. *J. Chem. Soc. Faraday Trans.* **1991**, *87*, 179–184. [CrossRef]
25. Gutfraind, R.; Sheintuch, M.; Avnir, D. Fractal and multifractal analysis of the sensitivity of catalytic reactions to catalyst structure. *J. Chem. Phys.* **1991**, *95*, 6100–6111. [CrossRef]
26. Sanders, L.M.; Ghaisas, S.V. Fractals and patterns in catalysis. *Phys. A* **1996**, *233*, 629–639. [CrossRef]
27. Che, M.; Bennett, C.O. The influence of particle size on the catalytic properties of supported metals. *Adv. Catal.* **1989**, *36*, 55–172.
28. Matsubu, J.; Zhang, S.; DeRita, L.; Marinkovic, N.S.; Chen, J.G.; Graham, G.W.; Pan, X.; Christopher, P. Adsorbate-mediated strong metal–support interactions in oxide-supported Rh catalysts. *Nat. Chem.* **2017**, *9*, 120–127.
29. Huang, R.; Wu, J.; Zhang, M.; Liu, B.; Zheng, Z.; Luo, D. Strategies to enhance photocatalytic activity of graphite carbon nitride-based photocatalysts. *Mater. Des.* **2021**, *210*, 110040.
30. Dobrescu, G.; Berger, D.; Papa, F.; Fangli, I.; Rusu, M. Fractal dimensions of lanthanum cobaltites samples by adsorption isotherm method. In *Interdisciplinary Applications of Fractal and Chaos Theory*; Dobrescu, R., Vasilescu, C., Eds.; Editura Academiei Romane: Bucuresti, Romania, 2004; pp. 295–302.
31. Dobrescu, G.; Berger, D.; Papa, F.; Ionescu, N.I. Fractal dimension of lanthanum ferrite samples by adsorption isotherm method. *Appl. Surf. Sci.* **2003**, *220*, 154–158. [CrossRef]
32. Dobrescu, G.; Berger, D.; Papa, F.; Ionescu, N.I.; Rusu, M. Fractal analysis of micrographs and adsorption isotherms of $La_{1-x}Sr_xCoO_3$ samples. *J. Optoelectron. Adv.* **2003**, *5*, 1433–1437.
33. Wojsk, R.; Terzyk, A.P. The structural parameters of microporous solid, including fractal dimension, on the basis of the potential theory of adsorption-the general solution. *Comput. Chem.* **1997**, *21*, 83–87.
34. Papa, F.; Berger, D.; Dobrescu, G.; State, R.; Ionescu, N.I. Correlation of the Sr-dopant content in $La_{1-x}Sr_xCoO_3$ with catalytic activity for hydrogen peroxide decomposition. *Rev. Roum. Chim.* **2018**, *63*, 447–453.
35. Aoa, M.; Pharma, G.H.; Sage, V.; Pareeka, V. Structure and activity of strontium substituted $LaCoO_3$ perovskite catalysts forsyngas conversion. *J. Mol. Catal. A Chem.* **2016**, *416*, 96–104. [CrossRef]
36. Prasad, D.H.; Park, S.Y.; Oh, E.O.; Ji, H.; Kim, H.R.; Yoon, K.J.; Yoon, J.; Son, J.W.; Lee, J.H. Synthesis of nano-crystalline $La_{1-x}Sr_xCoO_{3-\delta}$ perovskite oxides by EDTA-citrate complexing process and its catalytic activity for soot oxidation. *Appl. Catal. A Gen.* **2012**, *447*, 100–106. [CrossRef]
37. Trypolskyi, A.I.; Gurnyk, T.M.; Strizhak, P.E. Fractal dimension of zirconia nanopowders and their activity in the CO oxidation. *Catal. Commun.* **2011**, *12*, 766–771. [CrossRef]
38. Dobrescu, G.; Papa, F.; State, R.; Balint, I. Characterization of bimetallic nanoparticles by fractal analysis. *Powder Technol.* **2018**, *338*, 905–914. [CrossRef]
39. Miyazaki, A.; Matsuda, K.; Papa, F.; Scurtu, M.; Negrila, C.; Dobrescu, G.; Balint, I. Impact of particle size and metal-support interaction on denitration behavior of wee-defined Pt-Cu nanoparticles. *Catal. Sci. Technol.* **2015**, *5*, 492–503. [CrossRef]
40. Lanza, R.; Canu, P.; Jaras, S.G. Partial oxidation of methane over Pt-Ru bimetallic catalyst for syngas production. *Appl. Catal. A Gen.* **2008**, *348*, 221–228. [CrossRef]
41. Liao, P.C.; Carberry, J.J.; Fleisch, T.H.; Wolf, E.E. CO oxidation activity and XPS studies of $PtCu_Y$-Al_2O_3 bimetallic catalysts. *J. Catal.* **1982**, *74*, 307–316. [CrossRef]
42. Soares, O.S.G.P.; Órfão, J.J.M.; Pereira, M.F.R. Bimetallic catalysts supported on activated carbon for the nitrate reduction in water: Optimization of catalysts composition. *Appl. Catal. B Environ.* **2009**, *91*, 441–448. [CrossRef]
43. Baglio, V.; Stassi, A.; Di Blasi, A.; D'Urso, C.; Antonucci, V.; Arico, A.S. Investigation of bimetallic Pt-M/C as DMFC cathode catalysts. *Electrochim. Acta* **2007**, *53*, 1360. [CrossRef]

44. Papa, F.; Negrila, C.; Dobrescu, G.; Miyazaki, A.; Balint, I. Preparation, characterization and catalytic behavior of Pt-Cu nanoparticles in methane combustion. *J. Nat. Gas Chem.* **2011**, *20*, 537–542. [CrossRef]
45. Dobrescu, G.; Papa, F.; Atkinson, I.; Culita, D.; Balint, I. Correlation between the basicity and the fractal dimension of Rh-nanoparticles supported on Al_2O_3, TiO_2 and WO_3. *IOSR-JAC* **2021**, *14*, 11–25.
46. Jang, W.; Shim, J.; Kim, H.; Yoo, S.; Roh, H. A review on dry reforming of methane in aspect of catalytic properties. *Catal. Today* **2019**, *324*, 15–26. [CrossRef]
47. Wang, H.Y.; Ruckenstein, E. Carbon dioxide reforming of methane to synthesis gas over supported rhodium catalysts: The effect of support. *Appl. Catal. A Gen.* **2000**, *204*, 143–152. [CrossRef]
48. Curtis Conner, W.; Bennett, C.O. Are the pore and surface morphologies of real fractal catalysts? *J. Chem. Soc. Faraday Trans.* **1993**, *89*, 4109–4114. [CrossRef]
49. Rudzinski, W.; Lee, S.L.; Sanders yan, C.C.; Panczyk, T. A fractal approach to adsorption on heterogeneous solid surfaces. 1. The relationship between geometric and energetic surface heterogeneities. *J. Phys. Chem. B* **2001**, *105*, 10847–10856. [CrossRef]
50. Berger, D.; Matei, C.; Papa, F.; Voicu, G.; Fruth, V. Pure and doped $LaCoO_3$ obtained by combustion method. *Prog. Solid State Chem.* **2007**, *35*, 183–191. [CrossRef]
51. Dobrescu, G.; Berger, D.; Papa, F.; Fangli, I.; Sitaru, I. Fractal analysis of micrographs and adsorption isotherms of $LaCoO_3$ and $LaFeO_3$ samples. *Ann. West Univ. Timis. Ser. Chem.* **2003**, *12*, 1505–1512.
52. Papa, F.; Negrila, C.; Miyazaki, A.; Balint, I. Morphology and chemical state of PVP-protected Pt, Pt-Cu and Pt-Ag nanoparticles prepared by alkaline polyol method. *J. Nanopart. Res.* **2011**, *13*, 5057–5064. [CrossRef]
53. Schepers, H.E.; van Beek, J.H.G.M.; Bassingthwaighte, J.B. Four methods to estimate the fractal dimension from self-affine signals (medical application). *IEEE Eng. Med. Biol.* **1992**, *11*, 57–64. [CrossRef]
54. Botet, R.; Jullien, R. A theory of aggregating systems of particles: The clustering of clusters process. *Ann. Phys. Fr.* **1988**, *13*, 153–221. [CrossRef]
55. Family, F.; Vicsek, T. Scaling of the active zone in the Eden process on percolation networks and the ballistic deposition model. *J. Phys. A Math. Gen.* **1985**, *18*, L75–L81. [CrossRef]
56. Chauvy, P.F.; Madore, C.; Landolt, D. Variable length scale analysis of surface topography: Characterization of titanium surfaces for biomedical applications. *Surf. Coat. Technol.* **1998**, *110*, 48–56. [CrossRef]
57. Fripiat, I.J.; Gatineau, L.; van Damme, H. Multilayer physical adsorption on fractal surfaces. *Langmuir* **1986**, *2*, 562–567. [CrossRef]
58. Pfeifer, P.; Wu, Y.J.; Cole, M.W.; Krim, J. Multilayer adsorption on a fractally rough surface. *Phys. Rev. Lett.* **1989**, *62*, 1997–2000. [CrossRef] [PubMed]

Article

Enhanced Catalytic Hydrogen Peroxide Production from Hydroxylamine Oxidation on Modified Activated Carbon Fibers: The Role of Surface Chemistry

Wei Song [1,2,*], Ran Zhao [1], Lin Yu [1], Xiaowei Xie [1], Ming Sun [1] and Yongfeng Li [1]

[1] School of Chemical Engineering and Light Industry, Guangdong University of Technology, Guangzhou Higher Education Mega Centre, Guangzhou 510006, China; zr066524@mail2.gdut.edu.cn (R.Z.); gych@gdut.edu.cn (L.Y.); xwxie@gdut.edu.cn (X.X.); sunmgz@gdut.edu.cn (M.S.); gdliyf@gdut.edu.cn (Y.L.)

[2] Analysis and Test Center, Guangdong University of Technology, Guangzhou Higher Education Mega Centre, Guangzhou 510006, China

* Correspondence: songw@gdut.edu.cn

Abstract: Herein, direct production of hydrogen peroxide (H_2O_2) through hydroxylamine (NH_2OH) oxidation by molecular oxygen was greatly enhanced over modified activated carbon fiber (ACF) catalysts. We revealed that the higher content of pyrrolic/pyridone nitrogen (N5) and carboxyl-anhydride oxygen could effectively promote the higher selectivity and yield of H_2O_2. By changing the volume ratio of the concentrated H_2SO_4 and HNO_3, the content of N5 and surface oxygen containing groups on ACF were selectively tuned. The ACF catalyst with the highest N5 content and abundant carboxyl-anhydride oxygen containing groups was demonstrated to have the highest activity toward catalytic H_2O_2 production, enabling the selectivity of H_2O_2 over 99.3% and the concentration of H_2O_2 reaching 123 mmol/L. The crucial effects of nitrogen species were expounded by the correlation of the selectivity of H_2O_2 with the content of N5 from X-ray photoelectron spectroscopy (XPS). The possible reaction pathway over ACF catalysts promoted by N5 was also shown.

Keywords: activated carbon fiber; surface modification; hydrogen peroxide; hydroxylamine; pyrrolic/pyridone nitrogen

1. Introduction

Hydrogen peroxide (H_2O_2), as a green chemical, attracts research attention in both energy and environmental related fields because it has the highest content of reactive oxygen among the common oxidants and the green by-product [1]. It has been widely used as a bleach in the paper and textile industry, an energy carrier in fuel cells, an oxidant in chemical production, wastewater treatment, hydrometallurgy and electronics industry [2]. Notwithstanding, the current industrial production of H_2O_2 is mainly through the anthraquinone oxidation process [3], which involves multistep reactions, massive energy consumption and waste generation. Furthermore, the cost and safety problems are also raised ineluctably by the handle, transport and storage of high concentration H_2O_2. Nevertheless, in many practical applications, H_2O_2 with only a low concentration could satisfy the demand in the reactions, such as selective oxidation, on-site degradation of dye, sewage treatment and disinfection (<30 mM) [4–6]. In this context, research on alternative production methods of H_2O_2 and it's in situ use has been the research focus [5,7–15]. The direct generation of H_2O_2 by the reaction of molecular hydrogen (H_2) and oxygen (O_2) is considered the most promising method [16–19], but the industrial application is obscured by the dangers of the explosive reaction mixture and the insufficiency of catalysts with high selectivity without considering the reaction systems of O_2 and H_2 at high pressure [20–25]. In recent years, both photo- and electro-catalytic H_2O_2 production techniques are in the process of research, but the former is still suffered from a low selectivity and yield of H_2O_2

with affordable raw materials while the latter faces the problems of low efficiency and complicated devices accompanied with high cost [2,26,27].

Therefore, in order to avoid the risk of explosion, it would be an effective way to choose an appropriate hydrogen source replacing H_2 in the direct synthesis process of H_2O_2. Thus, hydroxylamine (NH_2OH) was viewed as an available alternative to H_2 since it could be transformed into H_2O_2 by O_2 in the conditions of room temperature and normal pressure in an aqueous solution ($2NH_2OH + O_2 = N_2 + 2H_2O + H_2O_2$) [28]. This reaction is a simple step and easy to handle procedure with two major kinds of catalytic systems. Homogeneous manganese complexes catalysts with high TOF values were firstly studied in this reaction [29–31], but they suffered from separating and recycling problems. Afterward, noble metal particles (Au and Pd) dispersed on different supports were reported to catalyze this system effectively [32–34], but the low concentration of H_2O_2 (0.05–0.1 wt.%) generation and the high cost of noble metal remain as the major impediment for the industrial applications.

Based on the research above, activated carbon (AC) was found to be an effective catalyst used in the direct H_2O_2 production process through NH_2OH oxidation by O_2 in our earlier research [35,36]. We found that the catalytic properties of AC were closely related to the surface oxygen-containing groups. In order to fulfill wider and higher demands in practical application, the selectivity and activity towards H_2O_2 formation still need to be further improved. Moreover, the reactivity usually originated from the structure of carbon materials. Considering this, nitrogen doping on carbon materials has been found to be one of the most effective ways to improve the selectivity of H_2O_2 in metal-free catalytic systems, although the selectivity usually depends on both oxygen- and nitrogen-doped atoms [13]. It is noteworthy that N-doping played a beneficial role for H_2O_2 selectivity only at a low active surface site density while becoming detrimental at higher contents in some reactions [37,38]. Meanwhile, the microporous volume content on carbon materials was also found to have proportionality with selectivity in many catalytic systems [37,39].

Considering the above factors, the easily available polyacrylonitrile-based (PAN-based) activated carbon fiber (ACF) would be an ideal candidate catalyst material. On the one hand, compared to the AC material with a large number of mesopores, PAN-based ACF has only microporous structures and the pore channels are directly open on the surface. More importantly, PAN-based ACF has an intrinsic nitrogen content without any further nitrogen-doped steps and the relatively simple surface modification process would be cost-effective [40]. In the present work, for the sake of comparative study of the generation and modification on the surface doped species, the improvements on the content and type of surface functional groups were intended by treating the PAN-based ACF with concentrated mixed acid in different volume ratios. The selectivity and yield of H_2O_2 in the reaction on modified ACF catalysts were well interconnected with the amounts of pyrrolic/pyridone nitrogen (N5) and desorbed carboxyl-anhydride groups from the ACF surface. The possible reaction pathway over ACF catalysts promoted by N5 was also shown.

2. Results and Discussion

2.1. Material Structure

Figure 1 shows the scanning electron microscopy (SEM) images of ACF-0 sample and the corresponding elemental mapping images, along with the SEM images of ACF samples before and after surface modification by different volume ratios of concentrated H_2SO_4 and HNO_3 (H_2SO_4/HNO_3 (v/v)). From Figure 1a–d, three main elements (C, N, O) were found on the ACF-0 sample which exhibited a long fiber feature with a diameter of about 15 µm. The surface roughness of the ACF samples gradually increased with the increase of H_2SO_4/HNO_3 (v/v) from 0.5 to 4, as shown from Figure 1e–h. For the ACF-Rw sample, the surface still kept smooth similar to the ACF-0 sample, while slight surface roughness was observed on the surface of ACF-R1. By increasing the value of

H$_2$SO$_4$/HNO$_3$ (v/v) to 2 and 4, some auricular-like sheet protrusions were both found on the surface of ACF-R2 and ACF-R4 samples. Consequently, the surface modification caused differently morphological changes on the ACF samples through the erosion of carbon surface by different H$_2$SO$_4$/HNO$_3$ (v/v), and some microporous structures on the ACF samples might be destroyed through mixed acid treatment with higher content of concentrated H$_2$SO$_4$.

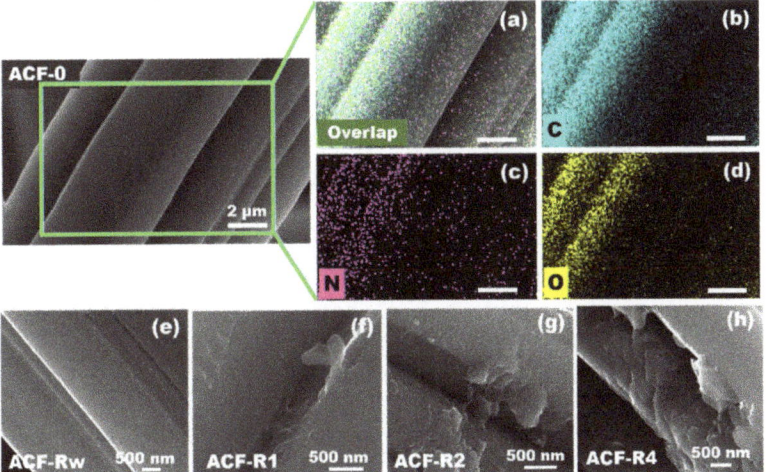

Figure 1. SEM images of the ACF-0 sample and the corresponding elemental mapping images for (**a**) overlap, (**b**) C, (**c**) N, (**d**) O. The SEM images of ACF samples after surface modification (**e**) ACF-Rw, (**f**) ACF-R1, (**g**) ACF-R2, (**h**) ACF-R4.

The pore size distribution of ACF-0 and ACF-R4 samples, and nitrogen adsorption-desorption isotherms of the ACF-0 sample without surface modification is shown in Figure 2. The detailed texture parameters for the ACF samples were shown in Table 1. Moreover, the ACF-0 sample displayed the features of type I isotherm, indicating the existence of micropores. There were two types of pores on the ACF-0 sample: micropore (1.41 nm and 1.13 nm) and supermicropore (0.78 nm and 0.57 nm) [41]. As for the ACF-R4 sample, the micropore content (1.14 nm) was greatly enhanced and some micropores enlarged to 1.69 nm while the supermicropore content was decreased compared with the ACF-0 sample. As displayed in Table 1, minor changes in the pore size and surface area were found between the ACF-0 and ACF-Rw samples but obviously decrease in micropore volume and surface area were observed on the ACF-R2 and ACF-R4 samples. As for the ACF-Rw sample, the surface area of micropores ($S_{mic.}$), the total surface area (S_{BET}) and the surface area of mesoporous ($S_{mes.}$) was 895, 934 and 39 m^2/g, respectively. Correspondingly, the micropore volume ($V_{mic.}$), the total pore volume (V_{total}) and the micropore width ($d_{pore.}$) of ACF-Rw were calculated to be 0.361, 0.395 cm^3/g and 0.85 nm. These results were similar to those of the ACF-0 sample. With the increase of H$_2$SO$_4$/HNO$_3$ (v/v), the values of V_{Total} and S_{BET} initially reduced to 0.323 cm^3/g and 686 m^2/g on the ACF-R1 sample, then declined to 0.229 cm^3/g and 481 m^2/g on the ACF-R2 sample. Moreover, the d_{pore} for ACF samples by mixed acid oxidation increased slightly from 0.87 to 0.95 nm, with increasing H$_2$SO$_4$/HNO$_3$ (v/v) from 0.5 to 4. Notably, the mesopores of the ACF samples were not altered much through the modification by mixed acid, whereas the micropores decreased remarkably by higher values of H$_2$SO$_4$/HNO$_3$ (v/v), especially in the ACF-R2 and ACF-R4 samples. This suggests that the higher content of H$_2$SO$_4$ caused severe destruction of the microporous structures while the higher content of HNO$_3$ preserved the textual characteristics of the ACF sample.

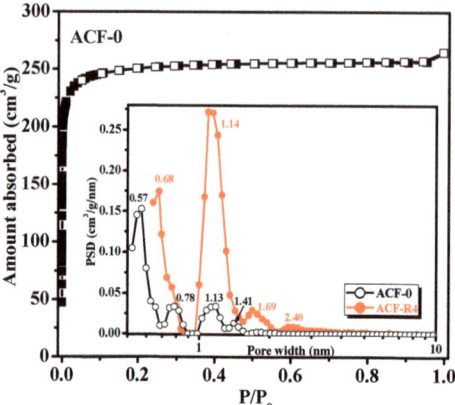

Figure 2. The pore size distribution (PSD) of ACF-0 and ACF-R4 samples obtained from the density functional theory method and N_2 adsorption-desorption isotherms of ACF-0 sample.

Table 1. Textural parameters of the ACF samples before and after surface modification.

Catalyst	S_{BET} [a] (m^2/g)	$S_{mic.}$ [b] (m^2/g)	$S_{mes.}$ [b] (m^2/g)	$V_{mic.}$ [b] (cm^3/g)	V_{Total} [c] (cm^3/g)	d_{pore} [d] (nm)
ACF-0	944	909	35	0.374	0.411	0.87
ACF-Rw	934	895	39	0.361	0.395	0.85
ACF-R1	686	601	85	0.261	0.323	0.94
ACF-R2	481	356	124	0.146	0.229	0.95
ACF-R4	368	322	46	0.132	0.160	0.95

[a] Multipoint Braunauer-Emmett-Teller (BET). [b] Calculated by the t-plot method. [c] Estimated from the amounts of gas adsorbed at a relative pressure of 0.994. [d] Average pore diameter calculated from $2V_{Total}/S_{BET}$ for slit pore.

2.2. Surface Properties

The Fourier transformation infrared (FTIR) spectra of ACF samples is shown in Figure 3a. According to the references, the peak at 1225 cm^{-1} is originated from the stretching mold of C–N and C–O in carboxylic anhydrides, ethers, lactones and phenols [42,43]. The peak at 1405 cm^{-1} and 1580 cm^{-1} is respectively related to the nitrogen groups and the double bond of C=C in quinoid structure. Meanwhile, the peak at 1730 cm^{-1} is owing to the stretching vibration of the C=O band in carboxyl and lactones groups attached to the aromatic rings, and the peak at 910 cm^{-1} is related to the anhydride groups [41,43,44]. After modification by different values of H_2SO_4/HNO_3 (v/v), the intensities of the above peaks were wholly enhanced to different extents, suggesting the formation of large quantities of oxygen-containing species on ACF surface. For ACF-R1, ACF-R2 and ACF-R4 samples, the intensities of the peaks at 910 cm^{-1}, 1225 cm^{-1}, 1580 cm^{-1} and 1730 cm^{-1} were all greatly increased, indicating the enlargement in phenols, quinones, lactones, carboxyls and anhydrides. The highest peak intensity was found on ACF-R2 and ACF-R4 samples, especially at the position of 1730 cm^{-1}, which confirms the further enrichment of anhydride and carboxylic groups by a higher content of H_2SO_4 in mixed acid.

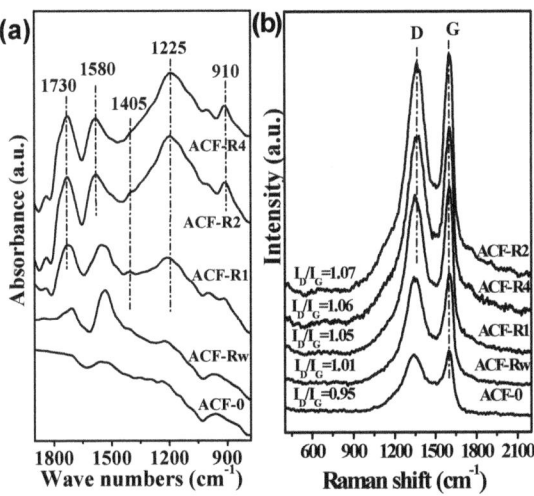

Figure 3. The (**a**) FTIR and (**b**) Raman spectra of ACF samples.

As shown in Figure 3b, Raman spectroscopy detection was conducted to investigate the defects on carbon structure of ACF samples with surface modification. Usually, carbon fiber mainly has two characteristic peaks, one of which is the D peak at the position of 1350–1375 cm^{-1}, and the other is the G peak at the position of 1580–1603 cm^{-1} [45]. The D peak is related to amorphous and defects of carbon structure while the G peak is related to graphite crystal structure. Generally, the calculation of I_D/I_G ratio from integral areas values of D and G peak was used to measure the structural defects of carbon materials [46]. It is widely known that I_D/I_G value increases with more structural defects generated on the carbon material. Obviously, the intensity of Raman spectra on ACF samples gradually increased, meanwhile the I_D/I_G values of all the ACF samples increased from 0.95 to 1.07 with increasing H_2SO_4/HNO_3 (v/v) from 0.5 to 2. Whereas the I_D/I_G values of ACF-R4 decreased to 1.06 with increasing H_2SO_4/HNO_3 (v/v) from 2 to 4. Therefore, it is believed that there were more surface defects and structural changes on the ACF carbon framework according to the surface modification with more H_2SO_4 contents in mixed acid. These results were well matched with the textural characteristics in ACF samples shown in Table 1.

The narrow scan of C 1s regions in X-ray photoelectron spectroscopy (XPS) of ACF samples is exhibited in Figure 4a. Moreover, the deconvolution results of the C 1s spectrum are given in Table 2. For modified carbon materials, the C 1s spectra usually involved graphitic carbon (C–graphite, Peak I), ether, alcohol or phenolic groups (C–O, Peak II), carbonyl or quinone groups (C=O, Peak III), carboxylic groups (–COO–, Peak IV) and Peak V for the satellite peak from the π–π* electron shake-up [47–49]. The intensities of peak I was decreased by the oxidation of mixed acid, whereas the intensities for the peaks attributed by C–O groups were increased [49]. However, the areas of peak II of all ACF samples by acid oxidation increased not so obviously compared to that of peak III or peak IV, which indicated phenolic groups may not be tailored much by adjusting different values of H_2SO_4/HNO_3 (v/v). Similarly, the integral area of peak IV for both ACF-Rw (4.5%) and ACF-R1 sample (5.4%) was more than two times larger than that of the ACF-0 (2.0%). Notably, the area of peak IV increased to 7.9% and 8.7% on the ACF-R2 and ACF-R4 samples respectively, clearly confirming the generation of a large amount of surface carboxylic groups by the higher content of H_2SO_4.

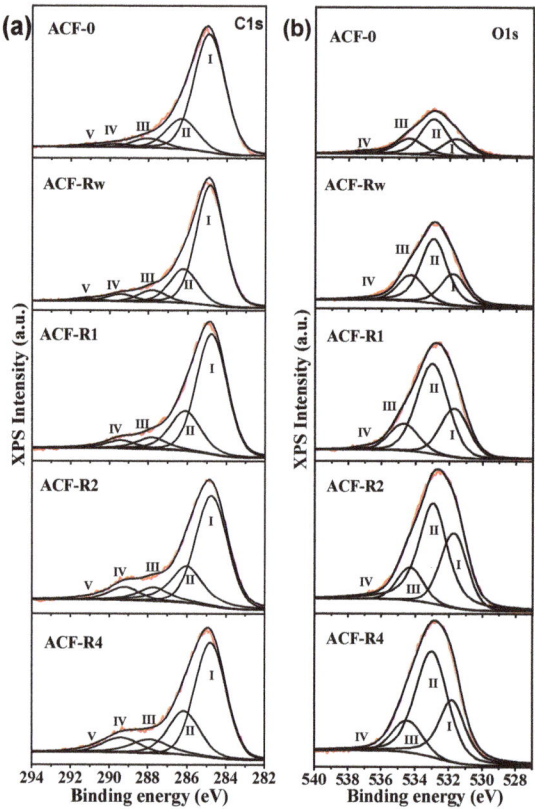

Figure 4. High resolution of XPS spectra in (**a**) C 1s (**b**) O 1s regions of ACF samples.

Table 2. Deconvolution results of the C 1s XPS spectra for the ACF samples, values given in % of total intensity.

Catalyst	Functional Groups/Binding Energy (eV)				
	Peak I C–graphite 284.7–284.8	Peak II C–O 286.0–286.3	Peak III C=O 287.7–288.1	Peak IV –COO– 289.2–289.6	Peak V $\pi-\pi^*$ 291.2
ACF-0	72.5	18.8	5.7	2.0	1.0
ACF-Rw	67.9	19.9	6.4	4.5	1.3
ACF-R1	63.4	23.4	7.8	5.4	—
ACF-R2	57.4	24.8	9.5	7.9	1.9
ACF-R4	56.5	23.7	10.3	8.7	0.8

Figure 4b exhibits the narrow scan of XPS spectra in O 1s regions of the ACF samples. Moreover, the deconvolution results of the O 1s spectrum are displayed in Table 3. As shown in Figure 4b, the O 1s XPS spectra can be deconvoluted into three main peaks, namely Peak I, Peak II and Peak III, which are associated with the C=O group, C–O group and adsorbed H_2O or O_2, respectively [42]. The adsorbed CO or CO_2 in the ACF surface can be attributed to the minor Peak IV, the binding energy of which was at 536.9–537.0 eV. Obviously, the intensities of both Peak III and Peak IV decreased by surface modification, whereas the peaks corresponding to C=O groups increased evidently. As for Peak I, the intensities increased from 25.5 to 30.5% by surface modification with increasing the content of H_2SO_4, and similar results were obtained on ACF-R2 and ACF-R4 samples.

Additionally, the atomic ratio of surface O/C in the ACF samples by acid oxidation was enhanced significantly from 21.3 to 32.9% with increasing the H_2SO_4/HNO_3 (v/v) from 0.5 to 4. The above results suggested that more carboxylic species were generated by mixed acid oxidation with higher content of H_2SO_4, being consistent with the results of FTIR measurement.

Table 3. Deconvolution results of the O 1s XPS spectra for the ACF samples, values given in % of total intensity.

Catalyst	Functional Groups/Binding Energy (eV)				O/C (%)
	Peak I C=O 531.6–531.8	Peak II C–O 532.9–533.0	Peak III H_2O_{ads}/ O_{2ads} 534.3–534.7	Peak IV CO_{2ads}/ CO_{ads} 536.9–537.0	
ACF-0	23.2	51.0	23.3	2.5	12.0
ACF-Rw	25.5	54.2	20.1	0.7	21.3
ACF-R1	27.6	56.3	15.3	0.9	24.9
ACF-R2	30.7	56.0	12.9	0.5	32.2
ACF-R4	30.5	55.8	13.5	0.2	32.9

For the sake of examining the crucial role of the intrinsic nitrogen doped in ACF samples, the deconvolution results of N 1s XPS profiles of ACF samples are exhibited in Figure 5. Moreover, the corresponding results of the deconvolution are displayed in detail in Table 4. According to the curve fitting results and references, five distinct types of nitrogen contained species were deconvoluted from the N 1s spectra: NX (-NO_2), N4 (pyridine-N oxide), NQ (quaternary N), N5 (pyrrolic/pyridone) and N6 (pyridine) [50–53]. It was evident that the content of N6 in ACF-0 was highest among all ACFs. Moreover, the content of N5 significantly increased to 33.2%, 43.4%, 50.5% and 46.3% for the ACF-Rw, ACF-R1, ACF-R2 and ACF-R4, respectively. As shown in Table 4, the content of N6 on the ACF-0 decreased from 15.6 to 5.2% corresponding to the ACF-R2. Moreover, the content of both NQ and N4 on the ACF sample decreased nearly one half by surface modification. It was reported that $-NO_2$ and pyridine were the main forms of the nitrogen introduced from HNO_3 oxidation and different forms of nitrogen can be transformed to each other [42,54]. The content of NX initially reached the maximum (32.3%) on ACF-Rw, then decreased to 22.3%, 15.3% and 19.0% on ACF-R1, ACF-R2 and ACF-R4, respectively. No content of NX can be observed on the ACF-0 sample without surface modification. Accordingly, when the nitrogen form predominated in the ACF sample was quaternary N, the mixed acid modification transformed them to $-NO_2$ with the higher content of HNO_3. Meanwhile, more pyrrolic nitrogen species were generated by a higher content of H_2SO_4. In addition, the atomic ratios of surface N/C on ACF samples were gradually enhanced from 1.5 to 2.4 with the increase of H_2SO_4/HNO_3 (v/v) from 0.5 to 2, whereas that on ACF-R4 sample decreased to 2.0 by increasing the value of H_2SO_4/HNO_3 (v/v) to 4. These results suggest that the surface N-containing groups could be effectively tuned by mixed acid oxidation with different volume ratios of concentrated H_2SO_4 and HNO_3.

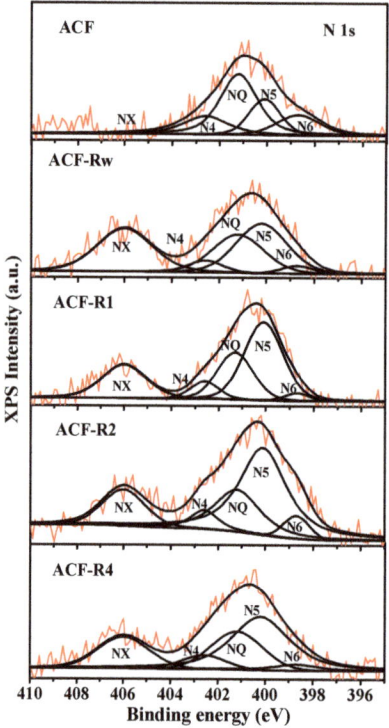

Figure 5. High resolution of XPS spectra in N 1s regions of ACF samples.

Table 4. Deconvolution results of the N 1s XPS spectra for the ACF samples, values given in % of total intensity.

Catalyst	Functional Groups/Binding Energy (eV)					N/C (%)
	N6 Pyridine 398.7	N5 Pyrrolic/Pyridone 400.1–400.2	NQ Quaternary N 401.2–401.3	N4 Pyridine-N-oxide 402.6	NX –NO_2 406.0	
ACF-0	15.6	20.2	46.8	17.5	—	1.4
ACF-Rw	3.2	33.2	26.0	5.4	32.3	1.5
ACF-R1	2.8	43.4	24.3	7.3	22.3	1.7
ACF-R2	5.2	50.5	22.8	6.2	15.3	2.4
ACF-R4	4.7	46.3	22.9	7.1	19.0	2.0

The temperature-programmed desorption (TPD) results of the ACF samples were shown in Figure 6. After being heated, carbon oxides were the main decomposition products of surface oxygen-containing functional groups [55–58]. As shown in Figure 7, the anhydrides and carboxylic acids usually decomposed into CO_2 at relatively lower temperatures while the lactones decomposed into CO_2 at higher temperatures. Meanwhile, the carboxylic anhydrides, ethers, phenols, carbonyl-quinones generally decomposed into CO [58]. Only little quantities of CO_x were obtained on the ACF-0 sample while significant quantities of CO_x were obtained on the other three ACF samples. For the ACF samples modified by mixed acid, the data of CO_x gradually rose with increasing the H_2SO_4/HNO_3 (v/v) from 0.5 to 4. Especially, the desorption quantity of CO from the ACF-R1 sample was almost five-fold larger than that of the ACF-0 sample, illustrating the formation of large quantities of phenol and carbonyl-quinone groups. On the flip side, the desorption amount of CO_2 from the ACF-R2 sample was almost more than 15 times greater than the

ACF-0 sample, primarily owing to the remarkable generation of lactones, anhydrides and carboxylic acids. The quantities of CO_x obtained from ACF-R4 were very similar to the ACF-R2 sample.

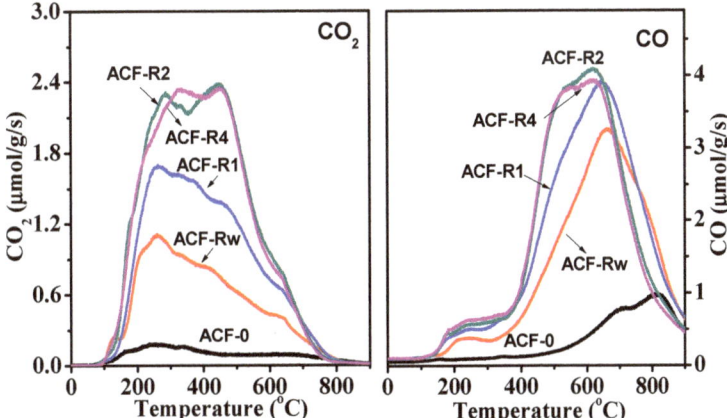

Figure 6. TPD profiles of ACF samples before and after surface modification.

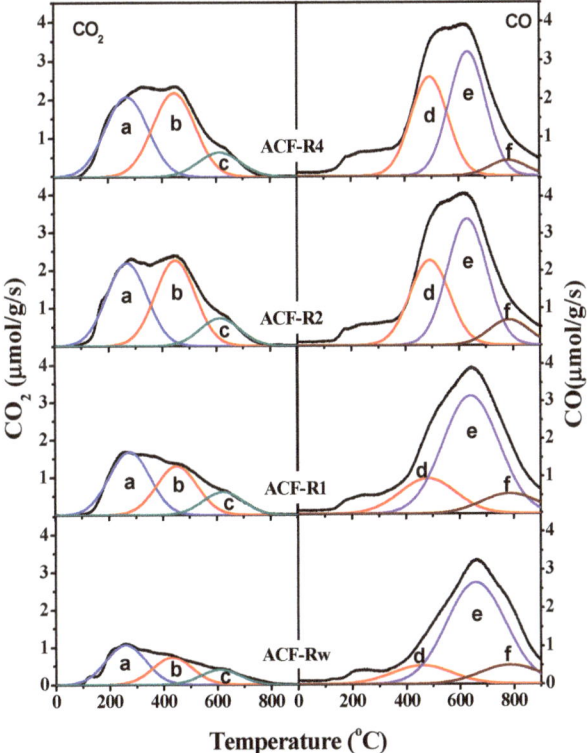

Figure 7. Deconvolution of the TPD profiles for ACF samples, in which peak a, peak b, peak c from the CO_2 desorption of the carboxyl, anhydride, lactone groups while peak d, peak e, peak f from the CO desorption of the anhydride, phenol, carbonyl groups on ACF samples.

Tables 5 and 6 show the detailed data of CO and CO_2 desorbed from specific surface groups on ACF samples. The desorption quantities of CO and CO_2 on ACF-0 sample were 244 μmol/g and 65 μmol/g, severally, and they were raised to 1008 μmol/g and 400 μmol/g on ACF-Rw sample. Upon increasing the H_2SO_4/HNO_3 (v/v) from 0.5 to 1, the desorption quantities of CO and CO_2 on the ACF-R1 sample remarkably raised to 1207 μmol/g and 678 μmol/g, respectively. Nevertheless, the amounts of CO desorbed from carbonyl-quinone groups on the ACF-R2 sample decreased to 125 μmol/g. While compared with ACF-Rw, more than two times larger amounts of CO_2 desorbed from carboxyl and anhydride groups were also found on the ACF-R2 sample. The desorption quantities of CO_2 and CO on the ACF-R4 sample were very similar to the ACF-R2 sample. Considering all these examinations, it could be deduced that the largest amounts of carboxyl (407 μmol/g) and anhydride (425 μmol/g) were obtained on the ACF-R2 and ACF-R4 samples while the most enrichment of phenol groups was detected on the ACF-R1 sample. This means that the moderate content of H_2SO_4 produced more phenol groups while the higher content of H_2SO_4 in the mixed acid created more carboxylic and anhydride groups. These results were consistent with the FTIR and XPS results of ACF samples.

Table 5. The desorption quantities of CO_2 from the ACF samples by the deconvolution of the TPD profiles.

Catalyst	CO_2 Desorption (μmol/g)			
	Carboxyl [a]	Anhydride [b]	Lactone [c]	Total
ACF-0	32	15	17	65
ACF-Rw	195	132	73	400
ACF-R1	317	245	116	678
ACF-R2	407	420	136	963
ACF-R4	406	425	122	953

Desorption temperatures: [a] 255–275 °C, [b] 430–451 °C, [c] 611–623 °C.

Table 6. The desorption quantities of CO from the ACF samples by the deconvolution of the TPD profiles.

Catalyst	CO Desorption (μmol/g)			
	Anhydride [d]	Phenol [e]	Carbonyl [f]	Total
ACF-0	15	83	146	244
ACF-Rw	132	739	137	1008
ACF-R1	245	823	139	1207
ACF-R2	420	627	125	1172
ACF-R4	425	551	67	1043

Desorption temperatures: [d] 458–491 °C, [e] 630–660 °C, [f] 785–812 °C.

2.3. H_2O_2 Production

Figure 8a shows the concentration of H_2O_2 production from NH_2OH oxidation by O_2 on the ACF catalysts. For the ACF-0 catalyst without surface oxidation, the cumulative concentration of H_2O_2 was very low and cannot be detected after reacting for 300 min. For the ACF-Rw catalyst, the concentration of H_2O_2 increased to 55.9 mmol/L at 420 min and then increased slightly. Similar trends plots were observed on the ACF-R1, ACF-R2 and ACF-R4 catalysts, on which the H_2O_2 concentration increased gradually with the increase of reaction time. When the reaction was conducted for 660 min, the concentration of H_2O_2 approached 88.6 mmol/L and 112 mmol/L on the ACF-R1 and ACF-R4 catalyst, respectively. With increasing the H_2SO_4/HNO_3 (v/v) from 1 to 2, the maximum concentration of H_2O_2 reached 123 mmol/L on the ACF-R2 catalyst, which clearly demonstrates more reactive species were generated on the ACF-R2 surface by an appropriately higher content of H_2SO_4 in mixed acid. In order to explore the stability of ACF catalysts, the recycling tests of ACF-R2 were performed as shown in Figure 8b. After three cycles, there was almost no decrease in the activity of the reused ACF-R2 catalyst with the yield of H_2O_2 about 49% (123 mmol/L) after reacting for 660 min.

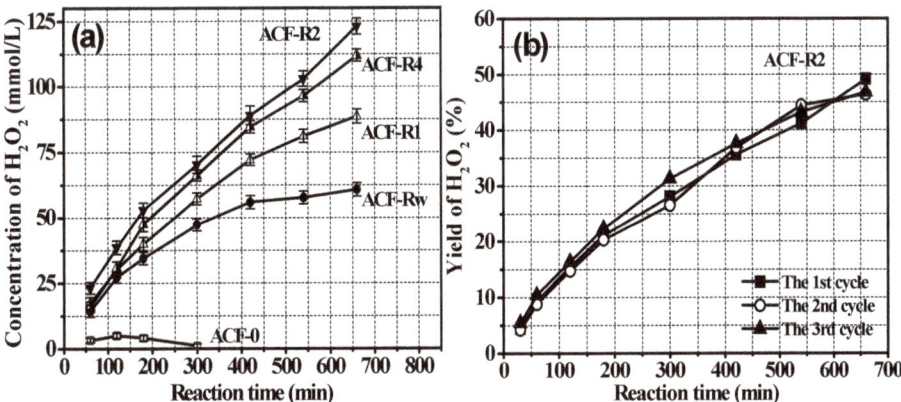

Figure 8. (a) The concentration of H_2O_2 formation on the ACF catalysts, (b) the yield of H_2O_2 for the recycling tests of the ACF-R2 catalyst (pH = 8.6, temp. = 25 °C, the error bar calculated by STDEV method).

The selectivity of H_2O_2 along with the NH_2OH conversion on the ACF catalysts at the reaction time of 180 min was shown in Figure 9a. The selectivity of H_2O_2 was only 46.0% on the ACF-Rw catalyst although the higher conversion of NH_2OH (30%) was observed on it, which was possibly induced by the higher surface area and more carbonyl-quinone groups generating through the surface modification. In view of similar conversion toward NH_2OH (~22%) consumption, the selectivity of H_2O_2 was 73.6% on the ACF-R1 catalyst prepared by an equal volume of H_2SO_4 and HNO_3. Whereas the selectivity of H_2O_2 was greatly enhanced to 99.3% on the ACF-R2 catalyst obtained by further increasing the content of H_2SO_4. However, the selectivity of H_2O_2 decreased to 87.6% on the ACF-R4 catalyst by the increase of the H_2SO_4/HNO_3 (v/v) from 2 to 4. Thus, the formation of more reactive nitrogen and oxygen containing groups on the ACF catalysts greatly enhanced the selectivity toward H_2O_2 formation.

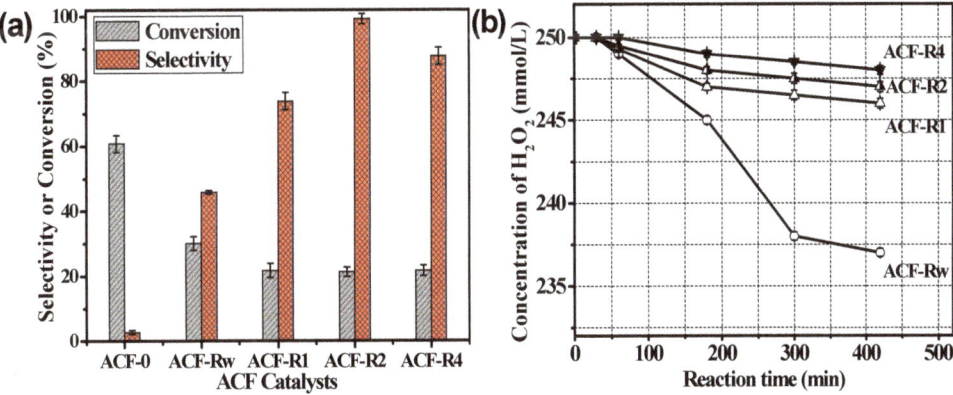

Figure 9. (a) The selectivity of H_2O_2 and conversion of NH_2OH on the ACF catalysts (pH = 8.6, temp. = 25 °C, time = 180 min.), (b) the decomposition of H_2O_2 over ACF catalysts (the error bar calculated by STDEV method).

The activity of H_2O_2 decomposition over ACF catalysts was shown in Figure 9b. According to the reference [59], the activity toward H_2O_2 decomposition was directly related to the basic sites (chromene groups) on the AC materials surface, while the formation of surface carboxylic groups (–COOH) will accordingly retard the catalytic decomposition of H_2O_2. It was also found that the acidic function groups of AC materials treated by HNO_3 would suppress the H_2O_2 decomposition rate. As shown in Figure 9b, almost no decomposition of H_2O_2 was detected on the ACF-R2 and ACF-R4 catalyst during the first 60 min. After reacting for 420 min, the concentration of H_2O_2 in the ACF-R1, ACF-R2 and ACF-R4 catalyst system only decreased to 246 mmol/L, 247 mmol/L and 248 mmol/L, respectively. As for the ACF-Rw catalyst, with the smallest amounts of carboxylic groups, the concentration of H_2O_2 quickly decreased to 245 mmol/L only within 180 min. Therefore, the modified ACF catalysts with large amounts of carboxylic groups by mixed acids retarded the catalytic decomposition of H_2O_2 and exhibited a higher activity of H_2O_2 generation.

The catalytic performance in the reaction of H_2O_2 production from NH_2OH oxidation over modified ACF catalysts was listed and compared to those of previously reported catalysts in Table 7. The modified ACF catalysts showed a higher formation concentration of H_2O_2 than the Au/MgO and Pd/Al_2O_3 system with a longer reaction time. The ACF-R2 and ACF-R4 catalysts exhibited similarly catalytic performance with the ACH system but with higher selectivity toward H_2O_2. As for the homogeneous Mn (II/III)-complex system, both the concentration and the yield of H_2O_2 were higher than all heterogeneous catalysts systems without considering their separating and recycling problems. Meanwhile, the concentration of H_2O_2 over ACF-R2 and ACF-R4 catalysts was higher than the most reactive carbon supported Au and Pd catalysts, which were used in the direct H_2O_2 production process from H_2 and O_2 at high pressure. Thus, compared with the Au-Pd/C catalyst, the ACF catalysts system had a longer reaction time (>9 h) while the supported Au and Pd catalysts system only took 0.5 h to obtain a similar concentration of H_2O_2. Considering the practical application, the reaction of the ACF catalysts system was easy to handle at atmospheric pressure whereas the high pressure was necessary for the supported Au and Pd catalysts system.

Table 7. Comparative performance in the production of H_2O_2 over ACF catalysts with reference catalysts (pH = 7.0–8.6, temp. = 2–27 °C).

Catalyst	[H_2O_2] (mmol/L)	Reaction Time (h)	Hydrogen Source	Conversion (%)	Selectivity (%)	Yield (%)	Reference
ACF-Rw	60.8	11.0	NH_2OH	72.9	33.3	24.3	This work
ACF-R1	88.5	11.0	NH_2OH	50.5	70.1	35.4	This work
ACF-R4	112	11.0	NH_2OH	47.4	94.5	44.8	This work
ACF-R2	123	11.0	NH_2OH	49.2	100	49.2	This work
ACF-R2 [a]	117	11.0	NH_2OH	56.0	83.6	46.8	This work
ACF-R2	53	3.0	NH_2OH	21.3	99.3	21.2	This work
ACH	114	7.0	NH_2OH	-	-	46.7	[35]
ACP	~50	3.0	NH_2OH	~23	~87	20.0	[36]
Mn^{2+}-Tiron	~225	6.0	NH_2OH	-	-	~90	[29]
Mn^{3+}-Complex [b]	~185	0.75	NH_2OH	-	-	~74	[30]
Au/MgO	32.6	1.0	NH_2OH	-	-	81.5	[32]
Pd/Al_2O_3	37.6	1.0	NH_2OH	-	-	94.0	[33]
Au-Pd/C [c]	69	0.5	H_2	-	-	-	[21]
AuPd/C [c]	62	0.5	H_2	-	-	-	[22]
Au-Pd Catalyst [c]	94	0.5	H_2	-	-	-	[23]
Au + Pd/C [c]	79	0.5	H_2	-	-	-	[24]
Pd{Au}/C [c]	77.5	0.5	H_2	-	-	-	[25]

[a] The third cycle of the ACF-R2 catalyst. [b] The $[Na]_5[Mn(3,5-(SO_3)_2-Cat)_2]\cdot 10H_2O$ complex with addition of Tiron as catalyst. [c] Reaction conditions: 10 mg catalyst in 5.6 g methanol and 2.9 g water solvent, 420 psi 5% H_2/CO_2 + 160 psi 25% O_2/CO_2, with stirring 1200 rpm at 2 °C.

2.4. Effect of Surface Nitrogen- and Oxygen-Containing Groups

Obviously, there was no direct correlation between the selectivity of H_2O_2 with the surface area or the microporous volume of ACF catalysts. That is, the H_2O_2 formation was affected little by the microporous structure. With the aim of exploring the reactivity and the surface chemistry of ACF catalysts, we correlated the selectivity of H_2O_2 with the percentage of N5 (pyrrolic/pyridone) from XPS spectra, and the concentration of H_2O_2 on the specific surface area of ACFs with the amounts of desorbed carboxyl-anhydride groups over the ACF catalysts from TPD results, as shown in Figure 10. Clearly, there was a perfectly positive correlation between the selectivity and the percentage of N5 on the ACF catalysts shown in Figure 10a. It has been considered that, for nitrogen doping, the wholeness of the π conjugate system was broken by the higher electronegativity on the N atom doped in the carbon basal framework of ACF. Moreover, this could induce charge redistribution, which changes the adsorption performance of the reactive intermediates over the carbon materials [57,58]. Thus, compared with pyridine, pyrrolic/pyridone structure in the carbon skeleton possessing more electronegativity was beneficial for the effective adsorption of reactants, which greatly enhanced the selectivity of H_2O_2 on ACF-R2 with higher content of N5.

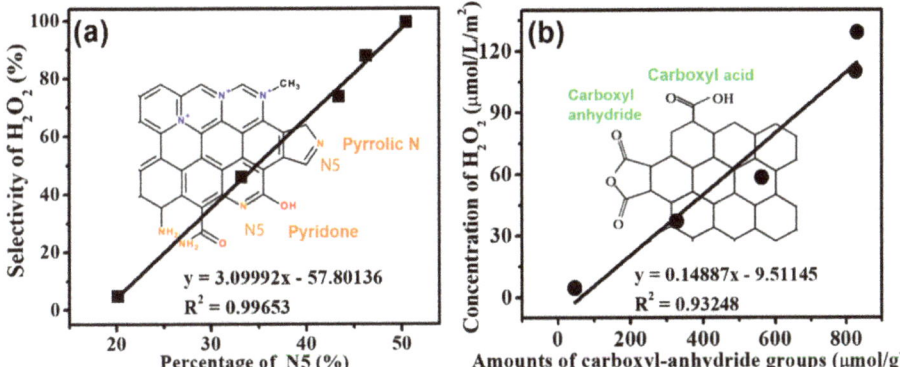

Figure 10. The relationship between (**a**) the selectivity of H_2O_2 and percentage of N5, (**b**) the concentration of H_2O_2 and the amounts of carboxyl-anhydride groups on the ACF catalysts (pH = 8.6, temp. = 25 °C, time = 180 min.).

On the other hand, the correlation between the concentration of H_2O_2 on a specific surface area of ACFs with the amounts of CO_2 desorbed from carboxyl-anhydride groups demonstrated that the yield of H_2O_2 increased in a positive correlation way with the increment of carboxyl-anhydride groups on ACF catalysts, as shown in Figure 10b. This could be ascribed to the more hydrophilic surface on ACFs induced by the formation of large quantities of carboxyl-anhydride species, which are in favor of both effective contact with the hydrophilic reactant and maintaining the existence of H_2O_2. Therefore, the highest selectivity of the ACF-R2 catalyst can be sensibly and directly ascribed to the great quantity of surface oxygen-containing groups and nitrogen-containing groups, particularly the pyrrolic/pyridone nitrogen groups.

For the sake of further clarifying the crucial function of the surface nitrogen, a possible promotion mechanism is proposed. Scheme 1 shows the possible reaction pathway of H_2O_2 production from NH_2OH and O_2 on ACF catalysts promoted by N5. Similar to the reaction mechanism proposed in our previous work [60], NH_2OH loses protons and electrons when contacted with the quinone species on the ACF surface, forming the HNO intermediate. Then the HNO reacts with NH_2OH, producing N_2 and H_2O. The quinoid groups subsequently transfer the protons and electrons to O_2 through the redox cycles of quinone and hydroquinone, completing a typical process of H_2O_2 formation. The role of N5 can be explained from two aspects, namely pyrrolic nitrogen and pyridone structure.

The pyrrolic nitrogen doped on a carbon structure with more electronegativity formed in the edges of the carbon basal plane on ACF, which promotes the electrons transfer between O_2 and NH_2OH. Thus, the adsorbed O_2 species on the ACF surface received the electrons transferred easily from the nitrogen species with extra electrons, and then formed $HO_2^\bullet$ intermediates [53]. For the pyridone structure, the NH group is considered a portion of the six-membered ring on the brink of an extended carbon basal plane. The electronic surrounding of the NH species is thought similar to that of pyrrole because the excess electrons of the N atom could be delocalized among the condensed aromatic system and entrapped at defects on the carbon basal layer [40]. Meanwhile, the pyridone structure is usually in presence of two tautomeric structures including 2-hydroxypyridine and α-pyridone. Usually, these two tautomeric forms are transformed to each other by the intramolecular proton transfer, which may facilitate the protons transfer to the $HO_2^\bullet$ intermediates, forming H_2O_2. Therefore, the higher selectivity of H_2O_2 can be attributed to the higher content of N5 on the ACF catalyst.

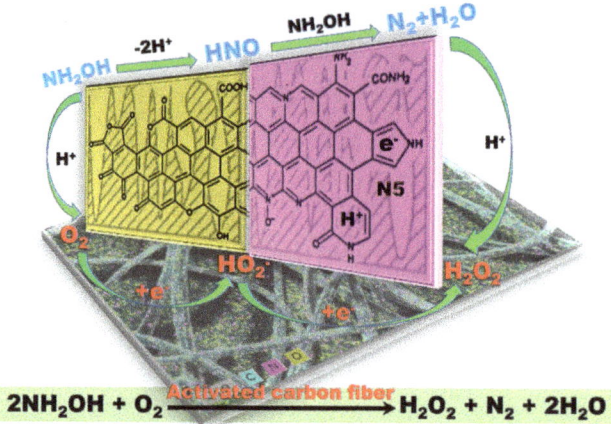

Scheme 1. The possible reaction pathway of H_2O_2 production from NH_2OH and O_2 on ACF catalysts promoted by N5.

3. Materials and Methods

3.1. Surface Modification of ACF

Ten grams of PAN-based ACF (Jilin, Jiyan high-tech Fibers) were put into 100 mL of concentrated hydrochloric acid (HCl, 37%) and mixed for removing the possible impurities including ashes or inorganic substances. The mixtures were firstly stirred for 3 h at ambient temperature, then Cl^- was thoroughly removed from the filtrate by washing with hot water (detected with $AgNO_3$). The obtained sample was put into a vacuum oven and dried at 80 °C overnight, which was christened ACF-0. Then, the ACF-0 (0.5 g) was mixed and stirred in 50 mL of concentrated sulfuric acid (H_2SO_4, 98%) and concentrated nitric acid (HNO_3, 68%) at 60 °C for one hour with a volume ratio of 0.5, 1, 2 and 4, respectively. The oxidized ACF was washed by hot water in order to obtain nearly neutral pH of the filtrate and put into a vacuum oven, then dried at 80 °C overnight. The samples as prepared thus were noted as ACF-Rw, ACF-R1, ACF-R2 and ACF-R4, respectively.

3.2. Characterization of the ACF Catalysts

Field-emission scanning electron microscopy (FE-SEM) images were recorded on a Philips Fei Quanta 200F instrument operating at 20 kV, while elemental mapping images of ACF-0 were obtained on a Hitachi SU8220 SEM instrument working at 15 kV. Nitrogen adsorption-desorption detection was measured by a Micrometrics ASAP 2460 instrument under −196 °C. Moreover, the ACF catalysts were outgassed at 250 °C overnight before the start of measurement. The multipoint Braunauer–Emmett–Teller (BET) analysis was used to calculate the specific surface area (S_{BET}). Fourier transformation infrared (FTIR) spectra of the ACF catalysts were conducted on an IR spectrometer (Bruker Vector 22) by making KBr pellets containing 0.5 wt.% of ACF. The Raman spectra of ACF catalysts were obtained on a Horiba LabRAM HR Evolution Raman spectrometer by using a 532 nm laser. The measurements of X-ray photoelectron spectroscopy (XPS) were carried out on an ES-CALAB MK-II spectrometer (VG Scientific Ltd., West Sussex, UK) with an Al Kα radiation source under an accelerated voltage of 20 kV. For correcting the charge effect, the binding energy (BE) of C1s was adjusted to 285.0 eV. The sensitivity factors and the peak areas of the elements were used to calculate the surface atomic ratio of O/C [61]. Temperature-programmed desorption (TPD) was accomplished in a quartz tubular reactor, which was linked to a quadrupole mass spectrometer (Omnistar, Balzers). After the ACF catalyst (40 mg) was filled in the reactor, the temperature was increased to 900 °C with a heating rate of 10 °C/min in helium flow of 30 mL/min. The mass spectrometer was used to monitor the outlet gas.

3.3. Catalyst Testing

The general reaction of NH_2OH with O_2 was performed in a 100 mL of jacketed glass reactor by stirring at room temperature under atmospheric pressure, as reported elsewhere [31]. In a typical reaction process, 0.15 g of ACF catalyst was put into the aqueous solution of reactant, which was made of hydroxylammonium chloride ($NH_2OH \bullet HCl$, 1.74 g) and 50 mL of deionized water. Before adding the ACF catalyst, the pH value of $NH_2OH \bullet HCl$ aqueous solution was regulated to 8.6 by the solution of 1 M NaOH. Moreover, O_2 was bubbled into the reaction mixtures at a constant flow rate of 25 mL/min, which was tailored by a mass flow controller. Samples of the reactants were taken out periodically in order to analyze the concentration of H_2O_2 by the colorimetric method, which was based on the titanium (IV) sulfate [62]. Similarly, the colorimetric method with the Fe (III)-1,10-phenanthroline complexes was used to detect the concentration of $NH_2OH \bullet HCl$ [63]. The recycling tests of ACF catalysts were performed with the same conditions mentioned above. For each cycle, the used ACF catalyst was washed with hot water and dried at 80 °C in a vacuum oven overnight. The tests of H_2O_2 decomposition were carried out in similar reaction conditions only without feeding $NH_2OH \bullet HCl$ and O_2. The initial concentration of H_2O_2 was 0.25 M without adjusting the pH value. The dosage of ACF catalyst for each decomposition test was 0.15 g. The yield toward H_2O_2 formation was calculated in accordance with the stoichiometric ratio of the reaction ($2NH_2OH + O_2 = H_2O_2 + 2H_2O + N_2$), as the following equation:

$$H_2O_2 \; Yield \; (\%) = 2 \times n(H_2O_2)/n(NH_2OH \bullet HCl) \times 100\% \tag{1}$$

where $n(H_2O_2)$ is the moles of H_2O_2 generated in the reaction, and $n(NH_2OH \bullet HCl)$ is the moles of $NH_2OH \bullet HCl$ in feed.

4. Conclusions

Proper tuning of the surface chemistry of ACFs with intrinsic nitrogen content could expeditiously promote the selectivity of H_2O_2 production through NH_2OH oxidation. Mixed acid oxidation of ACF under mild reaction conditions effectively increased the surface oxygen groups and tailored the pyrrolic/pyridone nitrogen doped on a carbon structure, which then accelerated the selectivity for H_2O_2 over 99.3% on ACF-R2 catalyst. The higher content of H_2SO_4 in the mixed acid created more pyrrolic/pyridone nitrogen, carboxyl and anhydride groups, enhancing the selectivity and yield toward H_2O_2 formation. In our present work, both an easy and low-priced synthetic process for H_2O_2 generation was described, while a new comprehension on the conception and mechanistic examination of metal-free N- and O-doped carbon materials were also provided.

Author Contributions: Conceptualization, W.S.; methodology, W.S. and L.Y.; formal analysis, R.Z.; data curation, X.X. and M.S.; writing-original draft preparation, W.S. and R.Z.; writing-review and editing, X.X. and M.S.; supervision, Y.L.; funding acquisition, W.S. and Y.L. All authors have read and agreed to the published version of the manuscript.

Funding: This research was funded by the National Natural Science Foundation of China (21603039, 51678160).

Conflicts of Interest: The authors declare no conflict of interest.

References

1. Campos-Martin, J.M.; Blanco-Brieva, G.; Fierro, J.L.G. Hydrogen peroxide synthesis: An out- look beyond the anthraquinone process. *Angew. Chem. Int. Ed.* **2006**, *45*, 6962–6984. [CrossRef]
2. Xue, Y.; Wang, Y.; Pan, Z.; Sayama, K. Electrochemical and photoelectrochemical water oxidation for hydrogen peroxide production. *Angew. Chem. Int. Ed.* **2021**, *60*, 2–14. [CrossRef] [PubMed]
3. Goor, G.; Kunkel, W.; Weiberg, O. Hydrogen peroxide. In *UllmannCs Encyclopedia of Industrial Chemistry*, 5th ed.; Elvers, B., Hawkins, S., Ravenscroft, M., Schulz, G., Eds.; Wiley-VCH: New York, NY, USA; Basel, Switzerland; Cambridge, UK; Weinheim, Germany, 1989; Volume A13, pp. 443–466.
4. Kholdeeva, O.; Maksimchuk, N. Metal-organic frameworks in oxidation catalysis with hydrogen peroxide. *Catalysts* **2021**, *11*, 283. [CrossRef]
5. Puértolas, B.; Hillb, A.K.; García, T.; Solson, B.; Torrente-Murciano, L. In-situ synthesis of hydrogen peroxide in tandem with selective oxidation reactions: A mini-review. *Catal. Today* **2015**, *248*, 115–127. [CrossRef]
6. Hu, X.; Zeng, X.; Liu, Y.; Lu, J.; Zhang, X. Carbon-based materials for photo- and electrocatalytic synthesis of hydrogen peroxide. *Nanoscale* **2020**, *12*, 16008–16027. [CrossRef] [PubMed]
7. Chen, Q.; Beckman, E.J. One-pot green synthesis of propylene oxide using in situ generated hydrogen peroxide in carbon dioxide. *Green Chem.* **2008**, *10*, 934–938. [CrossRef]
8. Miller, J.A.; Alexander, L.; Mori, D.I.; Ryabov, A.D.; Collins, T.J. In situ enzymatic generation of H_2O_2 from O_2 for use in oxidative bleaching and catalysis by TAML activators. *New J. Chem.* **2013**, *37*, 3488–3495. [CrossRef]
9. Asghar, A.; Raman, A.A.A.; Daud, W.M.A.W. Recent advances, challenges and prospects of in situ production of hydrogen peroxide for textile waste water treatment in microbial fuel cells. *J. Chem. Technol. Biotechnol.* **2014**, *89*, 1466–1480. [CrossRef]
10. Pan, Z.; Wang, K.; Wang, Y.; Tsiakaras, P.; Song, S. In-situ electrosynthesis of hydrogen peroxide and wastewater treatment application: A novel strategy for graphite felt activation. *Appl. Catal. B Environ.* **2018**, *237*, 392–400. [CrossRef]
11. Giorgianni, G.; Abate, S.; Centi, G.; Perathoner, S. Direct synthesis of H_2O_2 on Pd based catalysts: Modelling the particle size effects and the promoting role of polyvinyl alcohol. *ChemCatChem* **2019**, *11*, 550–559. [CrossRef]
12. Freakley, S.J.; Kochius, S.; van Marwijk, J.; Fenner, C.; Lewis, R.J.; Baldenius, K.; Marais, S.S.; Susan, D.J.O.; Harrison, T.L.; Alcalde, M.; et al. A chemo-enzymatic oxidation cascade to activate C-H bonds with in situ generated H_2O_2. *Nat. Commun.* **2019**, *10*, 4178. [CrossRef]
13. Van Schie, M.M.C.H.; Kaczmarek, A.T.; Tieves, F.; de Santos, P.G.; Paul, C.E.; Arends, I.W.C.E.; Alcalde, M.; Schwarz, G.; Hollmann, F. Selective oxyfunctionalisation reactions driven by sulfite oxidase-catalysed in situ generation of H_2O_2. *ChemCatChem* **2020**, *12*, 3186–3189. [CrossRef]
14. Lyu, J.; Niu, L.; Shen, F.; Wei, J.; Xiang, Y.; Yu, Z.; Zhang, G.; Ding, C.; Huang, Y.; Li, X. In situ hydrogen peroxide production for selective oxidation of benzyl alcohol over a Pd@hierarchical titanium silicalite catalyst. *ACS Omega* **2020**, *5*, 16865–16874. [CrossRef]
15. Liu, Y.; Zhao, Y.; Wang, J. Fenton/Fenton-like processes with in-situ production of hydrogen peroxide/hydroxyl radical for degradation of emerging contaminants: Advances and prospects. *J. Hazard. Mater.* **2021**, *404*, 124191–124210. [CrossRef]
16. Samanta, C. Direct synthesis of hydrogen peroxide from hydrogen and oxygen: An overview of recent developments in the process. *Appl. Catal. A Gen.* **2008**, *350*, 133–149. [CrossRef]
17. Edwards, J.K.; Freakley, S.J.; Carley, A.F.; Kiely, C.J.; Hutchings, G.J. Strategies for designing supported gold-palladium bimetallic catalysts for the direct synthesis of hydrogen peroxide. *Acc. Chem. Res.* **2014**, *47*, 845–854. [CrossRef] [PubMed]
18. Edwards, J.K.; Freakley, S.J.; Lewis, R.J.; Pritchard, J.C.; Hutchings, G.J. Advances in the direct synthesis of hydrogen peroxide from hydrogen and oxygen. *Catal. Today* **2015**, *248*, 3–9. [CrossRef]
19. Yi, Y.; Wang, L.; Li, G.; Guo, H. A review on research progress in the direct synthesis of hydrogen peroxide from hydrogen and oxygen: Noble-metal catalytic method, fuel-cell method and plasma method. *Catal. Sci. Technol.* **2016**, *6*, 1593–1610. [CrossRef]
20. Gao, G.; Tian, Y.; Gong, X.; Pan, Z.; Yang, K.; Zong, B. Advances in the production technology of hydrogen peroxide. *Chin. J. Catal.* **2020**, *41*, 1039–1047. [CrossRef]
21. Ntainjua, N.E.; Piccinini, M.; Pritchard, J.C.; Edwards, J.K.; Carley, A.F.; Moulijn, J.A.; Hutchings, G.J. Effect of halide and acid additives on the direct synthesis of hydrogen peroxide using supported gold–palladium catalysts. *ChemSusChem* **2009**, *2*, 575–580. [CrossRef]

22. Pritchard, J.C.; He, Q.; Ntainjua, E.N.; Piccinini, M.; Edwards, J.K.; Herzing, A.A.; Carley, A.F.; Moulijn, J.A.; Kiely, C.J.; Hutchings, G.J. The effect of catalyst preparation method on the performance of supported Au–Pd catalysts for the direct synthesis of hydrogen peroxide. *Green Chem.* **2010**, *12*, 915–921. [CrossRef]
23. Pritchard, J.; Kesavan, L.; Piccinini, M.; He, Q.; Tiruvalam, R.; Dimitratos, N.; Lopez-Sanchez, J.A.; Carley, A.F.; Edwards, J.K.; Kiely, C.J.; et al. Direct synthesis of hydrogen peroxide and benzyl alcohol oxidation using Au−Pd catalysts prepared by sol immobilization. *Langmuir* **2010**, *26*, 16568–16577. [CrossRef] [PubMed]
24. Tiruvalam, R.C.; Pritchard, J.C.; Dimitratos, N.; Lopez-Sanchez, J.A.; Edwards, J.K.; Carley, A.F.; Hutchings, G.J.; Kiely, C.J. Aberration corrected analytical electron microscopy studies of sol-immobilized Au + Pd, Au {Pd} and Pd {Au} catalysts used for benzyl alcohol oxidation and hydrogen peroxide production. *Faraday Discuss.* **2011**, *152*, 63–86. [CrossRef] [PubMed]
25. Pritchard, J.; Piccinini, M.; Tiruvalam, R.; He, Q.; Dimitratos, N.; Lopez-Sanchez, J.A.; Morgan, D.J.; Carley, A.F.; Edwards, J.K.; Kiely, C.J.; et al. Effect of heat treatment on Au–Pd catalysts synthesized by sol immobilisation for the direct synthesis of hydrogen peroxide and benzyl alcohol oxidation. *Catal. Sci. Technol.* **2013**, *3*, 308–317. [CrossRef]
26. Hou, H.; Zeng, X.; Zhang, X. Production of hydrogen peroxide by photocatalytic processes. *Angew. Chem. Int. Ed.* **2020**, *59*, 17356–17376. [CrossRef]
27. Song, H.; Wei, L.; Chen, L.; Zhang, H.; Su, J. Photocatalytic production of hydrogen peroxide over modifed semiconductor materials: A minireview. *Top. Catal.* **2020**, *63*, 895–912. [CrossRef]
28. Hughes, M.N.; Nicklin, H.G. Autoxidation of hydroxylamine in alkaline solutions. *J. Chem. Soc. A* **1971**, *1*, 164–168. [CrossRef]
29. Sheriff, T.S. Production of hydrogen peroxide from dioxygen and hydroxylamine or hydrazine catalysed by manganese complexes. *J. Chem. Soc. Dalton Trans.* **1992**, *6*, 1051–1058. [CrossRef]
30. Sheriff, T.S.; Carr, P.; Piggott, B. Manganese catalysed reduction of dioxygen to hydrogen peroxide: Structural studies on a manganese (III)–catecholate complex. *Inorg. Chim. Acta* **2003**, *348*, 115–122. [CrossRef]
31. Sheriff, T.S.; Carr, P.; Coles, S.J.; Hursthouse, M.B.; Lesin, J.; Light, M.E. Structural studies on manganese (III) and manganese (IV) complexes of tetrachlorocatechol and the catalytic reduction of dioxygen to hydrogen peroxide. *Inorg. Chim. Acta* **2004**, *357*, 2494–2502. [CrossRef]
32. Choudhary, V.R.; Jana, P.; Bhargava, S.K. Reduction of oxygen by hydroxylammonium salt or hydroxylamine over supported Au nanoparticles for in situ generation of hydrogen peroxide in aqueous or non-aqueous medium. *Catal. Commun.* **2007**, *8*, 811–816. [CrossRef]
33. Choudhary, V.R.; Jana, P. In situ generation of hydrogen peroxide from reaction of O_2 with hydroxylamine from hydroxylammonium salt in neutral aqueous or non-aqueous medium using reusable Pd/Al_2O_3 catalyst. *Catal. Commun.* **2007**, *8*, 1578–1582. [CrossRef]
34. Choudhary, V.R.; Jana, P. Factors influencing the in situ generation of hydrogen peroxide from the reduction of oxygen by hydroxylamine from hydroxylammonium sulfate over Pd/alumina. *Appl. Catal. A Gen.* **2008**, *335*, 95–102. [CrossRef]
35. Song, W.; Li, J.; Liu, J.; Shen, W. Production of hydrogen peroxide by the reaction of hydroxylamine and molecular oxygen over activated carbons. *Catal. Commun.* **2008**, *9*, 831–836. [CrossRef]
36. Song, W.; Yu, L.; Xie, X.; Hao, Z.; Sun, M.; Wen, H.; Li, Y. Effect of textual features and surface properties of activated carbon on the production of hydrogen peroxide from hydroxylamine oxidation. *RSC Adv.* **2017**, *7*, 25305–25313. [CrossRef]
37. Melchionna, M.; Fornasiero, P.; Prato, M. The rise of hydrogen peroxide as the main product by metal-free catalysis in oxygen reductions. *Adv. Mater.* **2019**, *31*, 1802920–1802924. [CrossRef]
38. Sun, Y.; Sinev, I.; Ju, W.; Bergmann, A.; Dresp, S.; Kühl, S.; Spöri, C.; Schmies, H.; Wang, H.; Bernsmeier, D.; et al. Efficient electrochemical hydrogen peroxide production from molecular oxygen on nitrogen-doped mesoporous carbon catalysts. *ACS Catal.* **2018**, *8*, 2844–2856. [CrossRef]
39. Raymundo-Piñero, E.; Cazorla-Amorós, D.; Linares-Solano, A. Temperature programmed desorption study on the mechanism of SO_2 oxidation by activated carbon and activated carbon fibres. *Carbon* **2001**, *39*, 231–242. [CrossRef]
40. Boehm, H.P. Catalytic properties of nitrogen-containing carbons. In *Carbon Materials for Catalysis*, 1st ed.; Serp, P., Figueiredo, J.L., Eds.; John Wiley & Sons, Inc.: Hoboken, NJ, USA, 2009; Chapter 7; pp. 219–238.
41. De la Puente, G.; Pis, J.J.; Menéndez, J.A.; Grange, P. Thermal stability of oxygenated functions in activated carbons. *J. Anal. Appl. Pyrolysis* **1997**, *43*, 125–138. [CrossRef]
42. Prahas, D.; Kartika, Y.; Indraswati, N.; Ismadji, S. Activated carbon from jackfruit peel waste by H_3PO_4 chemical activation: Pore structure and surface chemistry characterization. *Chem. Eng. J.* **2008**, *140*, 32–42. [CrossRef]
43. Yang, S.; Li, L.; Xiao, T.; Zheng, D.; Zhang, Y. Role of surface chemistry in modified ACF (activated carbon fiber)-catalyzed peroxymonosulfate oxidation. *Appl. Surf. Sci.* **2016**, *383*, 142–150. [CrossRef]
44. Macías-García, A.; Díaz-Díez, M.A.; Cuerda-Correa, E.M.; Olivares-Marín, M.; Gañan-Gómez, J. Study of the pore size distribution and fractal dimension of HNO_3-treated activated carbons. *Appl. Surf. Sci.* **2006**, *252*, 5972–5975. [CrossRef]
45. Kima, M.J.; Song, E.J.; Kim, K.H.; Choi, S.S.; Lee, Y.S. The textural and chemical changes in ACFs with e-beam and their influence on the detection of nerve agent simulant gases. *J. Ind. Eng. Chem.* **2019**, *79*, 465–472. [CrossRef]
46. Shi, M.; Bao, D.; Li, S.; Wulan, B.; Yan, J.; Jiang, Q. Anchoring PdCu amorphous nanocluster on graphene for electrochemical reduction of N_2 to NH_3 under ambient conditions in aqueous solution. *Adv. Energy Mater.* **2018**, *8*, 1800124–1800129. [CrossRef]

47. Terzyk, A.P. The influence of activated carbon surface chemical composition on the adsorption of acetaminophen (paracetamol) in vitro: Part II. TG, FTIR, and XPS analysis of carbons and the temperature dependence of adsorption kinetics at the neutral pH. *Colloids Surf. A* **2001**, *177*, 23–45. [CrossRef]
48. Swiatkowski, A.; Pakula, M.; Biniak, S.; Walczyk, M. Influence of the surface chemistry of modified activated carbon on its electrochemical behaviour in the presence of lead(II) ions. *Carbon* **2004**, *42*, 3057–3069. [CrossRef]
49. Brazhnyk, D.V.; Zaitsev, Y.P.; Bacherikova, I.V.; Zazhigalov, V.A.; Stoch, J.; Kowal, A. Oxidation of H_2S on activated carbon KAU and influence of the surface state. *Appl. Catal. B Environ.* **2007**, *70*, 557–566. [CrossRef]
50. Fels, J.R.; Kapteijn, F.; Moulijn, J.A.; Zhu, Q.; Thomas, K.M. Evolution of nitrogen functionalities in carbonaceous materials during pyrolysis. *Carbon* **1995**, *33*, 1641–1653.
51. Pietrzak, R. XPS study and physico-chemical properties of nitrogen-enriched microporous activated carbon from high volatile bituminous coal. *Fuel* **2009**, *88*, 1871–1877. [CrossRef]
52. Kundu, S.; Xia, W.; Busser, W.; Kundu, S.; Xia, W.; Busser, W.; Becker, M.; Schmidt, D.A.; Havenith, M.; Muhle, M. The formation of nitrogen-containing functional groups on carbon nanotube surfaces: A quantitative XPS and TPD study. *Phys. Chem. Chem. Phys.* **2010**, *12*, 4351–4359. [CrossRef]
53. Yang, G.; Chen, H.; Qin, H.; Yang, G.; Chen, H.; Qin, H.; Feng, Y. Amination of activated carbon for enhancing phenol adsorption: Effect of nitrogen-containing functional groups. *Appl. Surf. Sci.* **2014**, *293*, 299–305. [CrossRef]
54. Sun, H.; Kwan, C.; Wang, S.; Sun, H.; Kwan, C.K.; Suvorova, A.; Ang, H.M.; Tadé, M.O.; Wang, S. Catalytic oxidation of organic pollutants on pristine and surface nitrogen-modifified carbon nanotubes with sulfate radicals. *Appl. Catal. B Environ.* **2014**, *154–155*, 134–141. [CrossRef]
55. Boehm, H.P. Surface oxides on carbon and their analys is: A critical assessment. *Carbon* **2002**, *40*, 145–149. [CrossRef]
56. Zielke, U.; Hüttinger, K.J.; Hoffman, W.P. Surface-oxidized carbon fibers: I. Surface structure and chemistry. *Carbon* **1996**, *34*, 983–998. [CrossRef]
57. Zhang, J.; Zhang, H.; Cheng, M.; Lu, Q. Tailoring the electrochemical production of H_2O_2: Strategies for the rational design of high-performance electrocatalysts. *Small* **2020**, *16*, 1902845–1902861. [CrossRef]
58. Jiang, Y.; Ni, P.; Chen, C.; Lu, Y.; Yang, P.; Kong, B.; Fisher, A.; Wang, X. Selective electrochemical H_2O_2 production through two-electron oxygen electrochemistry. *Adv. Energy Mater.* **2018**, *8*, 1801909–1801933. [CrossRef]
59. Figueiredo, J.L.; Pereira, M.F.R. Carbon as Catalyst. In *Carbon Materials for Catalysis*, 1st ed.; Serp, P., Figueiredo, J.L., Eds.; John Wiley & Sons, Inc.: Hoboken, NJ, USA, 2009; Chapter 6; pp. 196–198.
60. Song, W.; Li, Y.; Guo, X.; Li, J.; Huang, X.; Shen, W. Selective surface modifification of activated carbon for enhancing the catalytic performance in hydrogen peroxide production by hydroxylamine oxidation. *J. Mol. Catal. Chem.* **2010**, *328*, 53–59. [CrossRef]
61. Figueiredo, J.L.; Pereira, M.F.R.; Freitas, M.M.A.; Órfão, J.J.M. Modification of the surface chemistry of activated carbons. *Carbon* **1999**, *37*, 1379–1389. [CrossRef]
62. Clapp, P.A.; Evans, D.F.; Sheriff, T.S. Spectrophotometric determination of hydrogen peroxide after extraction with ethyl acetate. *Anal. Chim. Acta* **1989**, *218*, 331–334. [CrossRef]
63. Yang, M. Hydroxylamine hydrochloride was determined by indirect spectrophotometry. *Chem. Ind. Eng. (China)* **1999**, *16*, 233–235.

Article

Improvement of Alkali Metal Resistance for NH₃-SCR Catalyst Cu/SSZ-13: Tune the Crystal Size

Zexiang Chen [1], Meiqing Shen [1,2], Chen Wang [1,3], Jianqiang Wang [1], Jun Wang [1,*] and Gurong Shen [4,*]

1. School of Chemical Engineering & Technology, Tianjin University, Tianjin 300072, China; tjuczx@tju.edu.cn (Z.C.); mqshen@tju.edu.cn (M.S.); chenwang87@tju.edu.cn (C.W.); jianqiangwang@tju.edu.cn (J.W.)
2. Collaborative Innovation Centre of Chemical Science and Engineering (Tianjin), Tianjin 300072, China
3. School of Environmental and Safety Engineering, North University of China, Taiyuan 030051, China
4. School of Materials Science and Engineering, Tianjin University, Tianjin 300350, China
* Correspondence: wangjun@tju.edu.cn (J.W.); gr_shen@tju.edu.cn (G.S.)

Abstract: To improve the alkali metal resistance of commercial catalyst Cu/SSZ-13 for ammonia selective catalytic reduction (NH$_3$-SCR) reaction, a simple method to synthesize Cu/SSZ-13 with a core–shell like structure was developed. Compared with smaller-sized counterparts, Cu/SSZ-13 with a crystal size of 2.3 µm exhibited excellent resistance to Na poisoning. To reveal the influence of the crystal size on Cu/SSZ-13, physical structure characterization (XRD, BET, SEM, NMR) and chemical acidic distribution (H$_2$-TPR, UV-Vis, Diethylamine-TPD, pyridine-DRIFTS, EDS) were investigated. It was found that the larger the crystal size of the molecular sieve, the more Cu is distributed in the crystal core, and the less likely it was to be replaced by Na to generate CuO. Therefore, a 2.3 µm sized Cu/SSZ-13 well-controlled the reactivity of the side reaction NH$_3$ oxidation and the generation of N$_2$O. The result was helpful to guide the extension of the service life of Cu/SSZ-13.

Keywords: alkali metal poison; Cu/SSZ-13; crystal size; NH$_3$-SCR

1. Introduction

For health and environmental considerations, the current community strongly demands to improve air quality. Environmental protection agencies around the world have introduced more and more stringent motor vehicle exhaust emission regulations on CO, nitrogen oxides (NO$_x$), unburned hydrocarbons and particulate matter [1]. The commercial technology of pollutants elimination is to adopt efficient catalytic converters. Cu/SSZ-13 (SSZ-13 is one framework type code of chabazite zeolite), one of the most excellent NO$_x$ removal catalysts, has been commercially used in the exhaust after-treatment system of medium and heavy diesel vehicles [2–6].

In the past decade, numerical theoretical and experimental studies have revealed the pore structure, NH$_3$-SCR catalytic active sites, high-temperature hydrothermal stability, sulfur poisoning mechanism and alkali metal poisoning mechanism of Cu/SSZ-13 [7–20]. Alkali metals, such as sodium (Na) and potassium (K) are present in conventional diesel fuel and biodiesel fuel, and have been implicated in internal injector deposits [21,22]. Besides, alkali metals also exist in automotive urea, such as diesel exhaust fluid (DEF) or AdBlue. Previous reports have shown that alkali metal ions are volatilized under high-temperature conditions and diffuse to the surface of the catalyst, causing the catalyst to be poisoned [23,24]. Fundamental researches have revealed the alkali metal poisoning mechanism of Cu/CHA catalysts [25,26]. Researchers suggest that alkali metal ions can destroy the active center Cu^{2+} and degrade the zeolite framework, causing irreversible deactivation. However, how to improve the tolerance of Cu/zeolite catalysts to alkali metals is rarely reported. Yan et al. reported that highly dispersed Cu$_4$AlOx mixed oxides have a good alkali metal resistance [27]. However, the multiple Cu complex cannot fit in the CHA's steric pore size and the inactive CuO clusters are easily formed in the CHA zeolites [28]. Du et al. compared the alkali resistance of conventional V_2O_5/WO_3-TiO_2 catalysts and Fe_2O_3/HY zeolite catalysts and found that the HY zeolites played an alkali-buffer role to retain the catalytic activity [29]. Zha et al. developed a hollandite Mn–Ti oxide promoted Cu-SAPO-34 (SAPO-34 is one framework type code of chabazite zeolite) catalysts (HMT@Cu-S). HMT layer can trap alkali metal ions to protect Cu/SAPO-34 [30]. The similarity of previous research lies in the use of outer shell materials to protect the inner structure. However, the synthesis of core–shell materials often requires fine control, and it is difficult to achieve industrial low-cost production. To meet industrial needs, we tried to develop a simple preparation process to obtain core–shell-like molecular sieves with good alkali metal resistance. In our previous study [31], it was found that different crystal sizes of the Cu/SAPO-34 can affect the spatial distribution of acid sites, which inspired us to adjust the acid sites on the outer surface of the molecular sieve by tuning the crystal size.

In this work, we report that Cu/SSZ-13 catalysts with different crystal sizes (0.4–2.3 μm) present distinct sodium metal ions resistance. Due to the similar deactivation effect of Na and K, we only choose Na to study the improvement of alkali metal resistance [25]. Na-poisoned Cu/SSZ-13 with crystal size of 2.3 μm showed much higher NH_3-SCR activity than ones with a size of 0.4 μm. The distribution of acidity and Cu ions is characterized by H_2-TPR, UV–vis, diethylamine-TPD, and pyridine-DRIFTs. By enlarging the crystal size of SSZ-13, the active center Cu ions are allowed to locate the subsurface or inner core, so that the outer layer of the zeolites buffer the foreign alkali metal poisoning. Adjusting the distribution of acid sites by modifying the crystal size provides a simple technical route to improve alkali metal poisoning.

2. Results and Discussions

The crystallization process of SSZ-13 zeolites is mainly divided into induction period and crystallization period [32]. During the induction period, the raw materials in the gel gradually gather to form crystal nuclei, and the crystal grows rapidly after entering the crystallization period. The crystal size of SSZ-13 can be affected by the crystallization time, crystallization temperature, gel pH and added amount of crystal seed. Through a series of trials, we found that the crystal size of SSZ-13 is most sensitive to the amount of seed crystal added. The as-prepared catalysts are visualized by SEM. As shown in Figure 1, the average crystal size of each sample is obtained based on the statistical results of ~100 grains. All samples present cubic shape and uniform particle size. The results show that the crystal sizes of the synthesized SSZ-13 are 0.4, 0.8, and 2.3 μm.

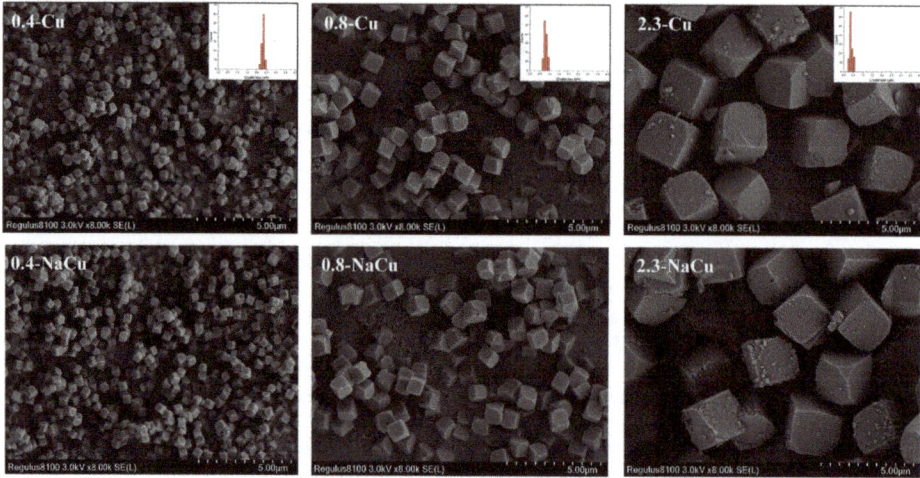

Figure 1. SEM images of Cu/SSZ-13 catalysts with different crystal sizes.

2.1. NH$_3$-SCR and NH$_3$ Oxidation Activity

Figure 2 presents the NH$_3$-SCR activity and side product N$_2$O concentration of catalysts before and after Na poisoning. Before Na poisoning, Cu/SSZ-13 with different crystal sizes present similar NH$_3$-SCR activities. It can be observed that between 350–400 °C, SCR activity increases with increasing crystal sizes. In this temperature range, the increasing temperature induces the active center Cu^{2+} to migrate from the solvated form (Cu^{II}(NH$_3$)$_n$(H$_2$O)$_m$) to the molecular sieve framework (Z$_2$Cu^{II} or ZCu^{II}OH) [33–35]. The change in reaction activity is mainly dominated by the number of isolated Cu ions. Such trend is consistent with the copper content in order: 2.3-Cu (1.44 wt %) > 0.8-Cu (1.38 wt %) > 0.4-Cu (1.05 wt %) (Table S1). Thereafter in the higher temperature region (>400 °C), all samples have decreased activities, which is caused by the side reaction NH$_3$ oxidation reaction. The NH$_3$ can be oxidized to N$_2$ by either NO$_x$ or O$_2$. The competition between SCR reaction and NH$_3$ oxidation by O$_2$ at high temperatures (>350 °C) is widely reported [6]. The amount of N$_2$O produced at a concentration of less than 11 ppm in the entire temperature window indicates that the catalysts have good N$_2$ selectivity (Figure 2c). NO$_x$ can be reduced to N$_2$O or N$_2$ by NH$_3$. Normally, 5–10 ppm of N$_2$O production for the state-of-the-art Cu-based catalysts is usually observed in the literature [2,3]. After Na poisoning (Figure 2b), the three samples show completely different NH$_3$-SCR reactivities, whose order is: 2.3-NaCu > 0.8-NaCu > 0.4-NaCu. The activity of 0.4-NaCu decreases most significantly, and the NO$_x$ conversion drops sharply from 98% to 30% at ~300 °C. Figure 2d shows the N$_2$O generation curve with temperature ramping. Surprisingly, the maximum amount of N$_2$O produced by 0.4-NaCu (~75 ppm) at 280 °C is 3 times that of 0.8-NaCu (~25 ppm) and 7 times that of 2.3-NaCu (~10 ppm). The results show that the geometric size of the molecular sieve greatly affects the NH$_3$-SCR catalytic activity and N$_2$ selectivity of the Na-poisoned CuSSZ-13 catalysts. In short, the Cu/SSZ-13 catalyst with a larger crystal size has better Na resistance.

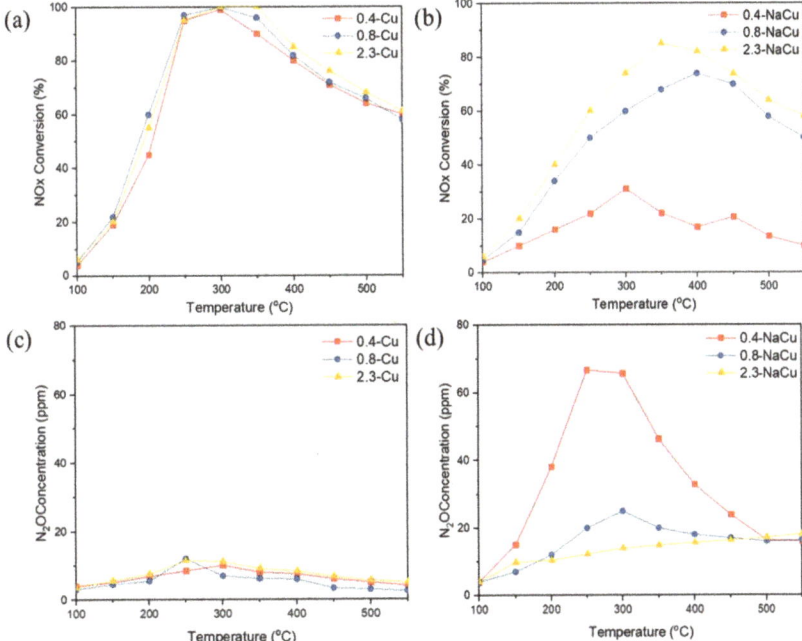

Figure 2. (**a**,**b**) NO$_x$ conversion and (**c**,**d**) N$_2$O formation of Cu/SSZ-13 before and after Na poisoning. The gas feed contains 500 ppm NO$_x$, 500 ppm NH$_3$, 10% O$_2$, 3% H$_2$O, balanced N$_2$. The gas hourly space velocity (GHSV) is 72,000 h^{-1}.

As a side reaction of NH$_3$-SCR, the NH$_3$ oxidation reaction was also tested (Figure 3). Before Na poisoning, the NH$_3$ oxidation performances of 0.4-Cu, 0.8-Cu and 2.3-Cu are almost the same. Temperature-dependent NO$_x$ and N$_2$O evolutions are recorded as shown in Figure S1. Same trends for all nitrogen oxides are observed. After Na poisoning, the NH$_3$ oxidation performance of all samples is improved. The active center of NH$_3$ oxidation can be isolated Cu ions and CuO$_x$ clusters. In the previous report [36], NH$_3$ oxidation activities on CuO$_x$ clusters are much higher than Cu ions. Therefore, the improvement of NH$_3$ oxidation is attributed to CuO$_x$ formation in the catalysts. It is noted that 0.4-NaCu has the largest increase in activity, reaching ~100% conversion at 250 °C. In other words, Cu/SSZ-13 with a larger particle size has relatively low ammonia oxidation activity and higher resistance to Na poisoning. This is consistent with the results of NH$_3$-SCR activity. More importantly, Figure S1 shows that there are two active centers in the oxidation process of NH$_3$ with temperature ramping. One active center is responsible for the formation of nitrogen oxides at 200–350 °C, and the other is responsible for above 350 °C. This trend of change is likely to be related to the migration of Cu with increasing temperature. As the crystal size increases, the former active center (might be Cu^{II}(NH$_3$)$_n$(H$_2$O)$_m$ [37]) gradually decreases. 2.3-NaCu can hardly see the NO$_x$ volcanic curve around 300 °C. This observation will be further discussed later.

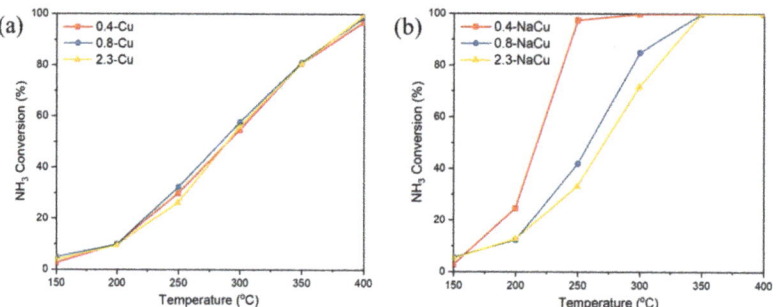

Figure 3. NH$_3$ oxidation activity of Cu/SSZ-13 (**a**) before and (**b**) after Na poisoning. The gas feed contains 500 ppm NH$_3$, 10% O$_2$ and balanced N$_2$. The gas hourly space velocity (GHSV) is 72,000 h^{-1}.

Combined with the results of NH$_3$-SCR activities, it is interesting to find that Na poisoned Cu/SSZ-13 with a smaller crystal size has a more serious decline in SCR activity and N$_2$ selectivity. Such reaction tread cannot be separable from the structure of the catalysts caused by different crystal growth paths. Therefore, the characterization of the texture properties and the active center were conducted.

2.2. Texture Properties

The crystal structure was characterized by X-ray diffraction experiments. All catalysts show uniform CHA-type X-ray diffraction patterns (9.7°, 13.1°, 16.3°, 18.0°, 20.9°, 25.4°, 26.3°, 31.1°), as shown in Figure S2. No characteristic peaks corresponding to CuO (35.6, 38.8°) or Na$_2$O (36.2, 46.1°) are observed, which suggests ICP (Table S1) determined ~1.4 wt % Cu and ~2 wt % Na (if Na poisoned) are well dispersed and the particle size of them are under the detection limit. Our previous work reported poisoned Na can induce desiliconization and form "H$_2$Si$_2$O$_5$" crystal phase (21.8°) [38]. As shown in Figure 4, it shows that the larger the crystal size of the molecular sieve is, the easier it is to desiliconize. However, the structural desiliconization occurs on the H/SSZ-13 but not Cu/SSZ-13, which shows that Cu ions play a role in protecting the Si framework. The nature of SSZ-13 determines the acid sites (Si-OH-Al) formation in the zeolites. Na poisoned samples presented the "H$_2$Si$_2$O$_5$" crystal phase (21.8°), which indicates desiliconization occurs. When Cu was loaded, such desiliconization was much weakened. The results show that desiliconization is likely to happen on the acid sites. Si-O(Cu^{II}OH)-Al or Al-O-Cu^{II}-O-Si-O-Al sites can be kept after Na poisoning.

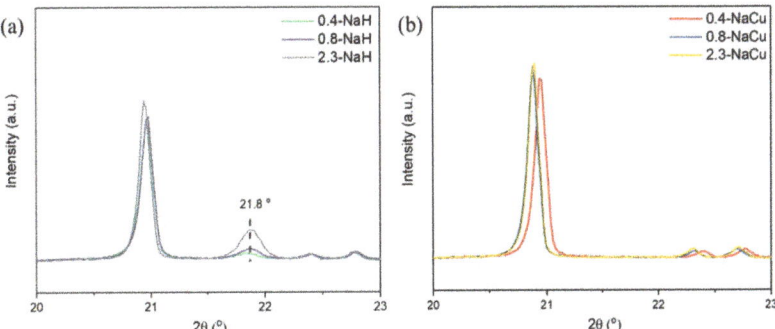

Figure 4. 20–23° XRD patterns of Na-poisoned (**a**) H/SSZ-13 supports and (**b**) Cu/SSZ-13 catalysts.

The N$_2$ adsorption and desorption experiments are used to characterize the pore structure of the catalysts. As shown in Table 1, the BET specific surface area of both SSZ-13 and CuSSZ-13 is greater than 800 m^2/g. However, the support SSZ-13 is easily damaged by Na poisoning, and the specific surface area is greatly reduced. This is consistent with the XRD results. The desiliconization is believed to cause such variation on the specific surface area. In contrast, the tolerance of Cu/SSZ-13 to Na is much improved. The pore distribution was also analyzed and found that the pore structure is rarely damaged by Na poisoning (Figure S3).

Table 1. BET Specific surface area of samples.

Samples	BET Specific Surface Area (m^2/g)	Samples	BET Specific Surface Area (m^2/g)
0.4-H	831	0.4-Cu	820
0.8-H	832	0.8-Cu	807
2.3-H	825	2.3-Cu	843
0.4-NaH	469	0.4-NaCu	795
0.8-NaH	500	0.8-NaCu	772
2.3-NaH	587	2.3-NaCu	833

The framework coordination structure of the catalyst was also characterized by NMR technology. Normalized ^{27}Al and ^{29}Si NMR spectra show that the coordination of Si and Al has little change before and after Na poisoning, independent of the crystal size of Cu/SSZ-13 (Figure S4). All samples present framework tetrahedral Al at 60 ppm. Corresponding feature of Si at −109, −103, and −99 ppm are attributed to Si(4Si, 0Al), Si(3Si, 1Al), and Si(2Si, 2Al), respectively.

The results of structural characterizations (XRD, BET, NMR) consistently indicate that even though Na poisoning may destroy the structure of the SSZ-13 supports, the loading of Cu greatly improves the tolerance to Na poisoning. The texture properties of Cu/SSZ-13 with different crystal sizes have little variation between each other.

2.3. Cu Distribution

Figure 5 presents hydrogen temperature-programmed reduction (H$_2$-TPR) profiles of CuSSZ-13 catalysts with different crystal sizes before and after Na poisoning. According to previous reports, it is generally believed that the H$_2$ reduction peak at 200 °C and 400 °C are attributed to the reduction of the eight-membered ring (8MR) Cu^{2+} ions and six-membered ring (8MR) Cu^{2+} ions to Cu$^+$, respectively. The reduction peak above 500 °C is the reduction of Cu$^+$ to Cu^0 [12,39]. As shown in Figure 5a, an H$_2$ reduction peak dominates at ~181 °C, while the one at ~400 °C is not significant. For Cu/SSZ-13 with Si/Al of ~25, Cu^{2+} is mostly in the form of $Z[Cu^{II}OH]$ (Z represents the negatively charged framework site), rather than Z_2Cu^{II}, because the probability of having two Al atoms in a six-membered ring (6MR) at the same time is relatively small. ^{29}Si NMR results also show that Si (3Si, 1Al) species dominate the framework structure (Figure S4). However, an unexpected result is that the larger the crystal size, the fewer Cu species can be reduced. This is in contradiction with the result that the total Cu content of 0.4-Cu, 0.8-Cu, and 2.3-Cu detected by ICP is 1.05 wt %, 1.38 wt %, and 1.44 wt %, respectively. Note that UV–vis characterization (Figure 6) confirms the ICP results, in which the charge transfer of O^{2-} to Cu^{2+} at 200 nm bands follows an order: 0.4-Cu < 0.8-Cu < 2.3-Cu. Why does the amount of reducible Cu decrease with increasing crystal size? What needs to be certain is that the kinetic radius of H_2 is small enough to pass through the molecular sieve and reach the Cu ion surface. Therefore, it is reasonable to believe that more stable Cu ions in the larger-sized Cu/SSZ-13 make them less reducible. Meanwhile, with the increasing crystal sizes, the shifting to a higher reduction temperature of Cu^+ to Cu^0 is observed. It also indicated that the interaction between Cu ions and supports becomes stronger for the Cu/SSZ-13 with a larger crystal size. After Na poisoning, the reduction peaks shift to lower temperatures, which was previously observed in the literature [13,40]. The peak at 150 °C is attributed to the reduction of $Z[Cu^{II}OH]$ to Cu^+, and the sharp one at 220 °C is attributed to the reduction of CuO to Cu^0. Here we classify the reducing species at ~220 °C as CuO, because UV–vis characterizes the formation of a large amount of CuO shown in Figure 6. The amount of produced CuO is proportional to the increase in NH_3 oxidation reaction activity. Besides, the amount of reducible Cu of different samples is still the same as that of fresh samples. The reason why the increase in the particle size leads to the stronger interaction between Cu and the supports is still unclear.

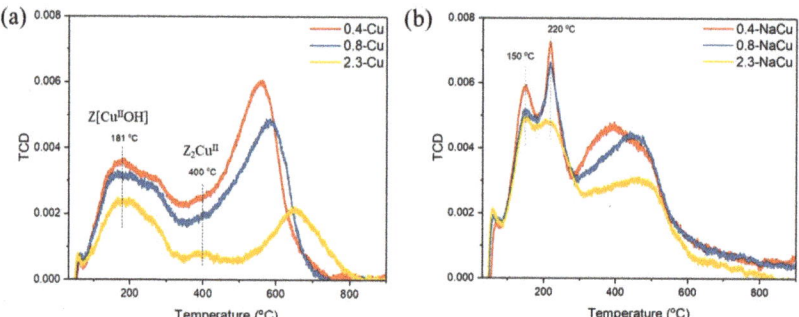

Figure 5. H_2-TPR profiles of Cu/SSZ-13 with different crystal sizes (**a**) before and (**b**) after Na poisoning.

2.4. Acidic Distribution

In the beginning, the target of designing the Cu/SSZ-13 with different crystal sizes is to tune the surface acid sites. The spatial acidic distribution of Cu/SSZ-13 was revealed by comparing the surface acid sites and bulky acid sites. Figure S5 shows the ammonia temperature-programmed desorption (NH_3-TPD) profiles. NH_3 molecules are small enough to penetrate the SSZ-13 zeolites and therefore can be used to characterize the bulky acid sites. The results show that Cu/SSZ-13 catalysts with different crystal sizes own a similar amount of bulky acid sites. Even Na-poisoned samples have a similar total acid amount, which is mainly dominated by weakly adsorbed acids.

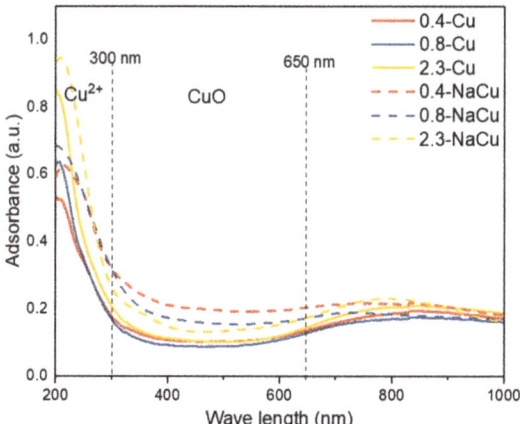

Figure 6. UV–vis spectra of the Cu/SSZ-13 with different crystal sizes.

Diethylamine is a probe molecule often used to investigate the surface acidity of microporous molecular sieves. Its kinetic diameter (4.5 Å) is larger than the pore size of the SSZ-13 molecular sieve (3.8 Å), which ensures that it can only bind to the acidic sites on the outer surface of the crystal. As shown in Figure 7a, only 0.4-Cu sample can adsorb diethylamine molecules. The results of diethylamine TPD for H/SSZ-13 catalysts in Figure 7c show that the smaller the crystal size, the more surface acid sites. It can be inferred that Cu is more present on the surface with a small crystal size, and as the crystal size increases, Cu is distributed in the subsurface or core, i.e., little diethylamine adsorption on the surface of 0.8-Na and 2.3-Na. For the Na-poisoned samples, two desorption peaks of diethylamine can be observed at ~200 °C and 300 °C in Figure 7b,d. Comparing the results of H/SSZ-13 and NaH/SSZ-13, it can be found that partial Na ions can substitute Brønsted acid sites and adsorb diethylamine on the surface of the molecular sieve. It is reasonable that there is almost no diethylamine adsorbing on the 2.3-NaH, because its surface acid sites are limited. Besides, with the decrease of the crystal size and the more acid sites on the surface, replaced Na ions remains on the outer surface. The acidic characteristics of the outer surface of Na-poisoned Cu/SSZ-13 are also in line with expectations. The Na substitution of Cu and the formation of CuO clusters are the basic characteristic of CuSSZ-13 alkali metal poisoning [25]. The newly generated acid sites on the outer surface of 0.8-Na and 2.3-Na are attributed to the formation of CuO clusters, which is consistent with the trends of CuO determined by UV–vis and ammonia oxidation activity.

The acidity of the outer surface of the catalyst was also quantitatively tested by pyridine infrared technology. Pyridine is a molecule with a larger kinetic radius of 5.7 Å, which can only interact with the acid sites on the outer surface of Cu/SSZ-13. According to previous studies [41–43], The IR bands at 1450 and 1593 cm^{-1} are attributed to the chemisorbed pyridine on the Lewis acid sites of the zeolites. The IR band at 1483 cm^{-1} is assigned to the pyridine bonded on both Lewis and Brønsted acid sites. The IR band at 1630 cm^{-1} is attributed to the Brønsted acid sites. As shown in Figure 8, the results show that the smaller the crystal size is, the more Brønsted acid and Lewis acid are on the surface. Although there is a certain deviation on surface quantification from the diethylamine-TPD, the trend is highly consistent.

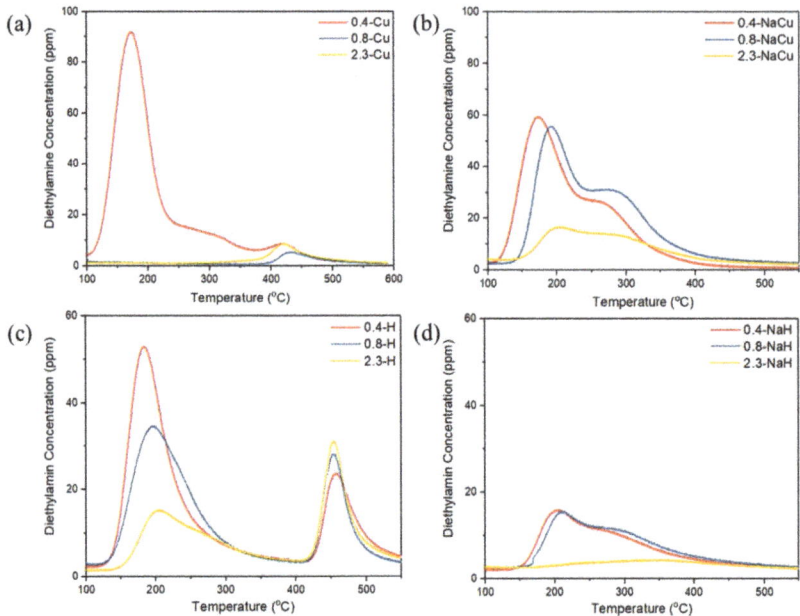

Figure 7. Diethylamine-TPD profiles of Cu/SSZ-13 (**a,b**) and H/SSZ-13 (**c,d**) before and after Na poisoning.

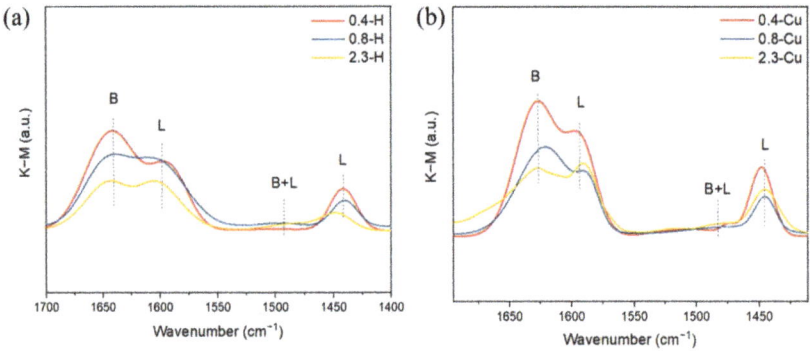

Figure 8. Pyridine-chemisorbed IR spectra of (a) H/SSZ-13 and (b) Cu/SSZ-13 with different crystal sizes before Na poisoning.

In addition to chemical adsorption methods, we expect to visually investigate the spatial distribution of Cu ions on SSZ-13. Combined with cross-section polishing technology, the Cu/SSZ-13 is cut and the Cu content is analyzed by EDS. As shown in Figure 9, point EDS analysis was conducted from the center to the edge of the Cu/SSZ-13 cross-section for four individual crystals. The detected Cu content is summarized in Table 2. The energy spectra are attached to the supporting information (Figure S6). The results show that the distribution of Cu in the molecular sieve forms a decreasing gradient from the center to the edge. Note that it is hard to detect Cu on the surface and subsurface. Such a feature provides sufficient outer surface space to buffer the corrosion of Na, thereby protecting the Cu ions in the core.

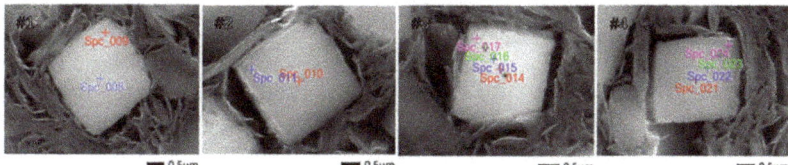

Figure 9. Point EDS analysis of 2.3-Cu sample.

Table 2. Cu content of 2.3-Cu cross-section from the center to the edge.

Detecting Number	Cu Content from Center to Edge/wt %
#1	0.87 ± 0.06, 0.10 ± 0.03
#2	0.76 ± 0.07, 0.08 ± 0.04
#3	0.80 ± 0.03, 0.62 ± 0.06, 0.21 ± 0.02, 0.09 ± 0.02
#4	0.67 ± 0.04, 0.55 ± 0.06, 0.32 ± 0.03, 0.09 ± 0.03

It is time to recall the changes in NH_3 oxidation performance of Cu/SSZ-13 with different crystal sizes before and after Na poisoning. According to our analysis, the side reaction of NH_3 oxidation of Na-poisoned Cu/SSZ-13 with larger crystal size is not significantly improved (Figure 3), and the number of active centers around 300 °C decreases with the increase of crystal sizes (Figure S1). This is rationalized by the difference in spatial acidity distribution. On the one hand, the smaller the crystal size, the more acidic the surface of Cu/SSZ-13 is, and the proportion of Cu ions on the surface is greater. This results in the surface Cu ions being susceptible to the Na corrosion to generate CuO species. On the other hand, the Cu ions distributed in the large crystal-sized molecular sieve have extremely strong stability. The interaction between Cu and framework O can be characterized by H_2-TPR. Because the interaction of Cu and O reflects how difficult the Cu^{2+} can be reduced (or O^{2-} can be split). From the H_2-TPR results, with increasing crystal sizes more Cu interacts with the framework O so strongly that it cannot be reduced by H_2. The chemical bond between the core Cu ions and framework O might be strong enough that NH_3 or H_2O cannot solvate Cu ions, which explains why the $Cu^{II}(NH_3)_n(H_2O)_m$ species are less distributed in the large-sized Cu/SSZ-13. This inference is based on the fact that the active center of NH_3 oxidation will change with increasing temperature. Chemically, isolated Cu (II) in the six-member ring of SSZ-13 is the active center. It can be solvated by H_2O and NH_3 molecules to form so-called hydrated or ammoniated Cu (II) species ($Cu^{II}(NH_3)_n(H_2O)_m$). Gao et al. reported that the activation energy of Cu/SSZ-13 in the low-temperature region (200–260 °C) and the high-temperature region (400–460 °C) of the NH_3 oxidation reaction are 64.1 kJ/mol and 150 kJ/mol, respectively [39]. Therefore, it is reasonable to believe that the reaction path is different in the two different temperature ranges, and the active center transforms from solvable $Cu^{II}(NH_3)_n(H_2O)_m$ to framework-anchored $ZCu^{II}OH$ as the temperature crosses ~300 °C.

3. Materials and Methods

3.1. Catalysts Preparation

Na/SSZ-13 (Si/Al = 25) was home-made according to previous reports [2,44,45]. The crystal size of SSZ-13 was controlled by adding different amounts of seed crystals [46]. During the hydrothermal synthesis process, other conditions remain the same, such as crystallization temperature (160 °C), crystallization time (72 h), silicon source (silica sol, Jiyida Co. Ltd., Qingdao, China), aluminum source (pseudo-boehmite, Chinalco, Zibo, China), template agent (1-Adamantanamine, Sigma-Aldrich, Beijing, China), etc. The templating agent was removed from as-prepared Na/SSZ-13 by calcinating at 650 °C for 4 h. Cu/SSZ-13 was obtained by the aqueous ion-exchange method. Briefly, following Na/SSZ-13 with the template removed, NH_4/SSZ-13 was obtained by twice ion-exchanging with 0.1M NH_4NO_3. Then NH_4/SSZ-13 was ion-exchanged by 0.1M $Cu(NO_3)_2$ to prepare

Cu/SSZ-13. H/SSZ-13 was also prepared by calcinating NH_4/SSZ-13 at 550 °C for 4 h. The samples were denoted by 'x-Cu' and 'x-H' to respectively represent Cu/SSZ-13 catalysts and H/SSZ-13 supports with different crystal sizes, where x is the crystal size (μm).

To simulate the alkali metal poisoning process, a combination of equal volume impregnation and high-temperature hydrothermal aging was carried out. In detail, it was impregnated with $NaNO_3$ on the samples and aged for 12 h at 600 °C under 10 vol % steam. 'x-NaCu' and 'x-NaH' were used to represent Na-poisoned samples.

3.2. Characterization

The morphology of the sample was pictured on a field emission electron microscopy (Regulus 8100, HITACHI, Tokyo, Japan) at 3 kV.

The powder X-ray diffraction (XRD) patterns of the catalysts were detected using a Rigaku Smartlab XRD (Tokyo, Japan) with Cu Kα radiation source. The X-ray emission tube was run at 40 kV and 40 mA. The diffraction angle (2 θ) range of 5–50° was chosen to identify the crystal phases.

The BET surface area was measured with a physisorption analyzer (ASAP 2460, Micromeritics) at 77K. Before testing, the samples were vacuumed under 300 °C for 3 h.

The NMR spectra were recorded using a JEOL JNM-ECZ600R spectrometer (Tokyo, Japan). The ^{29}Si and ^{27}Al NMR were carried out at a field strength of 14.1 T with a total scanning of 1200 and 300 times, relaxation delay of 10s and 2s, respectively. The tube diameter of 4 mm and 8 mm was chosen for testing ^{27}Al and ^{29}Si, respectively.

Hydrogen temperature-programmed reduction (H_2-TPR) was carried out in a Micromeritics AutoChem 2920 II system (Micromeritics). The catalysts were pretreated under 5% O_2/N_2 at 500 °C for 30 min. The H_2-TPR profile was recorded from ambient to 900 °C at a rate of 10 °C/min.

UV–vis spectra were collected in a SHIMADZU UV-2600 (Kyoto, Japan) equipped with an integrating sphere assembly. The sample powder was pressed into a cake in the button groove at ambient. A pressed cake of $BaSO_4$ was used as the reference.

Diethylamine temperature-programmed desorption measurement was conducted using a plug flow microreactor. The diethylamine was detected using an online Fourier transmission infrared spectrometer (MultiGas 6030, MKS). The sample was pretreated under 5% O_2/N_2 at 500 °C for 30 min, followed by diethylamine saturation at 100 °C. The weakly adsorbed diethylamine was purged by N_2 and then the sample was heated to 550 °C at a rate of 10 °C/min.

Operando pyridine titration experiment was monitored by a diffuse reflection infrared Fourier transform spectrometer (DRIFTS) (Nicolet 6700, Thermo). The sample was pretreated at 500 °C for 30 min under 5% O_2/N_2. Then it was cooled to 250 °C under N_2 and the background spectrum was collected. The in-situ time-resolved spectra were recorded when the N_2 carrier with pyridine mixed gas pass through the sample.

3.3. Catalytic Evaluation

The standard NH_3-SCR activity was tested with a plug flow microreactor. The powder sample was supported by quartz wool on both sides. Typically, 0.1 g catalyst was mixed with 0.9 g quartz sand to uniformly disperse catalysts powder. The simulated gas contains 500 ppm NH_3, 500 ppm NO, 10% O_2, 3% H_2O and balanced N_2. The gas hourly space velocity (GHSV) is 72,000 h^{-1}. The NO_x conversion was calculated as below,

$$NO_x \% = \frac{[NO_x]_{in} - [NO_x]_{out}}{[NO_x]_{in}} \times 100\% \quad (1)$$

The NH₃ oxidation activity was also investigated individually. The experiment setup is the same as that for the NH₃-SCR activity measurement. 500 ppm NH₃, 10% O₂ and balanced N₂ were used. The NH₃ conversion was calculated as below,

$$\mathrm{NH_3\ \%} = \frac{[\mathrm{NH_3}]_{in} - [\mathrm{NH_3}]_{out}}{[\mathrm{NH_3}]_{in}} \times 100\% \qquad (2)$$

The reagents and products were monitored by an online FTIR spectrometer (Multigas 6030, MKS). For each sample, the activity test included pre-treatment and steady-state activity testing. For example, the catalysts were firstly oxidized under 10% O₂/N₂ at 500 °C for 1 h. The ready catalyst was then fed with mixed gases of standard NH₃-SCR or NH₃ oxidation. The steady-state data were collected from 500 °C to 100 °C in an interval of 50 °C.

4. Conclusions

Cu/SSZ-13 catalysts with similar Cu loadings and different crystal sizes (0.4–2.3 μm) were controllably synthesized. Although the pore structure of H/SSZ-13 with a large crystal size is susceptible to Na poisoning, the loading of Cu ions has a strong protective effect on the structure. It was found that increasing the crystal size of Cu/SSZ-13 catalysts can effectively improve the Na resistance. Large crystal size is conducive to the distribution of acidic sites on the subsurface or inner core of the molecular sieve. The Cu ions distributed in the deep layer have strong redox stability and are not easily solvated by NH₃ or H₂O molecules. The richer Cu ions on the outer surface of Cu/SSZ-13 with smaller particle size is more susceptible to Na ion exchange to generate more CuO species. The decrease of Cu ions and the increase of CuO clusters are the main reasons for the decrease of NH₃-SCR activity and the increase of side reaction NH₃ oxidation activity. The results are helpful to guide industrial applications in heavy-duty diesel engines after treatment by improving the alkali metal tolerance of commercial Cu/SSZ-13 catalysts for NH₃-SCR reaction.

Supplementary Materials: The following are available online at https://www.mdpi.com/article/10.3390/catal11080979/s1, Figure S1: NO$_x$ and N₂O concentration evolution in the NH₃ oxidation; Figure S2. XRD patterns of Cu/SSZ-13 (a) before and (b) after Na poisoning; Figure S3. The micropore distribution of Cu/SSZ-13 (a) before and (b) after Na poisoning; Figure S4. (a–b) ²⁷Al NMR spectra and (c–d) ²⁹Si NMR spectra of CuSSZ-13 before and after Na poisoning; Figure S5. NH₃-TPD profiles of Cu/SSZ-13 with different crystal sizes (a) before and (b) after Na poisoning; Figure S6. Point EDS spectra corresponding to the results in Figure 9, Table S1. Element composition of Cu/SSZ-13 and Na-poisoned counterparts with different crystal sizes.

Author Contributions: Conceptualization, Z.C. and M.S.; Methodology, C.W.; Validation, Z.C., J.W. (Jianqiang Wang), and J.W. (Jun Wang); Formal analysis, Z.C.; Investigation, Z.C.; Writing—original draft preparation, Z.C.; Writing—review and editing, J.W. (Jianqiang Wang) and C.W.; Supervision, M.S. and G.S.; Project administration, J.W. (Jun Wang) and G.S.; Funding acquisition, J.W. (Jun Wang). All authors have read and agreed to the published version of the manuscript.

Funding: This research was funded by Foundation for Innovative Research Groups of the National Natural Science Foundation of China, grant number 51921004; Major Science and Technology Programs of Yunnan, grant number 202002AB080001-1; National Natural Science Foundation for Youth of China, grant number 21908207 and Postdoctoral Research Foundation of China, grant number 2020M670659; Shanxi Province Science Foundation for Youths, grant number 201901D211224.

Acknowledgments: We are grateful to all our funders.

Conflicts of Interest: The authors declare no conflict of interest.

References

1. Joshi, A. Review of Vehicle Engine Efficiency and Emissions. *SAE Tech. Pap.* **2021**. [CrossRef]
2. Kwak, J.H.; Tonkyn, R.G.; Kim, D.H.; Szanyi, J.; Peden, C.H.F. Excellent activity and selectivity of Cu-SSZ-13 in the selective catalytic reduction of NO$_x$ with NH₃. *J. Catal.* **2010**, *275*, 187–190. [CrossRef]
3. Ma, L.; Cheng, Y.; Cavataio, G.; McCabe, R.W.; Fu, L.; Li, J. Characterization of commercial Cu-SSZ-13 and Cu-SAPO-34 catalysts with hydrothermal treatment for NH₃-SCR of NO$_x$ in diesel exhaust. *Chem. Eng. J.* **2013**, *225*, 323–330. [CrossRef]

4. Fan, C.; Chen, Z.; Pang, L.; Ming, S.; Dong, C.; Brou Albert, K.; Liu, P.; Wang, J.; Zhu, D.; Chen, H.; et al. Steam and alkali resistant Cu-SSZ-13 catalyst for the selective catalytic reduction of NO_x in diesel exhaust. *Chem. Eng. J.* **2018**, *334*, 344–354. [CrossRef]
5. Jiang, H.; Guan, B.; Peng, X.; Zhan, R.; Lin, H.; Huang, Z. Influence of synthesis method on catalytic properties and hydrothermal stability of Cu/SSZ-13 for NH_3-SCR reaction. *Chem. Eng. J.* **2020**, *379*, 122358. [CrossRef]
6. Wang, J.; Peng, Z.; Chen, Y.; Bao, W.; Chang, L.; Feng, G. In-situ hydrothermal synthesis of Cu-SSZ-13/cordierite for the catalytic removal of NO_x from diesel vehicles by NH_3. *Chem. Eng. J.* **2015**, *263*, 9–19. [CrossRef]
7. Lomachenko, K.A.; Borfecchia, E.; Negri, C.; Berlier, G.; Lamberti, C.; Beato, P.; Falsig, H.; Bordiga, S. The Cu-CHA deNO$_x$ Catalyst in Action: Temperature-Dependent NH_3-Assisted Selective Catalytic Reduction Monitored by Operando XAS and XES. *J. Am. Chem. Soc.* **2016**, *138*, 12025–12028. [CrossRef] [PubMed]
8. Janssens, T.V.W.; Falsig, H.; Lundegaard, L.F.; Vennestrøm, P.N.R.; Rasmussen, S.B.; Moses, P.G.; Giordanino, F.; Borfecchia, E.; Lomachenko, K.A.; Lamberti, C.; et al. A Consistent Reaction Scheme for the Selective Catalytic Reduction of Nitrogen Oxides with Ammonia. *ACS Catal.* **2015**, *5*, 2832–2845. [CrossRef]
9. Bates, S.A.; Verma, A.A.; Paolucci, C.; Parekh, A.A.; Anggara, T.; Yezerets, A.; Schneider, W.F.; Miller, J.T.; Delgass, W.N.; Ribeiro, F.H. Identification of the active Cu site in standard selective catalytic reduction with ammonia on Cu-SSZ-13. *J. Catal.* **2014**, *312*, 87–97. [CrossRef]
10. Gunter, T.; Carvalho, H.W.; Doronkin, D.E.; Sheppard, T.; Glatzel, P.; Atkins, A.J.; Rudolph, J.; Jacob, C.R.; Casapu, M.; Grunwaldt, J.D. Structural snapshots of the SCR reaction mechanism on Cu-SSZ-13. *Chem. Commun.* **2015**, *51*, 9227–9230. [CrossRef] [PubMed]
11. Paolucci, C.; Khurana, I.; Parekh, A.A.; Li, S.; Shih, A.J.; Li, H.; Di Iorio, J.R.; Albarracin-Caballero, J.D.; Yezerets, A.; Miller, J.T.; et al. Dynamic multinuclear sites formed by mobilized copper ions in NO_x selective catalytic reduction. *Science* **2017**, *357*, 898–903. [CrossRef] [PubMed]
12. Hun Kwak, J.; Zhu, H.; Lee, J.H.; Peden, C.H.; Szanyi, J. Two different cationic positions in Cu-SSZ-13? *Chem. Commun.* **2012**, *48*, 4758–4760. [CrossRef] [PubMed]
13. Cui, Y.; Wang, Y.; Mei, D.; Walter, E.D.; Washton, N.M.; Holladay, J.D.; Wang, Y.; Szanyi, J.; Peden, C.H.F.; Gao, F. Revisiting effects of alkali metal and alkaline earth co-cation additives to Cu/SSZ-13 selective catalytic reduction catalysts. *J. Catal.* **2019**, *378*, 363–375. [CrossRef]
14. Jiang, H.; Guan, B.; Peng, X.; Wei, Y.; Zhan, R.; Lin, H.; Huang, Z. Effect of sulfur poisoning on the performance and active sites of Cu/SSZ-13 catalyst. *Chem. Eng. Sci.* **2020**, *226*, 115855. [CrossRef]
15. Olsson, L.; Wijayanti, K.; Leistner, K.; Kumar, A.; Joshi, S.Y.; Kamasamudram, K.; Currier, N.W.; Yezerets, A. A kinetic model for sulfur poisoning and regeneration of Cu/SSZ-13 used for NH_3-SCR. *Appl. Catal. B Environ.* **2016**, *183*, 394–406. [CrossRef]
16. Gao, F.; Wang, Y.; Washton, N.M.; Kollár, M.; Szanyi, J.; Peden, C.H.F. Effects of Alkali and Alkaline Earth Cocations on the Activity and Hydrothermal Stability of Cu/SSZ-13 NH_3-SCR Catalysts. *ACS Catal.* **2015**, *5*, 6780–6791. [CrossRef]
17. Deka, U.; Lezcano-Gonzalez, I.; Weckhuysen, B.M.; Beale, A.M. Local Environment and Nature of Cu Active Sites in Zeolite-Based Catalysts for the Selective Catalytic Reduction of NO_x. *ACS Catal.* **2013**, *3*, 413–427. [CrossRef]
18. Jangjou, Y.; Do, Q.; Gu, Y.; Lim, L.-G.; Sun, H.; Wang, D.; Kumar, A.; Li, J.; Grabow, L.C.; Epling, W.S. Nature of Cu Active Centers in Cu/SSZ-13 and Their Responses to SO_2 Exposure. *ACS Catal.* **2018**, *8*, 1325–1337. [CrossRef]
19. Song, J.; Wang, Y.; Walter, E.D.; Washton, N.M.; Mei, D.; Kovarik, L.; Engelhard, M.H.; Prodinger, S.; Wang, Y.; Peden, C.H.F.; et al. Toward Rational Design of Cu/SSZ-13 Selective Catalytic Reduction Catalysts: Implications from Atomic-Level Understanding of Hydrothermal Stability. *ACS Catal.* **2017**, *7*, 8214–8227. [CrossRef]
20. Lin, C.; Cao, Y.; Feng, X.; Lin, Q.; Xu, H.; Chen, Y. Effect of Si islands on low-temperature hydrothermal stability of Cu/SAPO-34 catalyst for NH_3-SCR. *J. Taiwan Inst. Chem. Eng.* **2017**, *81*, 288–294. [CrossRef]
21. Schwab, S.D.; Bennett, J.J.; Dell, S.J.; Galante-Fox, J.M.; Kulinowski, A.M.; Miller, K.T. Internal Injector Deposits in High-Pressure Common Rail Diesel Engines. *SAE Int. J. Fuels Lubr.* **2010**, *3*, 865–878. [CrossRef]
22. Liu, L.; Wu, X.; Ma, Y.; Zhang, X.; Ran, R.; Si, Z.; Weng, D. Potassium deactivation of Cu-SSZ-13 catalyst for NH_3-SCR: Evolution of salts, zeolite and copper species. *Chem. Eng. J.* **2020**, *383*, 123080. [CrossRef]
23. Williams, A.; McCormick, R.; Lance, M.; Xie, C.; Toops, T.; Brezny, R. Effect of Accelerated Aging Rate on the Capture of Fuel-Borne Metal Impurities by Emissions Control Devices. *SAE Int. J. Fuels Lubr.* **2014**, *7*, 471–479. [CrossRef]
24. Williams, A.; Burton, J.; McCormick, R.L.; Toops, T.; Wereszczak, A.A.; Fox, E.E.; Lance, M.J.; Cavataio, G.; Dobson, D.; Warner, J.; et al. Impact of Fuel Metal Impurities on the Durability of a Light-Duty Diesel Aftertreatment System. *SAE Tech. Pap.* **2013**. [CrossRef]
25. Wang, C.; Yan, W.; Wang, Z.; Chen, Z.; Wang, J.; Wang, J.; Wang, J.; Shen, M.; Kang, X. The role of alkali metal ions on hydrothermal stability of Cu/SSZ-13 NH_3-SCR catalysts. *Catal. Today* **2020**, *355*, 482–492. [CrossRef]
26. Wang, C.; Wang, C.; Wang, J.; Wang, J.; Shen, M.; Li, W. Effects of Na(+) on Cu/SAPO-34 for ammonia selective catalytic reduction. *J. Environ. Sci.* **2018**, *70*, 20–28. [CrossRef] [PubMed]
27. Yan, Q.; Nie, Y.; Yang, R.; Cui, Y.; O'Hare, D.; Wang, Q. Highly dispersed Cu$_y$AlO$_x$ mixed oxides as superior low-temperature alkali metal and SO_2 resistant NH_3-SCR catalysts. *Appl. Catal. A Gen.* **2017**, *538*, 37–50. [CrossRef]
28. Liu, A.; Liu, L.; Cao, Y.; Wang, J.; Si, R.; Gao, F.; Dong, L. Controlling Dynamic Structural Transformation of Atomically Dispersed CuO$_x$ Species and Influence on Their Catalytic Performances. *ACS Catal.* **2019**, *9*, 9840–9851. [CrossRef]

29. Du, Y.; Huang, Z.; Zhang, J.; Jing, G. Fe2O3/HY Catalyst: A Microporous Material with Zeolite-Type Framework Achieving Highly Improved Alkali Poisoning-Resistant Performance for Selective Reduction of NO_x with NH_3. *Environ. Sci. Technol.* **2020**, *54*, 7078–7087. [CrossRef] [PubMed]
30. Zha, K.; Kang, L.; Feng, C.; Han, L.; Li, H.; Yan, T.; Maitarad, P.; Shi, L.; Zhang, D. Improved NO_x reduction in the presence of alkali metals by using hollandite Mn–Ti oxide promoted Cu-SAPO-34 catalysts. *Environ. Sci. Nano.* **2018**, *5*, 1408–1419. [CrossRef]
31. Huang, S.; Wang, J.; Wang, J.; Wang, C.; Shen, M.; Li, W. The influence of crystallite size on the structural stability of Cu/SAPO-34 catalysts. *Appl. Catal. B Environ.* **2019**, *248*, 430–440. [CrossRef]
32. Kumar, M.; Luo, H.; Román-Leshkov, Y.; Rimer, J.D. SSZ-13 Crystallization by Particle Attachment and Deterministic Pathways to Crystal Size Control. *J. Am. Chem. Soc.* **2015**, *137*, 13007–13017. [CrossRef]
33. Gao, F.; Walter, E.D.; Kollar, M.; Wang, Y.; Szanyi, J.; Peden, C.H.F. Understanding ammonia selective catalytic reduction kinetics over Cu/SSZ-13 from motion of the Cu ions. *J. Catal.* **2014**, *319*, 1–14. [CrossRef]
34. Marberger, A.; Petrov, A.W.; Steiger, P.; Elsener, M.; Kröcher, O.; Nachtegaal, M.; Ferri, D. Time-resolved copper speciation during selective catalytic reduction of NO on Cu-SSZ-13. *Nat. Catal.* **2018**, *1*, 221–227. [CrossRef]
35. Gao, F.; Mei, D.; Wang, Y.; Szanyi, J.; Peden, C.H.F. Selective Catalytic Reduction over Cu/SSZ-13: Linking Homo- and Heterogeneous Catalysis. *J. Am. Chem. Soc.* **2017**, *139*, 4935–4942. [CrossRef] [PubMed]
36. Gao, F.; Kwak, J.H.; Szanyi, J.; Peden, C.H.F. Current Understanding of Cu-Exchanged Chabazite Molecular Sieves for Use as Commercial Diesel Engine $DeNO_x$ Catalysts. *Top. Catal.* **2013**, *56*, 1441–1459. [CrossRef]
37. Paolucci, C.; Parekh, A.A.; Khurana, I.; Di Iorio, J.R.; Li, H.; Albarracin Caballero, J.D.; Shih, A.J.; Anggara, T.; Delgass, W.N.; Miller, J.T.; et al. Catalysis in a Cage: Condition-Dependent Speciation and Dynamics of Exchanged Cu Cations in SSZ-13 Zeolites. *J. Am. Chem. Soc.* **2016**, *138*, 6028–6048. [CrossRef]
38. Chen, Z.; Wang, J.; Wang, J.; Wang, C.; Wang, J.; Li, W.; Shen, M. Disparate Essences of Residual, Ion-Exchanged, and Impregnated Na Ions on Topology Structure for Cu/SSZ-13 NH_3 Selective Catalytic Reduction Catalysts. *Ind. Eng. Chem. Res.* **2019**, *58*, 20610–20619. [CrossRef]
39. Gao, F.; Washton, N.M.; Wang, Y.; Kollár, M.; Szanyi, J.; Peden, C.H.F. Effects of Si/Al ratio on Cu/SSZ-13 NH_3-SCR catalysts: Implications for the active Cu species and the roles of Brønsted acidity. *J. Catal.* **2015**, *331*, 25–38. [CrossRef]
40. Zhao, Z.; Yu, R.; Zhao, R.; Shi, C.; Gies, H.; Xiao, F.-S.; De Vos, D.; Yokoi, T.; Bao, X.; Kolb, U.; et al. Cu-exchanged Al-rich SSZ-13 zeolite from organotemplate-free synthesis as NH_3-SCR catalyst: Effects of Na^+ ions on the activity and hydrothermal stability. *Appl. Catal. B Environ.* **2017**, *217*, 421–428. [CrossRef]
41. Wang, B.; Ma, L.; Han, L.; Feng, Y.; Hu, J.; Xie, W.; Bao, W.; Chang, L.; Huang, Z.; Wang, J. Assembly-reassembly of coal fly ash into Cu-SSZ-13 zeolite for NH_3-SCR of NO via interzeolite transformations. *Chem. Eng. Sci. X* **2021**, *10*, 100089.
42. Fan, J.; Ning, P.; Wang, Y.; Song, Z.; Liu, X.; Wang, H.; Wang, J.; Wang, L.; Zhang, Q. Significant promoting effect of Ce or La on the hydrothermal stability of Cu-SAPO-34 catalyst for NH_3-SCR reaction. *Chem. Eng. J.* **2019**, *369*, 908–919. [CrossRef]
43. Zhan, H.; Huang, S.; Li, Y.; Lv, J.; Wang, S.; Ma, X. Elucidating the nature and role of Cu species in enhanced catalytic carbonylation of dimethyl ether over Cu/H-MOR. *Catal. Sci. Technol.* **2015**, *5*, 4378–4389. [CrossRef]
44. Zones, S.I. Zeolite SSZ-13 and Its Method of Preparation. U.S. 4544538A, 1 October 1985.
45. Fickel, D.W.; Lobo, R.F. Copper coordination in Cu-SSZ-13 and Cu-SSZ-16 investigated by variable-temperature XRD. *J. Phys. Chem. C* **2010**, *114*, 1633–1640. [CrossRef]
46. Wang, J.; Shao, L.; Wang, C.; Wang, J.; Shen, M.; Li, W. Controllable preparation of various crystal size and nature of intra-crystalline diffusion in Cu/SSZ-13 NH_3-SCR catalysts. *J. Catal.* **2018**, *367*, 221–228. [CrossRef]

MDPI
St. Alban-Anlage 66
4052 Basel
Switzerland
Tel. +41 61 683 77 34
Fax +41 61 302 89 18
www.mdpi.com

Catalysts Editorial Office
E-mail: catalysts@mdpi.com
www.mdpi.com/journal/catalysts

www.ingramcontent.com/pod-product-compliance
Lightning Source LLC
LaVergne TN
LVHW070729100526
838202LV00013B/1198